Current Topics in Neuroendocrinology Volume 8

Editors

Detlev Ganten, Heidelberg · Donald Pfaff, New York

Current Topics in Neuroendocrinology

Vol. 1: Sleep. Clinical and Experimental Aspects
1982. 47 figures. VII, 129 pages. DM 95,–. ISBN 3-540-11125-5
Neurobiology of REM Sleep. A Possible Role for Dopamine · Endocrine and Peptide Functions in the Sleep-Waking Cycle · Sleep Regulation: Circadian Rhythm and Homeostasis · Haemodynamic Changes During Sleep in Man

Vol. 2: Adrenal Actions on Brain
1982. 25 figures. V, 153 pages. DM 95,–. ISBN 3-540-11126-3
Glucocorticoids and Hippocampus: Receptors in Search of a Function · Glucocorticoids and the Developing Nervous System · Feedback Actions of Adrenal Steroid Hormones · Electrophysiologic and Clinical Aspects of Glucocorticoids on Certain Neural Systems · Adrenal Steroids and Behavioral Adaptation: Relationship to Brain Corticoid Receptors

Vol. 3: Central Cardiovascular Control. Basic and Clinical Aspects
1983. 71 figures. V, 192 pages. DM 106,–. ISBN 3-540-11350-9
Functional and Anatomic Aspects of Central Nervous Cardiovascular Regulation · Autonomic Nervous System and Blood Pressure Control in Normotensive and Hypertensive Conditions · Reflex Control of Circulation in Normotensive and Hypertensive Humans · Corticotropin-Releasing Factor: Central Nervous System Effects on the Sympathetic Nervous System and Cardiovascular Regulation · Neuropeptides and Central Blood Pressure Regulation · Centrally Acting Drugs as a Tool to Study Central Mechanisms of Blood Pressure Control · The Blood-Brain Barrier and its Role in the Control of Circulating Hormone Effects on the Brain

Vol. 4: Neurobiology of Vasopressin
1985. 53 figures. V, 203 pages. DM 98,–. ISBN 3-540-11351-7
Biosynthesis of Vasopressin · Electrophysiological Studies of the Magnocellular Neurons · Volume Regulation of Antidiuretic Hormone Secretion · Vasopressin, Cardiovascular Regulation and Hypertension · Neuroanatomical Pathways Related to Vasopressin

Vol. 5: Actions of Progesterone on the Brain
1985. 61 figures. V, 216 pages. DM 98,–. ISBN 3-540-13433-6
Antiprogestins · Progesterone Receptors in Brain and Hypophysis · Effects of Estrogen and Progesterone as Revealed by Neurophysiological Methods · Progesterone Receptors in CNS Correlated with Reproductive Behavior · Estradiol-Progesterone Interactions in the Reproductive Behavior of Female Rats · Behavioral Effects on Humans of Progesterone-Related Compounds During Development and in the Adult

Vol. 6: Neurobiology of Oxytocin
1986. 38 figures. V, 175 pages. DM 98,–. ISBN 3-540-15341-1
Biosynthesis of Oxytocin in the Brain and Peripheral Organs · Regulation of Oxytocin Release · Proteolytic Conversion of Oxytocin, Vasopressin, and Related Peptides in the Brain · Oxytocin and Behavior · Oxytocin as an Ovarian Hormone · Oxytocin and the Milk-Ejection Reflex

Vol. 7: Morphology of Hypothalamus and Its Connections
1986. 94 figures. VI, 314 pages. DM 148,–. ISBN 3-540-16919-9
The Pituitary Portal System · Functional Ultrastructure of Gonadotropes: a Review · Ultrastructure of Anterior Pituitary Cells · Catecholamine-Peptide Interactions in the Hypothalamus · Neuroendocrine Projections to the Median Eminence · Afferents onto Neuroendocrine Cells · Interconnectedness of Steroid Hormone-Binding Neurons: Existence and Implications · Ultrastructure of Regulatory Neuroendocrine Neurons and Functionally Related Structures · Synaptogenesis and Neuronal Plasticity to Gonadal Steroids: Implications for the Development of Sexual Dimorphism in the Neuroendocrine Brain

D. Ganten and D. Pfaff (Eds.)

Neuroendocrinology of Mood

Coeditor

K. Fuxe

Contributors

L. F. Agnati, H. Ågren, M. Aronsson, M. S. Bauer, G. P. Chrousos
A. Cintra, T. J. Crow, M. A. Demitrack, M. J. Devlin, I. N. Ferrier
K. Fuxe, P. W. Gold, R. N. Golden, J.-Å. Gustafsson
A. Härfstrand, D. S. Janowsky, K. Kalogeras, M. A. Kling
B. Levant, P. Linkowski, D. L. Loriaux, N. Matussek
H. Y. Meltzer, J. Mendlewicz, J. F. Nash Jr., C. B. Nemeroff
R. M. Post, S. C. Risch, D. R. Rubinow, L. Terenius
L. Träskman-Bendz, B. T. Walsh, S. R. B. Weiss, H. Whitfield
P. C. Whybrow, F. A. Wiesel, M. Zoli

With 80 Figures

Springer-Verlag Berlin Heidelberg New York
London Paris Tokyo

Editors

Dr. Detlev Ganten, M.D., Ph.D.
Pharmakologisches Institut
Universität Heidelberg
Im Neuenheimer Feld 366
6900 Heidelberg/FRG

Dr. Donald Pfaff, Ph.D.
Rockefeller University
York Avenue, and 66th Street
New York, NY 10021/USA

Coeditor

Dr. Kjell Fuxe, M.D.
Karolinska Institute
P.O. Box 60400
10401 Stockholm, Sweden

The picture on the cover has been taken from Nieuwenhuys R., Voogd J., van Huijzen Chr.: The Human Central Nervous System. 2nd Edition. Springer-Verlag Berlin Heidelberg New York 1981

ISBN-13: 978-3-642-72740-5 e-ISBN-13: 978-3-642-72738-2

DOI: 10.1007/978-3-642-72738-2

Library of Congress Cataloging in Publication Data. Neuroendocrinology of mood. (Current topics in neuroendocrinology; v. 8) Includes bibliographies and index. 1. Mood (Psychology) – Physiological aspects. 2. Neuroendocrinology. 3. Neurotransmitters. 4. Affective disorders – Physiological aspects. I. Ganten, D. (Detlev), 1941-. II. Pfaff, Donald W., 1939-. III. Fuxe, Kjell. IV. Agnati,Luigi Francesco. V. Series. QP401.N37 1988 616.89'071 88-4897

Softcover reprint of the hardcover 1st edition 1988

Typesetting, printing and binding: Brühlsche Universitätsdruckerei, Giessen
2121/3145-543210

Table of Contents

Principles for the Hormone Regulation of Wiring Transmission and Volume Transmission in the Central Nervous System *

K. Fuxe [1], L. F. Agnati [2], A. Härfstrand [1], A. Cintra [1], M. Aronsson [1]
M. Zoli [2] and J.-Å. Gustafsson [3]

Contents

1 Introduction

We have recently suggested the existence in the central nervous system of two types of electrochemical transmission, namely wiring transmission (WT) and volume transmission (VT) (see Agnati et al. 1986a, b). The concepts are summarized in Tables 1 and 2. VT is a humoral type of chemical transmission. However, it

* This work has been supported by a grant (04X-715) from the Swedish Medical Research Council, a grant (MH25504) from the NIH, a grant from the Wallenberg Foundation, and CNR-I, MDI grants
1 Department of Histology and Neurobiology, Karolinska Institute, P.O. Box 60400, S-10401 Stockholm, Sweden
2 Department of Human Physiology, University of Modena, Modena, Italy
3 Department of Medical Nutrition, Huddinge Hospital, Huddinge, Sweden

Current Topics in Neuroendocrinology, Vol. 8

Table 1. Main features of morphofunctional roles of glial cells and neurons in wiring transmission and volume transmission

		Glial functions		Neuron functions	
		Physical	Biochemical	Physical	Biochemical
Wiring transmission	Electrical synapse	Synapse segregation	Modulation of coupling via ECF	Location of synapse	Modulation of coupling via ICF
	Chemical synapse	Synapse segregation	Modulation of synaptic cleft microenvironment	Location of Synapse	Control of the efficacy of transmission lines and their interplay
Volume transmission	Electrotonic signals	Shaping of ECF pathways for signal diffusion	Control of ECF ion composition	Location of sources and sinks for signals	Control of sources and sinks for signals
	Humoral signals	Shaping of ECF pathways for signal diffusion	Release, uptake and metabolism of signals	Location of release and recognition sites for signals	Release, uptake and metabolism of signals

ECF/ICF, extracellular/intracellular fluid

Table 2. The major characteristics of wiring transmission versus volume transmission

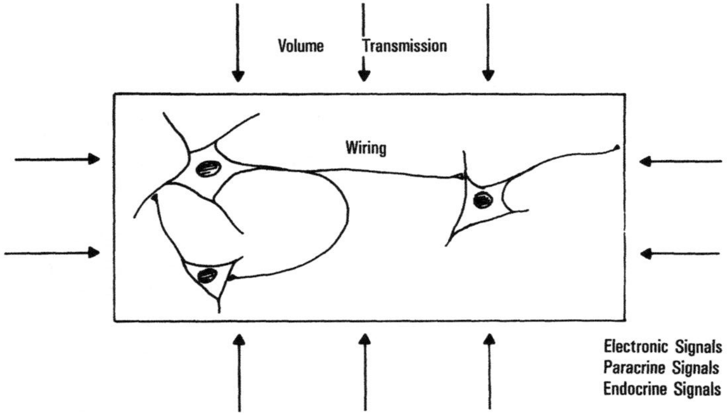

	Speed of transmission	Degree of divergence	Segregation ("safety" of the transmission)	Plasticity	Preferential information processing
Wiring transmission: Neuron-linked electrochemical transmission	High	Low to moderate	High	Low to moderate	Elementary elaboration Short-term action
Volume transmission: Humoral ("open") electrochemical transmission	Low	High to very high	Low	High to very high	Holistic elaboration Long-term action

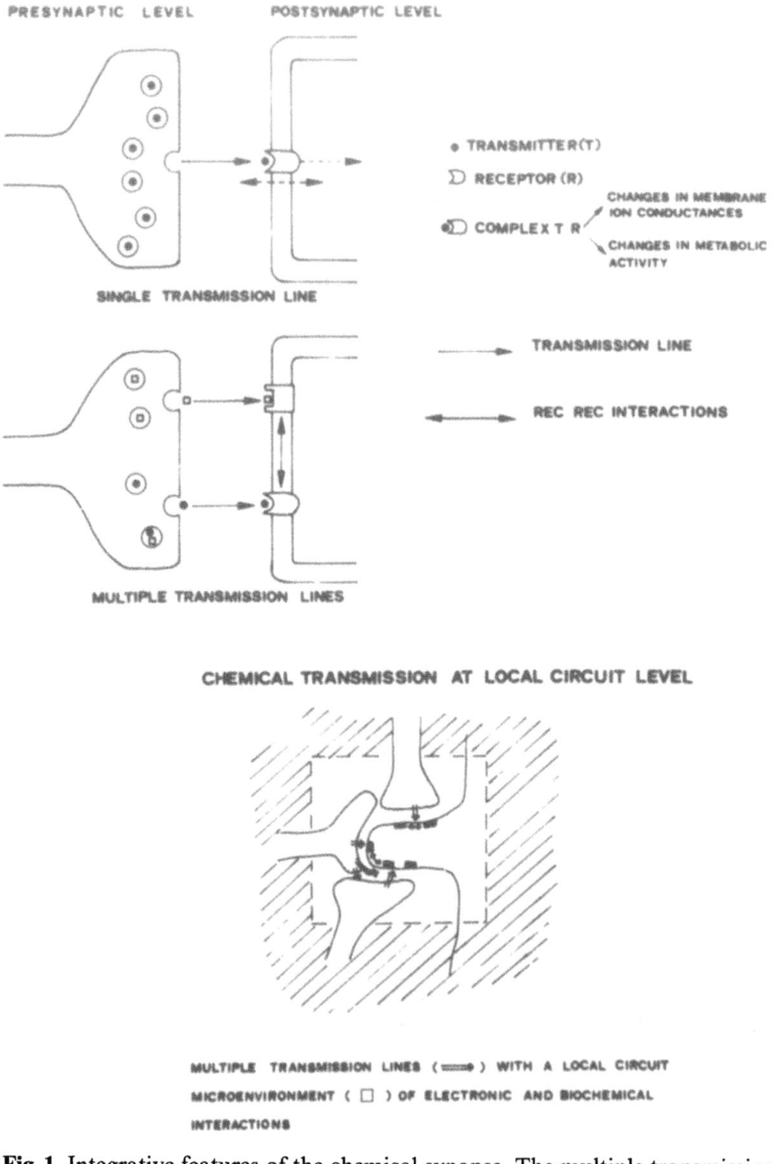

Fig. 1. Integrative features of the chemical synapse. The multiple transmission lines of the intramembrane receptor-receptor interactions are illustrated. The local circuit is shown as an electrometabolic unit

consists not only of humoral and paracrine signals, diffusing in the extracellular fluid to reach the appropriate receptors, but also of electrotonic signals, which also operate in the extracellular fluid. In fact, the extracellular space of the brain constitutes a restricted microenvironment. Thus, ion fluxes across cellular membranes can induce substantial changes in the ion composition. These ionic fluctuations in the extracellular fluid and the ionic fluxes from sources to sinks may represent

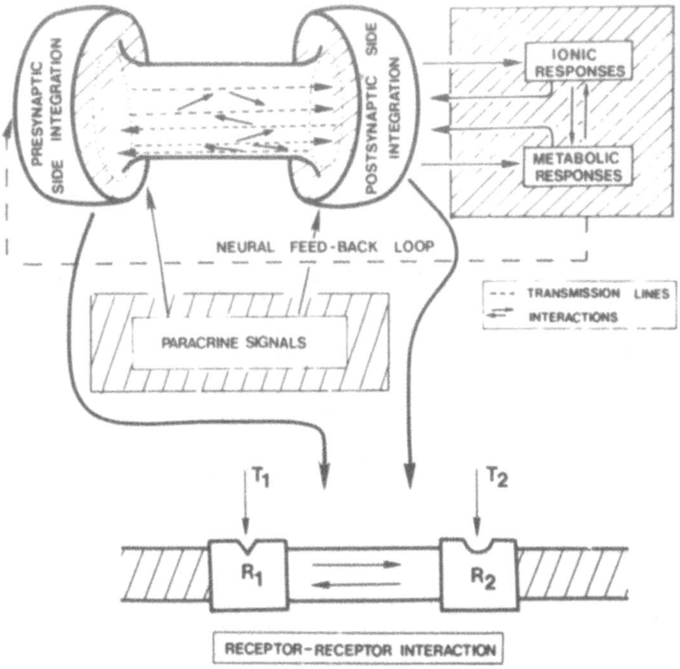

Fig. 2. The n-dimensional representation of synaptic transmission

signals for communication between neural groups (Nicholson 1980). In Table 1 the possible role of glia and neurons in WT and VT is summarized. In VT the glial cells control the extracellular fluid ion composition and the shaping of the extracellular fluid pathways (i.e., the communication channels between neural groups) for signal diffusion as well as the release, uptake and metabolism of humoral and paracrine signals. With regard to the function of neurons in VT they represent the location of sources and sinks for electrotonic signals and the sites of release and recognition of humoral and paracrine signals. From a biochemical standpoint the neurons control the sources and sinks of electrotonic signals and are involved in the uptake, release and metabolism of humoral and paracrine signals. When we focus our attention on chemical signals in WT and VT it is possible to recognize some main differential features. Thus, as seen in Table 2, the VT is characterized by low speed and long-term action, a high degree of divergence and plasticity and low safety of the transmission process. WT is the classical type of transmission which is neuron-linked and operates with high speed and safety and short-term actions, the divergency and plasticity being low. It seems clear that the integrative capability of the central and peripheral nervous system is increased by the presence of VT, which is not submitted to neuroanatomical constraints and may affect the computing charateristics of the neuronal networks.

In order to understand the actions of hormonal and paracrine signals on WT it is important to emphasize that the synapse is now regarded as a highly complex

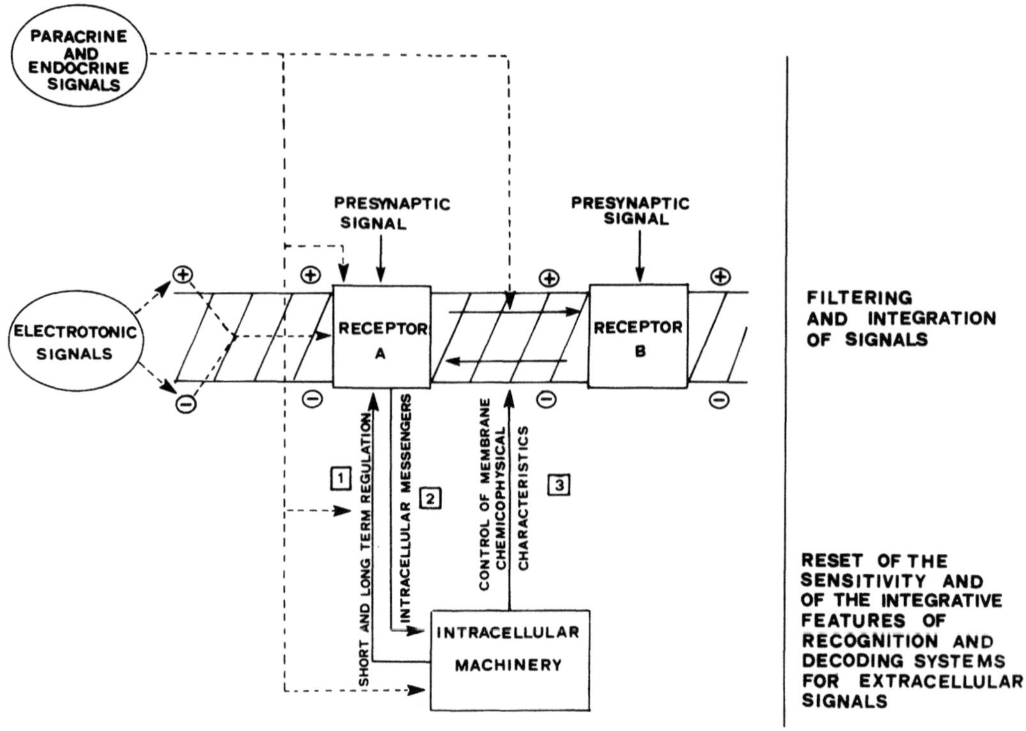

Fig. 3. Possible interaction between volume transmission and wiring transmission at membrane as well as at intracellular levels. At the intracellular level different targets for the modulatory interactions can be considered, such as [1] phosphorylation/dephosphorylation of the receptor complex; [2] cyclic nucleotide cascades; [3] lipid composition and polarization of the membrane

electrometabolic integrative unit (see Fig. 1). It consists of multiple transmission lines (Agnati et al. 1983, 1984a; Fuxe et al. 1984a), which interact with one another at the pre- and postsynaptic membrane via intramembrane receptor-receptor interactions (Fuxe and Agnati 1985) and via postreceptor intracellular signals. Furthermore, the ionic and metabolic responses participate in the integrative activity of the synaptic membranes. In Fig. 2 the complexity of the integration is illustrated at the membrane level, where filtration and integration signals take place. The intracytoplasmatic mechanisms control the recognition sites and the decoding mechanisms in the membrane for extracellular fluid signals, in this way, for example, resetting sensitivity in the integrative capacity of the receptor mechanisms. Hormonal and paracrine signals can directly modulate the receptor characteristics or the receptor-receptor interactions in the membrane (Fig. 3). The intracellular machinery and its short- and long-term regulation of the receptor mechanisms is also influenced, probably mainly via nuclear actions, at least as far as steroid and thyroid hormones are concerned (see Fuxe et al. 1981b). The electrotonic signals in the VT control the membrane polarization, in this way influencing the opening and closure of ionic channels. Changes in membrane polarization probably lead to allosteric changes in intramembrane proteins of the receptor complex and ion pumps. In this way electrical information can be transformed into chemical information. Thus, the two languages of the brain, i.e. the electrical and chemical signals, can be interconverted and the information coded into these two languages effectively integrated.

The above concept on WT and VT makes it easier to understand how mood can be affected by hormones such as steroid and thyroid hormones. We have, for example, observed that glucocorticoid receptor (Fuxe et al. 1985 d) immunoreactivity exist in very large numbers of nerve cells all over the cortical hemispheres (archi-, paleo- and neocortex), with predominant nuclear location. Thus, WT in the cortical areas of the brain controlling mood can be massively influenced by these hormones, which represent important signals in VT. Furthermore, these hormones also influence VT on the cortical networks subserving mood, since they can, by direct nuclear actions, regulate the synthesis and release of paracrine signals such as peptides from the cortical nerve cells to reach distant receptor populations in the cerebral cortex (see Fuxe et al. 1985 a; Agnati et al. 1986 b). Finally, in the frame of VT we can surmise that psychoactive drugs, even if they work on the wiring transmission (e.g. at synaptic level) reach their targets according to a VT mode. This gives further evidence that that endogenous signals may also effect WT by diffusing in the extracellular fluid of the brain. Mood control can be considered as the concerted result of a large number of endogenous and exogenous signals affecting the networks subserving mood via actions on WT and VT.

2 Humoral Modulation of Wiring Transmission

Firstly, it must be considered that there exists a blood-brain barrier, so that the central nervous system will not receive a number of peripheral signals. However, there also exist chemical and physical "windows" through which the brain receives and delivers lipophobic messages. Chemical windows consist of facilitated transport, active transports etc. The physical windows are represented by brain areas devoid of the blood-brain barrier, such as the area postrema, the median eminence and the subfornical organ. Also the neuronal inputs represent a part of the physical window (see Fig. 4).

3 Actions of Gonadal Steroids on Wiring Transmission

3.1 General Aspects

By means of autoradiographic and steroid receptor binding techniques nerve cells concentrating steroid sex hormone have been demonstrated in the central nervous system and been found to be concentrated in the limbic forebrain, the medial preoptic area and the hypothalamus, especially the medial part (see Cottingham and

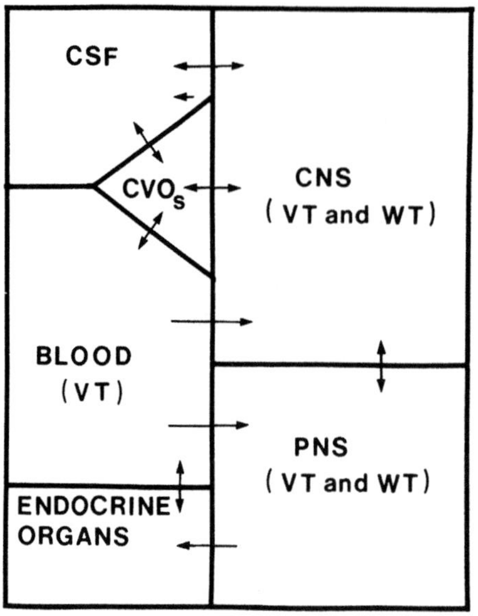

Fig. 4. Schematic illustration of interactions between central nervous system (*CNS*), peripheral nervous system (*PNS*) and endocrine organs via volume transmission (*VT*) and wiring transmission (*WT*). The role of circumventricular organs (*CVOs*) as "physical" windows for blood signals not passing the blood-brain barrier is indicated (e.g., peptide hormones). *CSF*, cerebrospinal fluid

Pfaff 1986; Stumpf 1968; Stumpf and Sar 1975b,c, 1978, 1981; Sar and Stumpf 1973, 1977). Recently, Cintra et al. (1986) have demonstrated, by means of a rat monoclonal antibody directed against the human estrogen receptor, purified from MCF-7 human breast cancer cells, that estrogen receptor immunoreactive (IR) nerve cells exist in the limbic forebrain areas, in the hypothalamus and the preoptic area with the same distribution as the estrogen-accumulating nerve cells (Figs. 5 and 6). Of substantial interest was the observation that the estrogen receptor immunoreactivity was exclusively present within the nuclei and that no translocation of the IR material took place in the cytoplasm following castration. These results strongly support the importance of the genomic actions of estrogens, inducing changes in the decoding of various types of proteins.

By means of combined autoradiography and immunohistochemistry it has been possible to demonstrate the accumulation of steroid hormones in transmitter-identified neurons such as dopamine (DA), noradrenaline (NA), vasopressin, β-endorphin, γ-aminobutyric acid (GABA) and somatotastin nerve cells (see Heritage et al. 1977; Stumpf and Sar 1981; Wuttke et al. 1981). As pointed out

Fig. 6. Camera lucida drawing of distribution of estrogen receptor immunoreactivity present in nuclei of nerve cells of the preoptic area and adjacent regions in a coronal section of the normal male rat brain. *BSTL*, bed nucleus striae terminalis, lateral part; *BSTM*, bed nucleus striae terminalis, medial part; *f*, fornix; *GP*, globus pallidus; *HDB*, nucleus of the horizontal limb of the diagonal band; *ICj*, island of Calleja; *LPO*, lateral preoptic area; *MPN*, median preoptic nucleus; *MPO*, medial preoptic area; *ox*, optic chiasm; *PVHap*, paraventricular nucleus, anterior parvocellular part; *PvPO*, periventricular preoptic nucleus; *sm*, striae medullaris of the thalamus; *SO*, supraoptic hypothalamic nucleus; *Tu*, olfactory tubercle; *VP*, ventral pallidum; *III V*, third ventricle

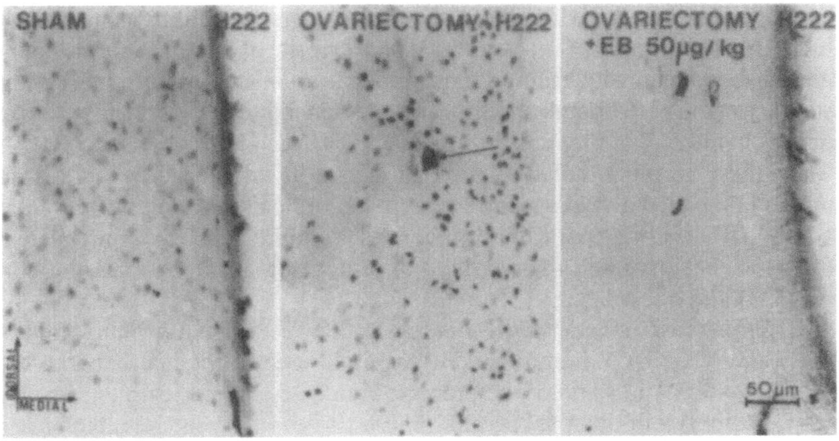

Fig. 5. Estrogen receptor immmunoreactivity in sham-operated, ovariectomized and estrogen benzoate-treated ovariectomized female rats. The figure shows the periventricular preoptic area (see hatched area in Fig. 6) using the indirect biotin avidine immunoperoxidase procedure. The primary antiserum was diluted 1/1000. The orientation is shown in the bottom left corner. The staining of the ependymal cells is nonreproducible and not modified by ovariectomy with or without estradiol benzoate treatment. The *arrow* points to an unspecific artifact. *Arrowheads* indicate specific estrogen receptor immunoreactive nuclei of nerve cells

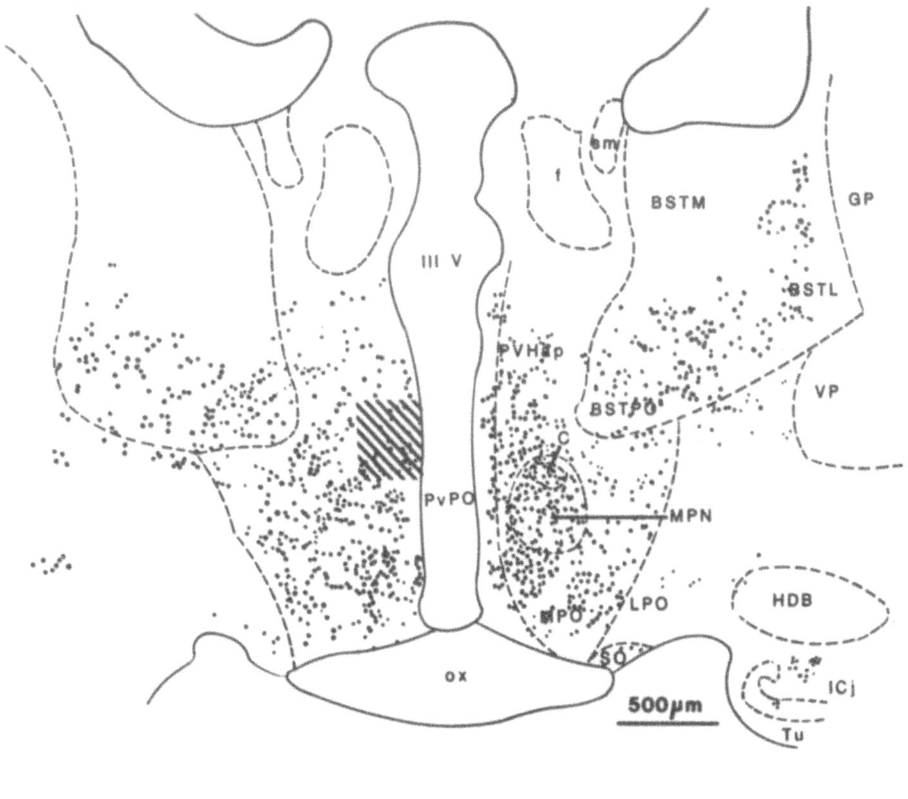

by Cottingham and Pfaff (1986) sex steroid hormone-binding neurons exhibit a high degree of interconnectedness. A network of steroid hormone-binding neurons is created which allows the amplification of steroid hormone actions and stability in the performance of the hormone-dependent network, as well as appropriate channeling of the inputs to this network (see Cottingham and Pfaff 1986). Thus, these hormone-dependent networks can be influenced by sex steroid hormones in the medial hypothalamus, in the preoptic area and in the limbic system, additionally enabling coordination of activity in the networks controlling reproductive behaviours and the secretion of luteinizing hormone-releasing hormone (LHRH) and of prolactin.

It should also be mentioned that regional levels of β-endorphin immunoreactivity and enkephalin immunoreactivity are altered in the neuroendocrine areas of the hypothalamus and of the preoptic area by estrogen treatment (see Dupont et al. 1981). It will be of substantial interest to evaluate the possible presence of estrogen immunoreactivity within these neurons of the mediobasal hypothalamus and of the preoptic area.

In agreement with the importance of genomic actions of estrogen it has been found that estrogens can regulate tyrosine hydroxylase gene transcription in the arcuate nucleus of the rat hypothalamus (Blum et al. 1985). Estrogens can also regulate proopiomelanocortin (POMC) gene expression in the rat hypothalamus. A decrease has been seen in POMC mRNA levels after estrogen treatment which is due at least in part to a decrease in the synthesis of POMC mRNA (see Roberts et al. 1985). As a matter of fact it seems likely that the majority of sex steroid actions on the presynaptic properties of transmitter-identified neurons, such as effects on synthesis mechanisms and release mechanisms for neurotransmitters, e.g. monoamines, and GABA are secondary to primary actions on the decoding of the genome of the estrogen IR neurons of the brain (for review, see McEwen et al. 1981).

3.2 Studies on Presynaptic Features of Monoamine Neurons

Estradiol-17β, progesterone and androgens have all been found to induce discrete changes in DA, NA and adrenaline (A) levels and utilization in the hypothalamus and in the preoptic area in male and female rats in various types of endocrine states (see Fuxe et al. 1981a; Andersson et al. 1981; Wuttke et al. 1981; Löfström and Beckström 1981; Fuxe et al. 1977a). Estradiol-17β appears to produce its central inhibitory feedback action on LHRH secretion at least in part by a direct action on the hypothalamus, leading to an activation of the lateral tuberoinfundibular DA pathway, which in turn, by axoaxonic influence and/or by effects on tanycytes, may inhibit the secretion of LHRH from the median eminence (see Quimet et al. 1984). This action of estradiol-17β may be mediated partly by an increase in the secretion of prolactin, which has the ability of increasing DA turnover in the lateral palisade zone of the median eminence, where the LHRH and DA terminals interact (see Fuxe et al. 1984b). However, it must also be considered that estrogen receptors are probably located in many of the DA nerve cell bodies of the mediobasal hypothalamus (see Heritage et al. 1977) and that estrogen treatment of hypophysectomized and castrated animals leads to a marked increase of DA utilization in both the lateral and the medial palisade zones of the median eminence (Fuxe et al. 1981 a).

The central facilitatory feedback action of estradiol-17β on LHRH secretion instead appears to involve an increase of NA utilization in the medial preoptic area (see Fuxe et al. 1977 a; Löfström and Beckström 1981; Wuttke et al. 1981). It has been postulated that this action of estrogen involves an inhibition of GABA interneurons in the medial preoptic area (see Wuttke et al. 1981). Thus, muscimol, when given intraventricularly, can reduce NA turnover in the anterior hypothalamus (see Fuxe et al. 1979 a). Intraventricular injection of GABA also reduces NA turnover in the medial preoptic area (see Wuttke et al. 1981). Thus, the central facilitatory feedback action of estrogen appears to involve increased NA release in the medial preoptic area mediated at least partly via an action on estrogen-sensitive GABAergic neurons in this region, leading to a loss of presynaptic inhibition of NA release (see Wuttke et al. 1981).

Evidence has also been obtained that androgenic steroids can produce discrete change of DA and NA utilization in the hypothalamus and preoptic area by activation of androgenic steroid receptors (see Andersson et al. 1981). Thus, evidence was obtained that the activity in noradrenergic mechanisms of the preoptic area can be turned off by the androgenic steroid R1881, while the inhibitory dopaminergic mechanism in the median eminence is turned on by this agent. As reported above, similar results have been obtained following treatment of castrated female rats with estrogens such that estrogen produces its central inhibitory feedback action on LHRH secretion.

The above results taken together indicate that sex steroids, by means of changes in genetic transcription via their nuclear actions, alter the formation of regulatory proteins controlling the chemical transmission in the steroid target cells. By axoaxonic contacts the steroid target cells in the local circuits of various regions of hypothalamus and preoptic area will influence the various DA and NA nerve terminal systems in a discrete way, as observed in the above-mentioned experiments (see Fuxe et al. 1979 b).

However, estrogens not only influence catecholamine turnover in regions where estrogen IR nerve cells exist but also influence DA utilization in parts of the forebrain where few estrogen IR nerve cells are found. In the hypophysectomized and castrated female rat it has, for example, been found that estradiol-17β can markedly reduce DA utilization in various parts of the striatum and of the nucleus accumbens. These results may be induced by activation of estrogen receptors of neural groups within the preoptic area and the hypothalamus which project to the ascending DA neurons and thus indirectly regulate DA utilization in the mesostriatal and mesolimbic systems. These results clearly indicate that estrogens can also modulate motor functions and mental activities such as mood (see Fuxe et al. 1981 b). When discussing the actions of estrogens on the striatal mechanisms, behavioural and neurochemical studies have also indicated antidopaminergic actions, which may be mediated at least in part via the pituitary gland through increases in the secretion of prolactin (see Euvrard and Boissier 1981). Estrogens are known to alleviate extrapyramidal symptoms such as hyperkinesias in patients receiving neuroleptic drugs (see Bedard et al. 1981).

Of special interest are our recent observations that steroid hormones can influence the coexistence of peptides and monoamines (Hökfelt et al. 1980) in the monoamine neurons (Fuxe et al. 1985 a), probably mainly via an influence on the synthesis of the peptide comodulator (see Sect. 4.2. paragraph on glucocorticoid receptors).

3.3 Studies on Monoamine Receptor Mechanisms

In a number of papers, estrogen, progesterone and other sex steroids have been found to modulate the binding characteristics of central α- and β-adrenergic receptors (Fuxe et al. 1979 b, c; Wilkinson 1978; Wilkinson et al. 1979 a, b, 1981). Our results have demonstrated that the α- and β-adrenergic receptors are sensitive to combined treatment with estrogen and progesterone resulting in the induction of sexual behaviour. The changes induced in the binding characteristics are ligand and region specific. In contrast, the gonadal steroids have been found to exert little effect on opiate receptor binding parameters (see Wilkinson et al. 1981; for review see McEwen et al 1970). Estrogen treatment has also been found to influence the 5-HT1 receptors and D2 receptors in the striatum. The effects of estrogen on striatal D2 receptors are complex. It was found at an early stage that estrogen can produce an increase in the number of D2 receptors in striatal membranes (see Bedard et al. 1981; Fuxe et al. 1979 c). These biochemical signs of DA receptor hypersensitivity may be related to the conversion of estrogens to catecholestrogens and/or to reduced DA release (Fuxe et al. 1981 a; Gordon 1985). However, the acute actions of estradiol on DA receptors produce a direct desensitization or an uncoupling of the receptor-effector mechanisms, characterized by a decrease in the proportion of D2 receptors in the high-affinity agonist state (Gordon 1985).

It seems likely that most of the effects of estrogen and progesterone treatment on monoamine receptors in steroid receptor-rich areas of the brain are produced via actions on the estrogen and progesterone receptors present in these areas. The activation of the steroid receptors may in turn produce changes in genetic transcription which can affect various aspects of biochemical signals regulating monoamine and other types of receptors (see Agnati et al. 1981; McEwen et al. 1981). This hyothesis is supported by the ability of in vivo estrogen administration to produce sustained and delayed increases in the receptor density values of transmitter receptors in regions where estrogen receptor immunoreactivity exists. Nongenomic actions of sex steroids are probably also involved, however, since changes in the binding characteristics of monoamine receptors can also be demonstrated in membrane preparations upon in vitro addition of the steroid. This may be the case in the above-mentioned estrogenic modulation of striatal DA receptor sensitivity. Membrane actions have also been demonstrated in a number of electrophysiological and biochemical experiments (see Baulieu 1981; Moss and Dudley 1985). Thus, within seconds estradiol-17α-butyric acid inhibits the firing rate of nerve cells applied directly to the membrane of these cells (Carette et al. 1979). These estrogen derivatives cannot cross the cell membrane. Furthermore, progesterone on the surface of maturing *Xenopus laevis* oocytes promotes oocyte

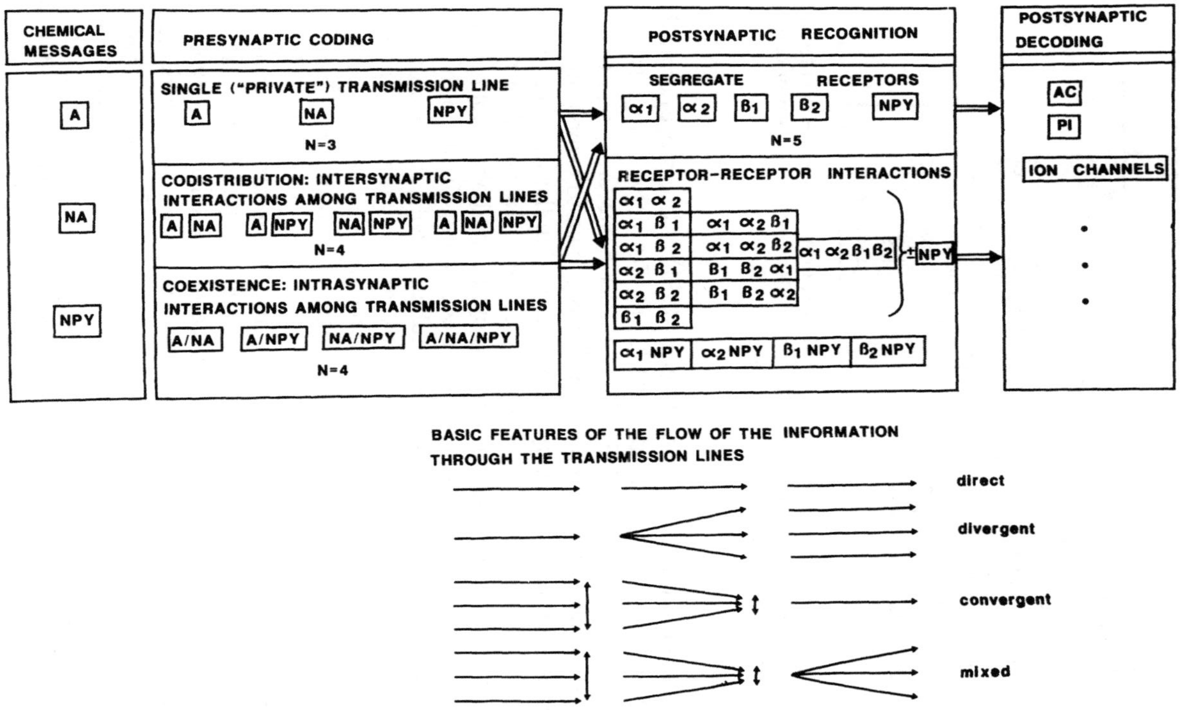

Fig. 7. Illustration of how a peptide (neuropeptide Y, *NPY*) and noradrenaline (*NA*) and adrenaline (*A*) can interact in synaptic transmission via either intersynaptic (codistribution) or intrasynaptic (coexistence) interaction. In both cases a large number of different types of receptor-receptor interactions are possible. These mechanisms allow a convergence of information flow at both the presynaptic and the postsynaptic level, which subsequently may converge into an action on one type of biological effector or diverge to several biological effectors

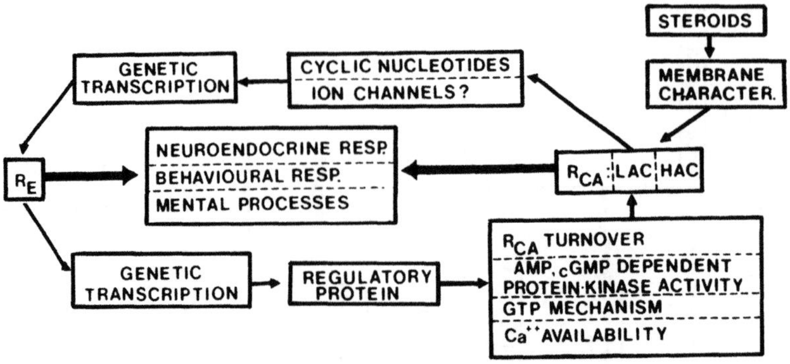

Fig. 8. Schematic illustration of the possible interrelationships between catecholamine (R_{CA}) and estrogen (R_E) receptors in target nerve cells. *LAC*, low-affinity component; *HAC*, high-affinity component

meiosis without entering the cell (see Baulieu 1981). Also, progesterone has recently been found in vitro to activate LHRH and DA nerve terminals by a nongenomic mechanism (Ramirez 1985). These effects, however, appear to be mediated via the metabolite pregnenolone, which has shown to exhibit highly potent activation of DA terminals in the hypothalamus and in the striatum. It was suggested that membrane actions of steroids may lead to changes in the transduction mechanisms, resulting in changes in adenylate cyclase and phospholipase C activity. In this way a chain of events may be started, leading finally to the release of LHRH and DA demonstrated in vitro on incubation with progesterone or pregnenolone. Thus, it seems likely that sex steroids can regulate membrane excitability, possibly via a special membrane site, since the action of estrogen on the membrane appears to be antagonized by antiestrogens (see Moss and Dudley 1985). An effect of membrane fluidity should, however, also be considered. These membrane changes could also affect the interactions among intramembrane macromolecular complexes, such as ion pumps, ion channels, etc. In particular, it seems likely that the membrane actions of gonodal steroids could also modulate the receptor-receptor interactions (see the paragraph below on glucocorticoid receptors). This steroid action could thus have a role on information handling by neural networks. In fact, as discussed above, intramembrane receptor-receptor interactions represent an important integrative mechanism in synaptic transmission allowing for a divergence and convergence of information flow in the synapses (see Fig. 7). They also increase the number of information signals which can be produced.

It should also be mentioned that gonadal steroids such as estrogen influence the transduction mechanisms; this is illustrated by a reduction of histamine-stimulated adenylate cyclase following 7 days of treatment with estrogen and reduction in isoproterenol-stimulated adenylate cyclase upon chronic estrogen treatment (see McEwen et al. 1981).

It must be emphasized, however, that not only can VT influence WT, but WT can also influence VT, as illustrated in Fig. 8. Thus, indications exist that catecholamine and serotonin (5-HT) receptors also can regulate steroid receptor synthesis (see Ginsburg et al. 1977; Kitayama et al., unpublished observations). Thus, the networks of the brain regulate their own sensitivity to hormonal signals by controlling the amount of steroid receptors present in the nerve cells. Of special interest in this regard is the fact that unsaturated fatty acids can affect sex steroid hormone receptors in the brain (see Kato 1985). Thus, transmitter receptors such as catecholamine receptors may, via regulation of phospholipase C, control the formation of unsaturated fatty acids, such as arachidonic acid, which when reaching cytoplasmatic steroidal receptors inhibit the binding capability of these receptors (see Kato 1985).

4 Actions of Glucocorticoids on Wiring Transmission

4.1 General Aspects

The existence of specific receptors for glucocorticoid steroids in the central nervous system was first provided in biochemical studies by McEwen et al. (1969, 1970). The nuclear concentration of ^{3}H-corticosterone, as seen using autoradiography , indicated that the receptors for glucocorticoid hormones are mainly nuclear in location. They were found principally within hippocampal formation , the septal area and the amygdaloid cortex. Corticosterone target neurons were also observed in the thalamus, but not within the hypothalamus, and in the preoptic area (see Stumpf and Sar, 1975a, 1981; see McEwen 1982; see Ganten and Pfaff 1982). However, by means of monoclonal antibodies against the rat liver glucocorticoid receptor (GR) in combination with the indirect immunoperoxidase technique, we have been able to demonstrate GR IR nerve and glial cells all over the brain and spinal cord of the male rat (Fig. 9) (Fuxe et al. 1985b–d, 1987; Agnati et al. 1985). The GR immunoreactivity in the nerve cells was found mainly in the nucleus, but also in the cytoplasm (Figs. 10 and 11). Following 2–4 day adrenalectomy the nuclear GR immunoreactivity of the nerve cells was found to disappear but the weak cytoplasmatic GR immunoreactivity remained. Following 4 h of treatment with corticosterone the GR immunoreactivity reappeared in the cell nuclei (Fig. 11). GR IR glial cells were mainly found in the white matter (Fig. 9), where they formed bands of cubic-like structures between the fiber bundles, probably mainly representing glial cells of the oligodendroglia type.

High densities of strongly GR IR nerve cell nuclei were present in the parvocellular part of the paraventricular hypothalamic nucleus rich in corticotropin-releasing factor (CRF) IR nerve cell bodies (Fig. 12), in the arcuate and periarcuate nucleus and in the CA1 and CA2 subregions of the hippocampal formation (Fig. 13). High densities of moderately GR IR nerve cells were present all over the cortical hemispheres and in the vast majority of the thalamic nuclei (Fig. 14–18). High densities of weakly to moderately GR IR nerve cells were observed over the entire nucleus caudatus putamen (Fig. 17, 19). These results illustrate the widespread distribution of GR IR nerve cells in the entire telencephalon and diencephalon. It should be mentioned that the majority of the hypothalamic and preoptic nuclei and several septal nuclei contained high densities of moderately GR IR nerve cells.

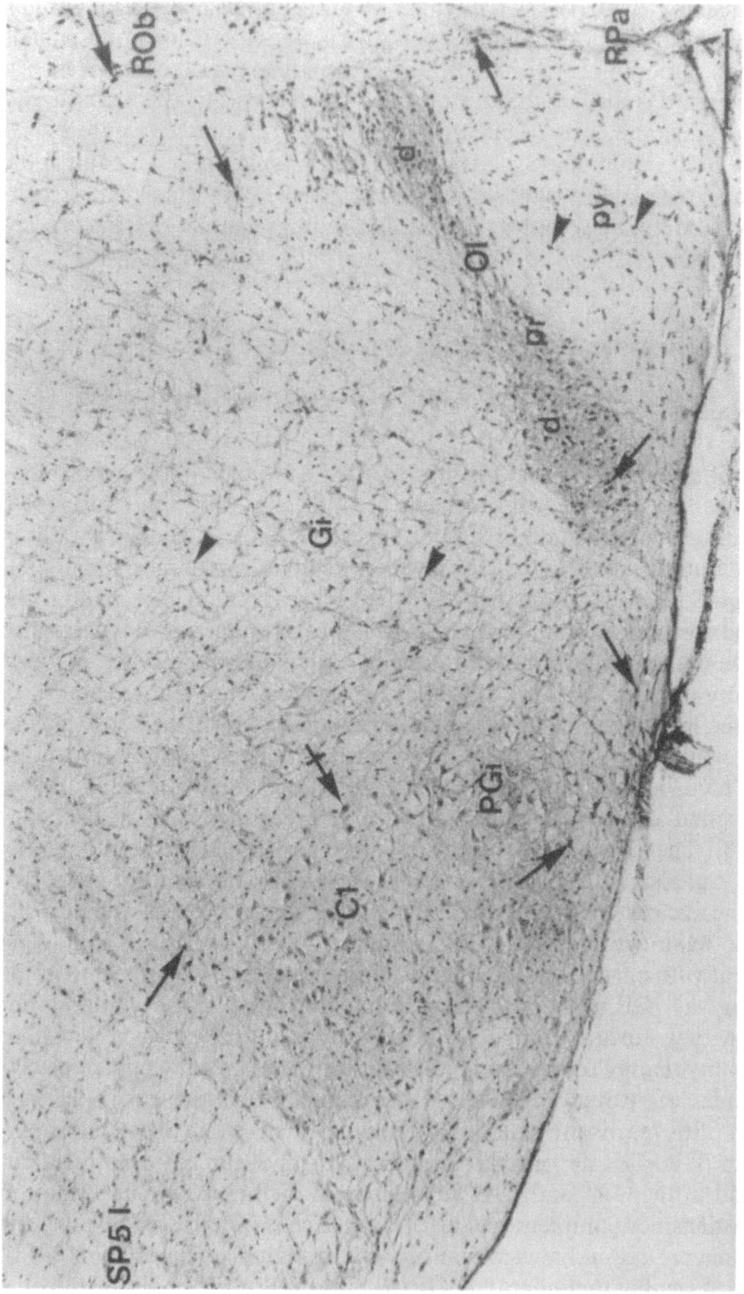

Fig. 9. Distribution of glucocorticoid receptor (GR) immunoreactivity in a coronal cryotome section (20 μm) of the ventral medulla oblongata (2 mm rostral to the obex). Some strongly GR immunoreactive (IR) nerve cells in the *C1* area, raphe obscurus (*ROb*), nucleus raphe pallidus (*RPa*) and within the dorsal subnuclei (*d*) of the inferior olive (*OI*) are indicated by *arrows*. The *arrow with a bar* shows a nerve cell where the cytoplasm is also GR IR. *Arrowheads* point to glial cells in the pyramidal tract (*py*) and in the nucleus reticularis gigantocellularis (*Gi*). SP5I-nucleus of the spinal tract of the trigeminal nerve, interpositus part. *Bar*, 250 μm

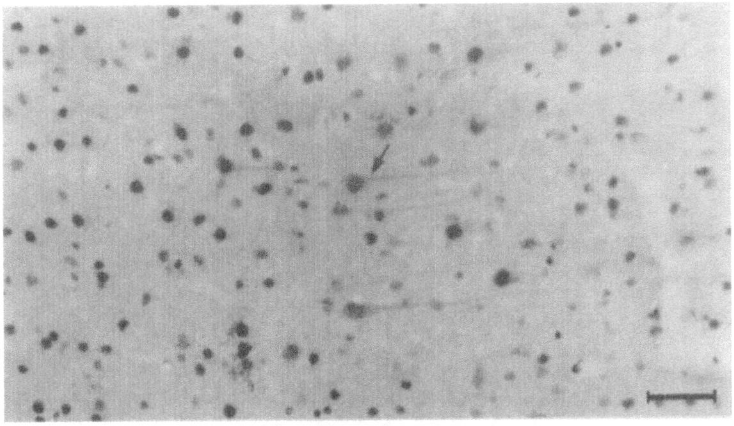

Fig. 10. Cellular localization of glucocorticoid receptor (GR) immunoreactivity (GR IR) in nerve cells of the frontoparietal cortex of the male rat. In the pyramidal nerve cells the GR immunoreactivity is seen not only in the nucleus but also in the pericaryon and in the apical dendrite. *Bar*, 200 µm

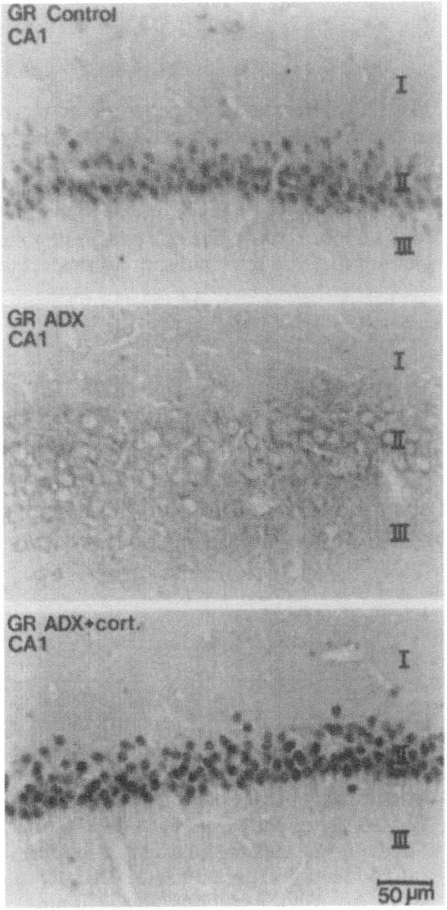

Fig. 11. Localization of glucocorticoid receptor (*GR*) immunoreactivity in pyramidal nerve cell nuclei and surrounding cytoplasm in the hippocampal subregion CA1 and its modulation by adrenalectomy with or without subsequent corticosterone treatment (4 h, 10 mg/kg, i.p.). Transverse sections. After adrenalectomy (*ADX*), the GR immunoreactivity is observed exclusively in the pericarya of the pyramidal nerve cells. After corticosterone (*cort*) treatment GR immunoreactivity even stronger than that present in the control animals is found in the nerve cell nuclei. König-Klippel level A4100 µm

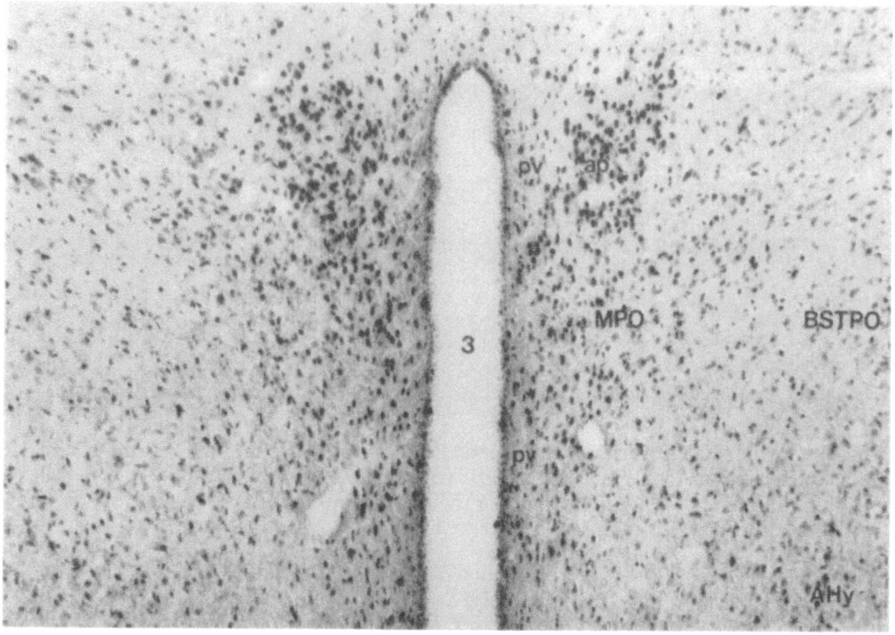

Fig. 12. Distribution of glucocorticoid receptor (GR) immunoreactivity in the anterior and dorsal periventricular part of the of the hypothalamus of the male rat. Densely packed strongly GR immunoreactive (IR) nerve cell nuclei are found in the anterior parvocellular nucleus of the paraventricular hypothalamic nucleus (*ap*) and in the periventricular hypothalamic nucleus (*pv*). Weakly to moderately GR IR nerve cell nuclei are found in the anterior hypothalamic nucleus (*AHy*), the medial preoptic nucleus (*MPO*) and the bed nucleus of the stria terminalis, preoptic part (*BSTPO*). *3*, third ventricle

In agreement with our observations showing a widespread distribution of GR IR neurons in the central nervous system, an in vitro quantitative autoradiographic analysis of adrenal steroid binding sites showed a widespread distribution of glucocorticoid binding sites in the rat central nervous system. The highest concentrations were found in the dentate gyrus, the lateral septum, the nucleus tractus solitarius, the nucleus paraventricularis and the amygdaloid complex (see de Kloet 1985; Rostene et al. 1985).

Fig. 14. Distribution of glucocorticoid receptor (GR) immunoreactivity in the frontoparietal cortex. The highest densities of GR immunoreactive nuclei are observed in layers 2, 3 and 6 of the cortex and in the nucleus caudatus putamen. DAB-nickel combination was used. Level: 1.3 mm behind Bregma. *FrPaM*, frontoparietal cortex, motor area; *FrPaSS*, frontoparietal cortex, somatosensory area; *cc*, corpus callosum; *ec*, external capsule; *cg*, cingulum; *LV*, lateral ventricle; *fi*, fimbria of hippocampus; *CPu*, caudate putamen; *AD*, anterodorsal thalamic nucleus; *DG*, dentate gyrus; *PCg*, posterior cingulate cortex; *CA3*, field CA3 of hippocampus. *Bar*, 250 μm

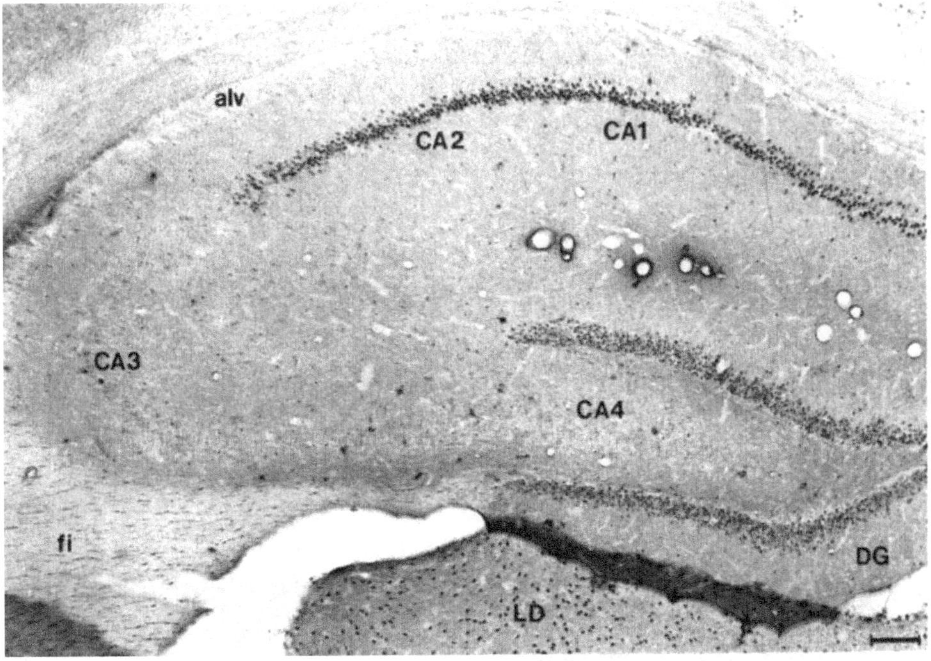

Fig. 13. Distribution of glucocorticoid receptor (GR) immunoreactivity by means of immunocytochemistry in the hippocampus of the male rat. Note the lack of GR immunoreactivity in the CA3 and CA4 areas. DAB-nickel combination was used. Approximate level: 2.8 mm behind Bregma. *CA1–CA4*, fields of hippocampus; *DG*, dentate gyrus; *alv*, alveus of hippocampus; *LD*; laterodorsal thalamic nucleus; *fi*, fimbria of hippocampus. *Bar*, 250 μm

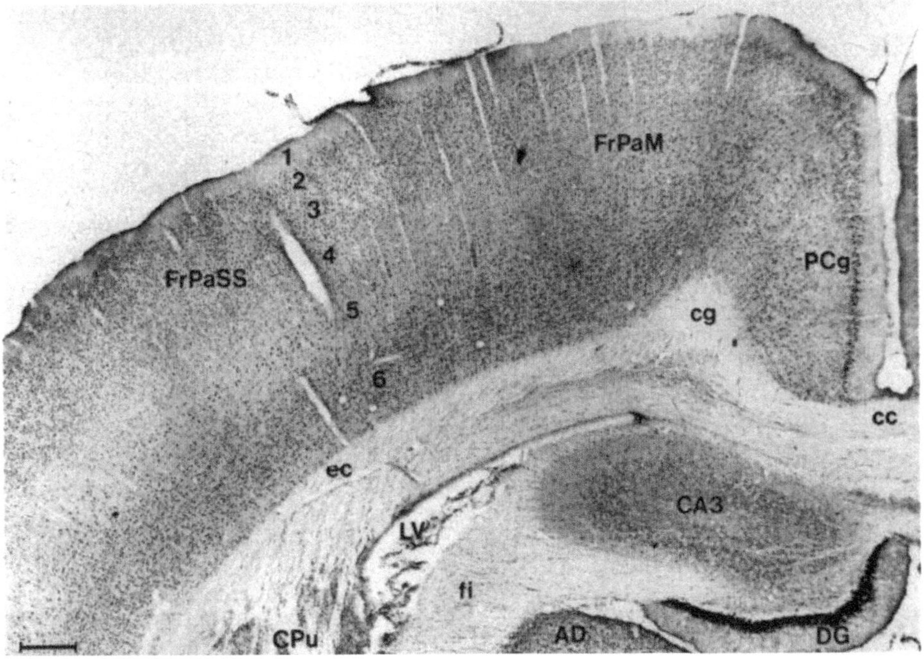

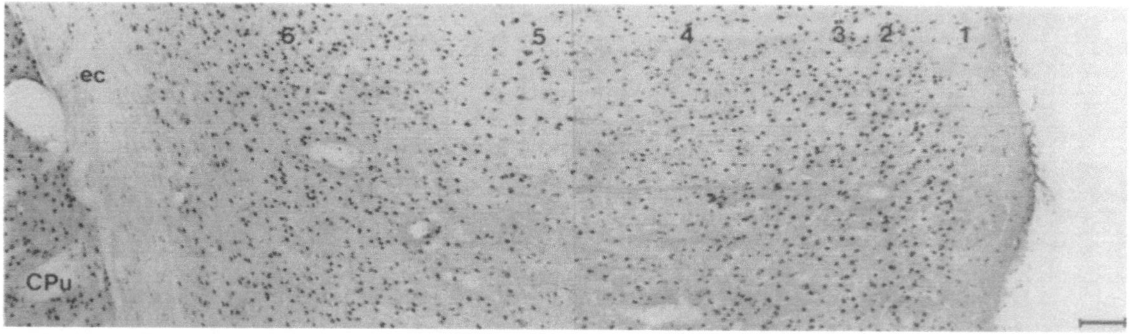

Fig. 15. Visualization of glucocorticoid receptor (GR) immunoreactive nerve cells in the frontoparietal cortex, somatosensory area. The incubation with the monoclonal anti-GR antibody was performed without Triton X-100. DAB-nickel combination was used. Approximate level: 1.3 mm below Bregma. *CPu*, caudate putamen; *ec*, external capsule. *Bar*, 200 μm

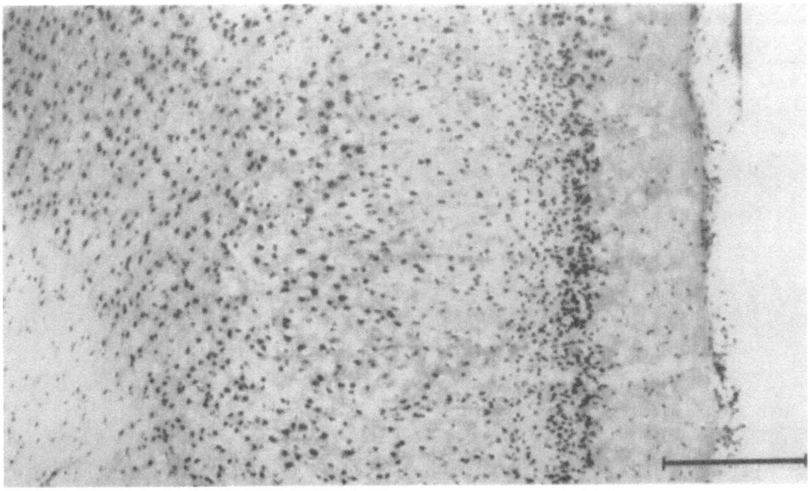

Fig. 16. Distribution of glucocorticoid receptor immmunoreactive (GR IR) nerve cell nuclei in the anterior cingulate cortex. High densities of strongly GR IR nerve cells are found especially in layers 2 and 3. *Bar*, 250 μm

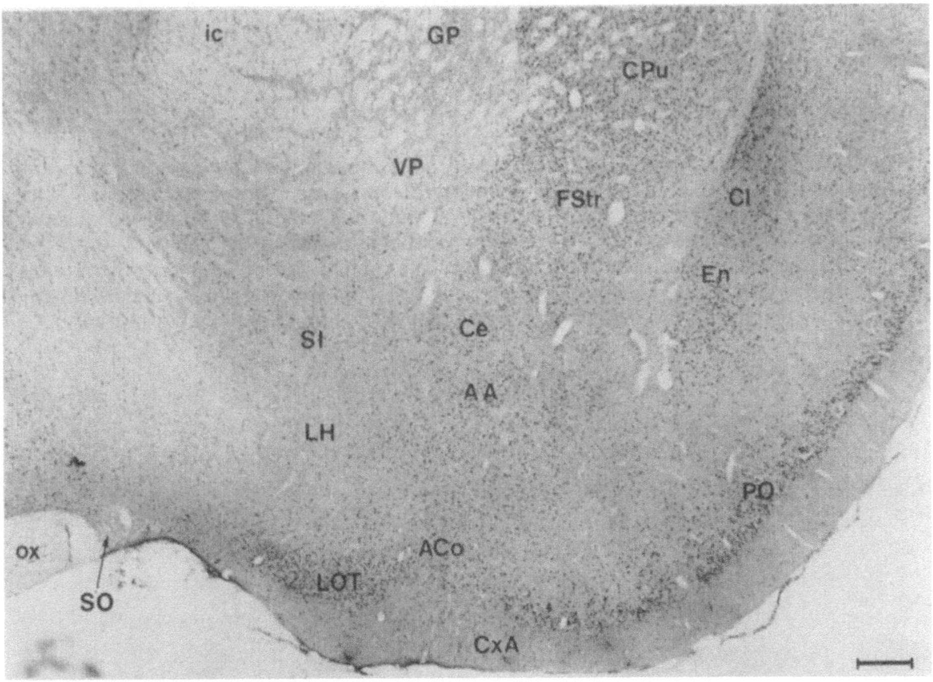

Fig. 17. Distribution of glucocorticoid (GR) immunoreactivity in the hypothalamus, dorsal striatum and piriform cortex. Note the low number of weak GR immunoreactive (IR) profiles in the globus pallidus, ventral pallidum and substantia innominata. High densities of strongly GR IR profiles are localized in the claustrum, in the endopiriform nucleus and in the olfactory and piriform cortex. Approximate level: 1.3 mm below bregma. DAB-nickel combination was used. *ic*, internal capsule; *ox*, optic chiasm; *LH*, lateral hypothalamic area; *SO*, supraoptic hypothalamic nucleus; *GP*, global pallidus; *CPu*, caudate putamen; *FStr*, fundus striati; *VP*, ventral pallidum; *SI*, substantia innominata; *AA*, anterior amygdaloid area; *CxA*, cortex-amygdala transition zone; *Cl*, claustrum; *En*, endopiriform nucleus; *LOT*, nucleus of the lateral olfactory tract; *PO*, primary olfactory cortex; *Ce* central amygdaloid nucleus; *Bar*, 250 μm

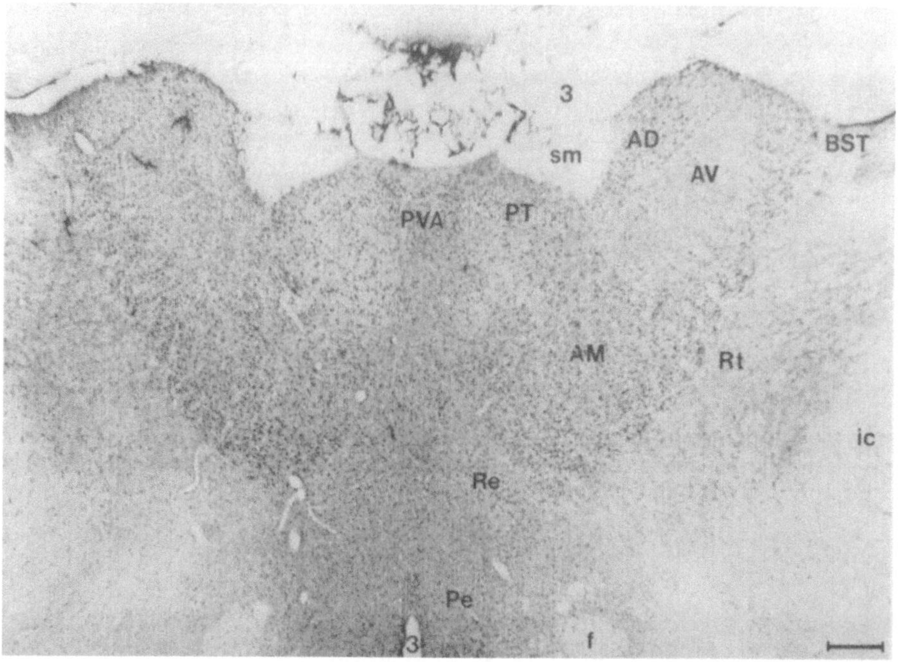

Fig. 18. Distribution of glucocorticoid receptor immunoreactivity in nerve cells of various anterior thalamic nuclei. DAB-nickel combination was used. Approximate level: 1.3 mm behind bregma. *3*, third ventricle; *sm*, striae medullaris thalamus; *PVA*, paraventricular thalamic nucleus, anterior part; *PT*, paratenial thalamic nucleus; *AD*, anterodorsal thalamic nucleus; *AV*, anteroventral thalamic nucleus; *BST*, bed nucleus striae terminalis; *AM*, anteromedial thalamic nucleus; *Rt*, reticular thalamic nucleus; *Re*, reuniens thalamic nucleus; *ic*, internal capsule; *f*, fornix; *Pe*, periventricular hypothalamic nucleus. *Bar*, 250 μm

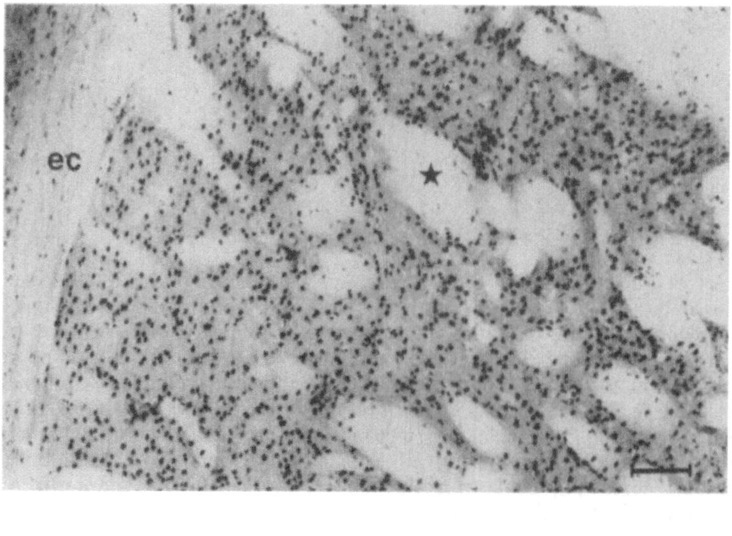

Table 3. Sites of coexistence of GR IR, monoaminergic and peptidergic neurons

Neurons	Approximate degree of GR coexistence (%)
Dopamine	
A8–A10	50
A11	0
A12–A14	100
A13	75
A15	0
Noradrenaline	
A1–A7	100
Adrenaline	
C1–C3	100
Dorsal strip	50
5-Hydroxytryptamine	
B1–B9	100
Corticotropin – releasing factor IR	
Paraventricular hypothalamic nucleus, parvocellular part	100
Paraventricular hypothalamic nucleus, magnocellular part	0
Bed nucleus of striae terminalis	50
Central amygdaloid nucleus	50
Growth hormone – releasing factor IR	
Arcuate nucleus	100
Somatostatin IR	
Periventricular hypothalamic nucleus	50
Neuropeptide Y IR	
C1–C3 area	100
A1 area	100
Arcuate nucleus	100
α-Melanocyte – stimulating hormone IR	
Arcuate nucleus	50
Perifornical nucleus	50
Central amygdaloid nucleus	50
Cholecystokinin IR	
Paraventricular hypothalamic nucleus	50
Hippocampus	100
Periventricular hypothalamic nucleus	75
β-Endorphin IR	
Arcuate nucleus	100

Nomenclature of monoaminergic neurons according to Dahlström and Fuxe (1964)
GR, glucocorticoid receptor; IR, immunoreactive

←

Fig. 19. High densities of glucocorticoid receptor immunoreactive nerve cell nuclei are found in the nucleus caudatus putamen DAB-nickel combination was used. *Star*, fibrae capsulae internae; *ec*, external capsule. *Bar*, 200 μm

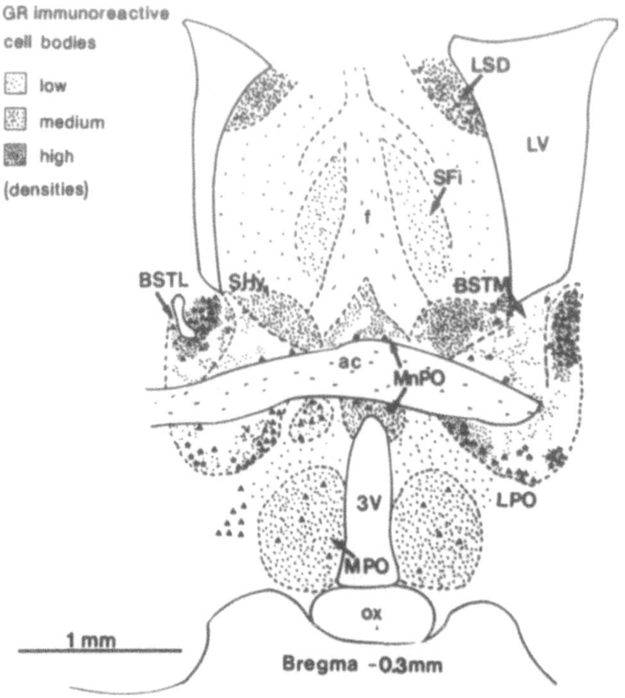

Fig. 20. Camera lucida drawing of the distribution of glucocorticoid receptor (GR) and corticotropin-releasing factor (CRF) immunoreactivities in the septal and preoptic area of the male rat, based on two-colour immunocytochemistry (see Härfstrand et al. 1986). Double-labelled nerve cells are represented by *triangles*. *Stars* indicate CRF immunoreactive cells without any demonstratable GR immunoreactivity. *ac*, anterior commisure; *ox*, optic chiasm; *LV*, lateral ventricle; *LSD*, laterodorsal part of the septal area; *SFi*, nucleus fimbrialis septalis; *SHy*, nucleus septohypothalamicus; *MnPo*, nucleus preopticus medianus; *BSTM*, *BSTL*, medial and lateral components of the bed nucleus of striae terminalis; *LPO*, lateral preoptic nucleus; *MPO*, medial preoptic nucleus

By means of two-color immunocytochemistry we have demonstrated that the vast majority of the NA, A and 5-HT neurons of the lower brain stem contain strong GR immunoreactivity (see Fuxe et al. 1985b; Härfstrand et al. 1986). These results indicate that glucocorticoids can in part control brain function via modulation of the synthesis and release of NA, A and 5-HT and their respective peptide comodulators. In view of the known function of NA, A and 5-HT pathways it is now possible to better understand how glucocorticoids can regulate cardiovascular and neuroendocrine function, the sleep-wakefulness cycle, attention and mood in man.

The various DA cell groups of the brain showed variability with regard to the number of DA nerve cell bodies positive for glucocorticoid receptors. Approximately 50% of the DA cells of groups A8, A9 and A10 in the ventral midbrain contain a weak to moderate degree of GR immunoreactivity. As seen in Table 3, the arcuate DA cell bodies all contain GR immunoreactivity, while the A11 DA

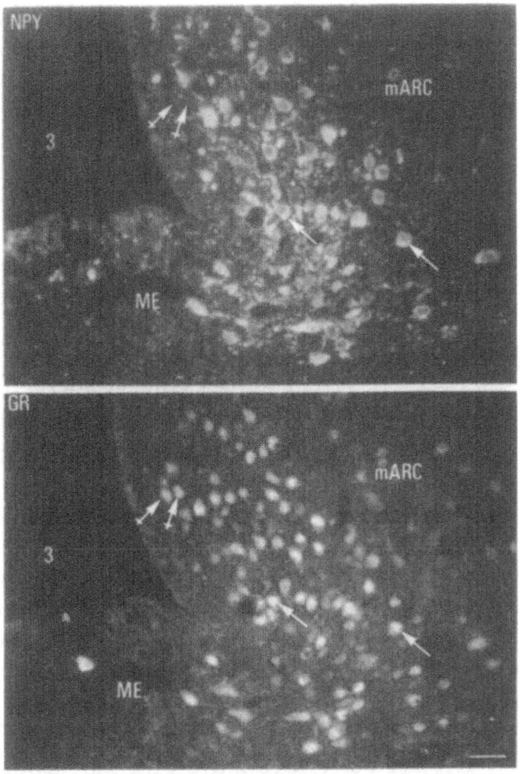

Fig. 21. Demonstration of nuclear glucocorticoid receptor immunoreactivity by means of FITC fluorophor in all neuropeptide Y (NPY) immunoreactive nerve cells of the medial parvocellular part of the arcuate nucleus (*mARC*). NPY immunoreactivity is located in the cytoplasm and demonstrated by rhodamine fluorophor. Two-colour immunofluorescence methodology. *Arrows* point to some of the double-labelled cells. *Arrows with bars* indicate glucocorticoid receptor immunoreactive cells lacking NPY immunoreactivity. *ME*, median eminence; *3*, third ventricle. *Bar*, 100 µm

cell group of the posterior hypothalamus showed no GR immunoreactivity in any of its cell bodies (see Härfstrand et al. 1986).

As seen in Table 3, all the CRF IR cells of the paraventricular hypothalamic nucleus projecting into the median eminence, as well as all the arcuate growth hormone-releasing factor (GRF) IR neurons projecting into the median eminence, contain GR immunoreactivity (Cintra et al. 1987). Many of the CRF IR cells of the preoptic nuclei and of the bed nucleus of the nucleus striae terminalis are also GR immunoreactive (Fig. 20). Thus, glucocorticoids can directly control the CRF and GRF neurons by an action at the nuclear level, probably representing the mechanisms underlying the central feedback action of glucocorticoids on GRF and CRF synthesis leading to an inhibition of GRF and CRF secretion (the delayed feedback). Also the neuropeptide Y (NPY) neurons of the parvocellular part of the arcuate nucleus all contain GR immunoreactivity (Fig. 21). This GR

IR is as strong as the one demonstrated in the CRF IR neurons. About 50% of the α-melanocyte-stimulating hormone (α-MSH) IR neurons of the arcuate nucleus and of the perifornical area exhibited GR immunoreactivity. Furthermore, all the β-endorphin IR neurons of the arcuate nucleus were GR immunoreactive (Table 3). Moreover it was discovered that α-MSH immunoreactivity exists in cortical pyramidal cells, which are also GR immunoreactive.

4.2 Morphometric and Microdensitometric Analysis of GR Immunoreactivity in the Central Nervous System

These analyses have been carried out using a computerized system for image analysis (IBAS II, Zeiss Kontron Munich, FRG). For details on the methodology, see Agnati et al. (1984b, 1985). A semiquantitative method to assess the relative amount of antigen has been developed: discrimination is performed at a level capable of excluding the background and the field area, (FAC)o is measured. The same procedure is repeated at different levels of discrimination allowing the selection of higher and higher grey tone values. After each discrimination procedure the corresponding field area, (FAC)i, is measured. Thus, the percent rations Yi = (FAC)i/(FAC)o can be calculated. By considering the plot of the Yi values as a function of the respective levels of discrimination, a curve can be obtained (Fig. 22) which expresses the relative content of antigen per area in the sampled field. The ED_{25}, ED_{50} and ED_{75} of the curve represent indices of low, medium and high content of antigen, respectively (see Zoli et al. 1986).

It is also possible to convert the exponential decay of the curve into a straight line. This linear transformation is useful, since the slope of the straight line can be used as a relative quantitative index of antigen content, which is more sensitive than the ED values (see below).

One result from the morphometric analysis is demonstrated in Fig. 23, which shows a density map of the GR IR neurons within the area of the paraventricular hypothalamic nucleus. The number of GR IR nerve cells is given per unit of square, in relation to the various subnuclei of this region. It is seen that the largest amounts of GR IR neurons are observed within the parvocellular part (FP). High densities are also found within the periventricular hypothalamic nucleus. Scattered GR IR cells are found within the magnocellular part, and none are found at this rostrocaudal level within the dorsal parvocellular subnuclei. A density map illustrating the coexistence and distribution of the tyrosine hydroxylase (TH) and GR immunoreactivity in the arcuate nucleus is shown in Fig. 24.

A microdensitometric analysis has been performed in a number of regions of the telencephalon and diencephalon, namely the caudate putamen, nucleus amygdaloideus medialis, somatosensory frontoparietal cortex, periventricular hypothalamic nucleus and the parvocellular part of the paraventricular hypothalamic nucleus. As seen in Fig. 22, the ED_{50} index shows similar amounts of GR immunoreactivity in the striatum, the amygdaloid cortex and the neocortex. All the indices are higher for the parvocellular part of the paraventricular hypothalamic nucleus, giving evidence for higher amounts of immunoreactivity in this region than in the other regions analysed. It should be noticed, however, that the

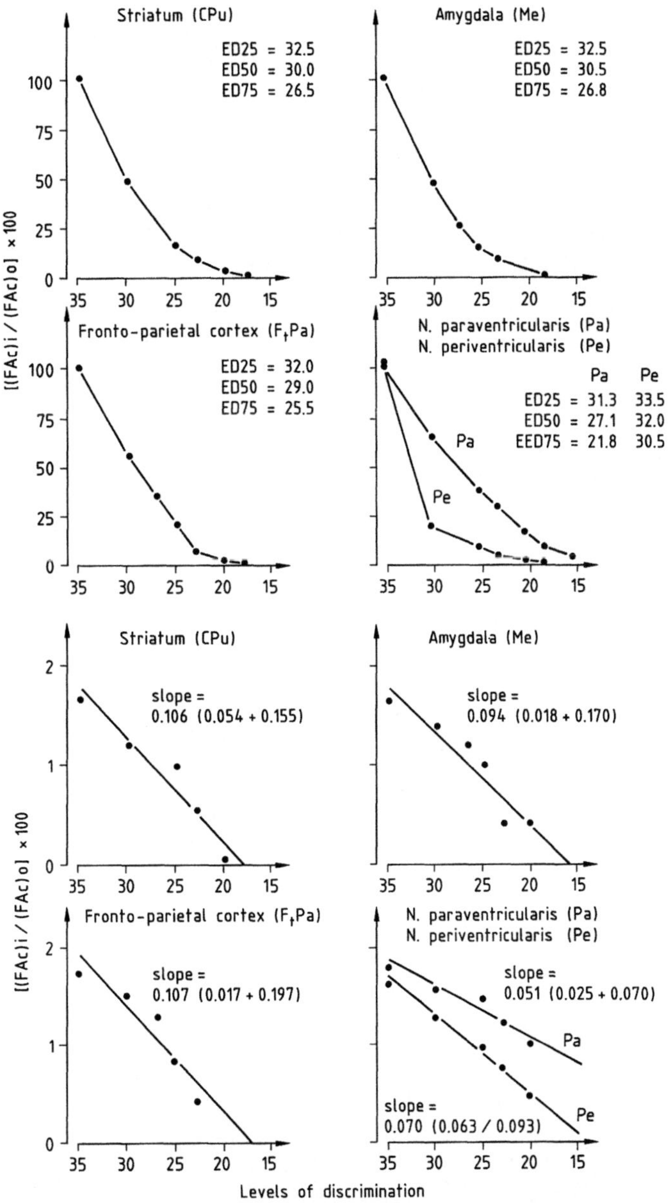

Fig. 22. Schematic representation of the procedure introduced in immunocytological studies to obtain a relative quantitative estimation of the antigen content. In the figure the effects of subsequent discriminations are illustrated. After each discrimination the field area of positive immunoreactivity, (FAC)i, is measured. The percent ratios Yi = ((FAC)i/(FAC)o) × 100 are calculated and by means of these values the estimation of the antigen content is performed. The Yi values are plotted as a function of the different levels of discrimination. A characterization of the curve is obtained by means of the ED values (*upper panels*). A more precise characterization of the curve is obtained after linear transformation by considering the slope values (*lower panels*), which are given with the respective 95% confidence interval

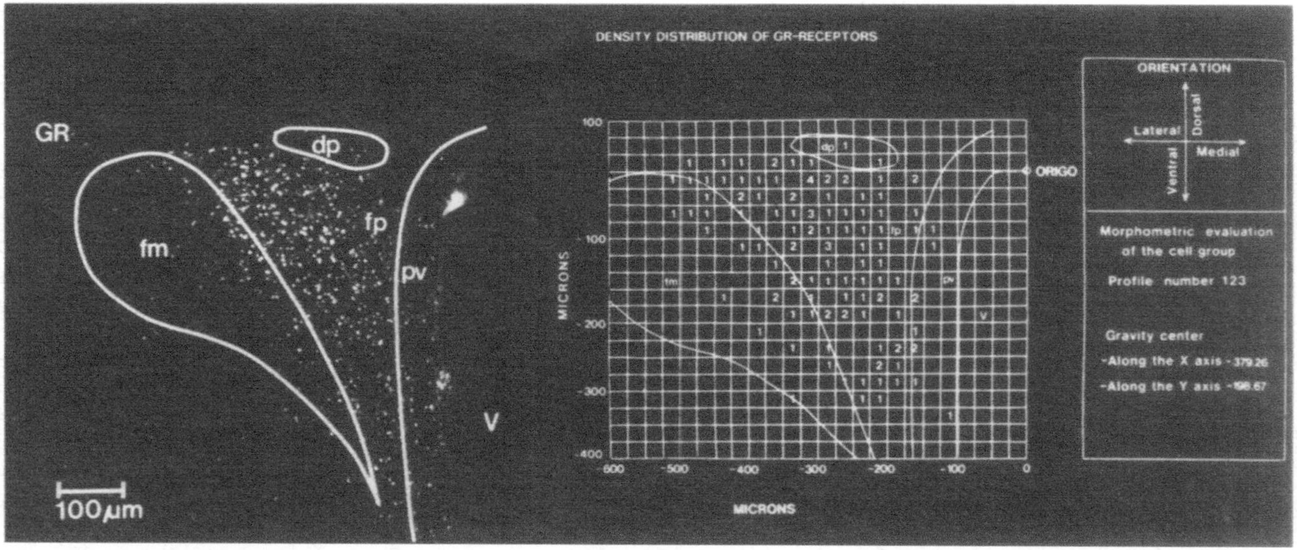

Fig. 23. Density distribution plot of glucocorticoid receptor immunoreactivity profiles in the periventricular and paraventricular hypothalamic nucleus. As origo the most dorsal part of the third ventricle (*v*) was chosen. The analysis was performed by means of a semiautomatic image analysis system within the parvocellular part (*fp*) of the paraventricular hypothalamic nucleus. *fm* magnocellular part of the paraventricular hypothalamic nucleus; *dp* dorsal parvocellular part of the paraventricular hypothalamic nucleus

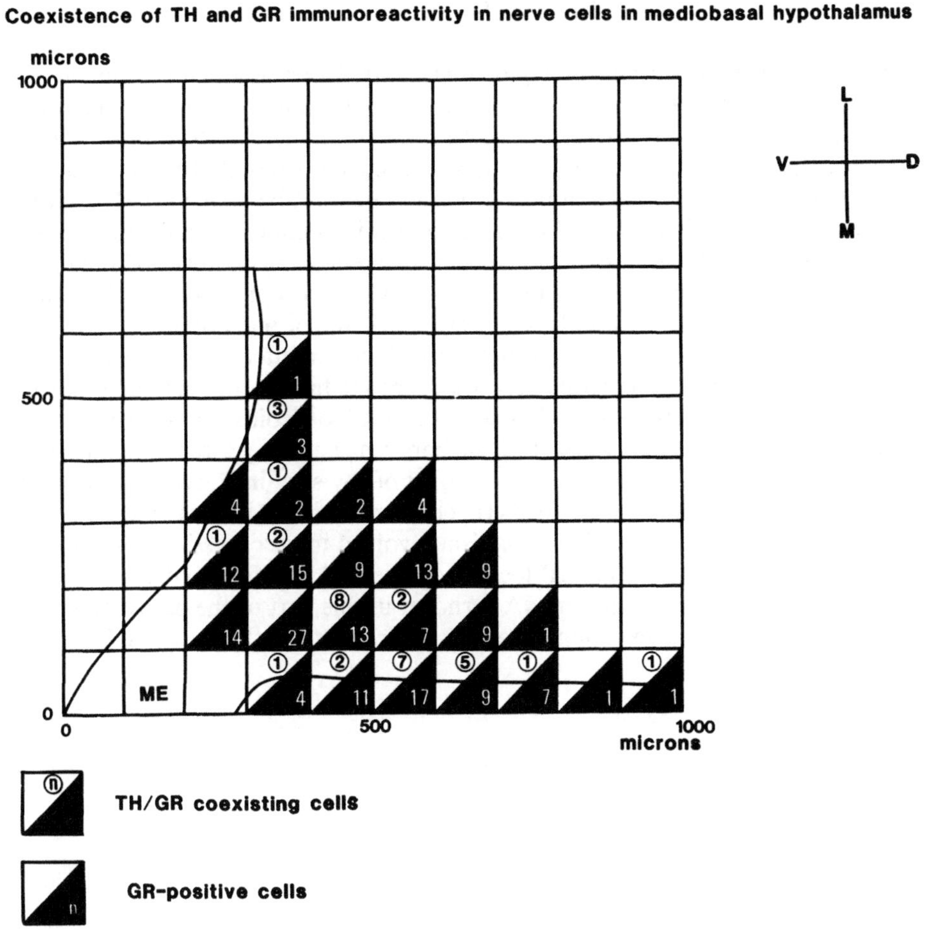

Fig. 24. Density map illustrating the coexistence of tyrosine hydroxylase (*TH*) and glucocorticoid receptor (*GR*) immunoreactive (IR) nerve cell bodies of the arcuate nucleus. The number of positive cells is shown in each square. In the *black triangle* is indicated the number of GR IR nerve cell bodies and in the *white triangle* the number of TH IR nerve cell bodies. The *circles* around the numbers in the white triangles indicate that the TH IR nerve cells have a nuclear GR immunoreactivity (coexistence). The orientation is indicated in the upper right part of the picture. The outline of the ventral surface and of the third ventricle is also given in the density map. *ME*, median eminence; *M*, medial; *V*, ventral; *D*, dorsal; *L*, lateral

curve within the periventricular hypothalamic nucleus is highly skewed, and therefore a certain amount of high grey tones may have escaped in this type of evaluation. By considering the slope values of the respective straight lines (see Fig. 22) an improved estimation of GR immunoreactivity may be obtained. Again, the parvocellular part of the paraventricular hypothalamic nucleus has the highest index as inferred from the low slope value. Using this type of semiquantitative method it is possible to study changes in the GR immunoreactivity in highly discrete neuronal populations of the central nervous system.

4.3 Studies on Presynaptic Features of Monoamine Neurons

The effects of corticosterone on regional DA and NA levels and utilization in the hypothalamus depend upon the endocrine state of the animal. However, a consistent finding has been the ability of corticosterone to increase NA utilization in the median eminence, even in hypophysectomized and adrenalectomized rats (see Andersson et al. 1981). In addition, corticosterone was able to increase NA levels selectively in the subependymal layer of the median eminence, indicating an activation also of NA synthesis. These results may reflect the existence of an inhibitory noradrenergic mechanism in the median eminence, controlling CRF release via an axoaxonic modulation. It is suggested that the primary action is on a GR IR local circuit neuron in the hypothalamus which can interact via an axoaxonic influence with the noradrenergic network in the median eminence (Andersson et al. 1981). Indications were also obtained that glucocorticoids can reduce NA utilization in the magnocellular part of the paraventricular hypothalamic nucleus. NA utilization in this region was also turned off by sustained hypersecretion of adrenocorticotropic hormone (ACTH). These results may indicate the existence of a facilitatory noradrenergic mechanism involved in vasopressin regulation. It is again suggested that the GR IR local circuit neurons are involved in producing this highly selective influence on NA turnover in this part of the paraventricular hypothalamic nucleus involving axoaxonic interactions in that nucleus. We have

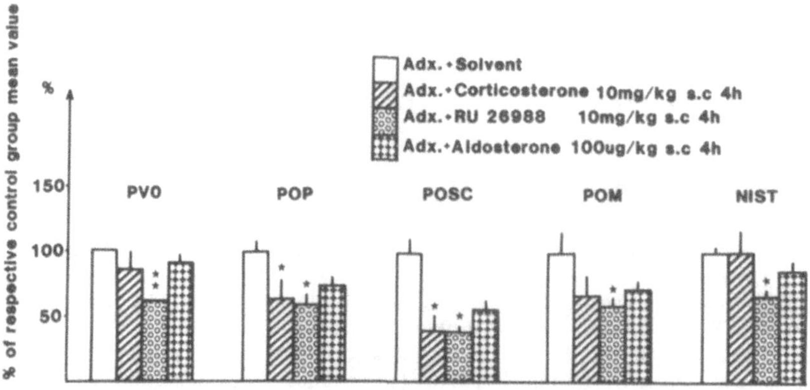

Figs. 25–27. Effects of corticosterone, RU 26988 and aldosterone on regional noradrenaline (NA) levels in discrete nuclei of the preoptic area (**Fig. 25**) and of the hypothalamus (**Fig. 26**) and on cathecolamine (CA) levels of the median eminence (**Fig. 27**) of adrenalectomized male rats (4 days after adrenalectomy). Means + SEM ($n=6$). Significances refer to rats treated with solvent alone: *$p<0.05$; **$p<0.01$ (Dunn test). *PVO*, area anterior to the anterior periventricular hypothalamic nucleus; *PV II*, posterior periventricular hypothalamic region; *POP*, *POSC*, *POM*, periventricular, suprachiasmatic and medial parts of the preoptic area. *NIST*, ventral part of the bed nucleus of striae terminalis. *PAFM*, *PAFP*, magno- and parvocellular parts of the paraventricular hypothalamic nucleus; *DM*, dorsomedial hypothalamic nucleus; *SEL*, subependymal layer of the median eminence; *MPZ*, medial palisade zone of the median eminence; *LPZ*, lateral palisade zone of the median eminence; *BZ*, border zone, an area located at the border of the medial and lateral hypothalamus between the fornix and the ventral surface of the brain; *PV I*, anterior periventricular hypothalamic region

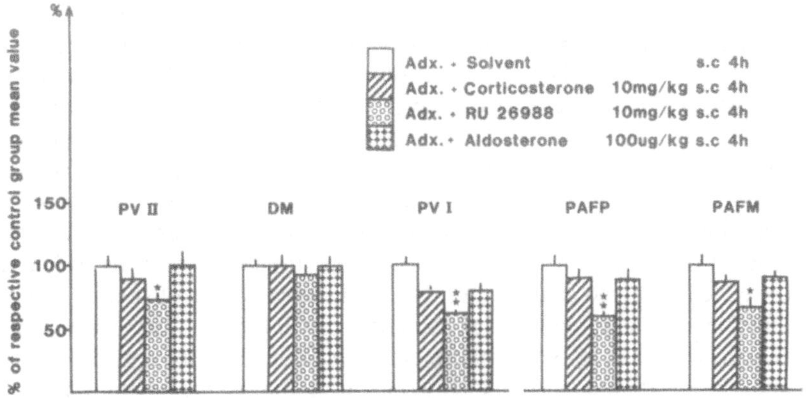

Fig. 26

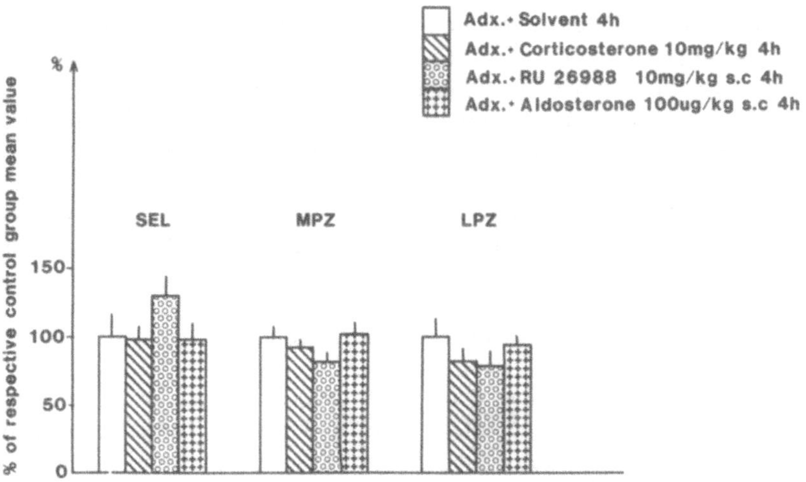

Fig. 27

administered the selective GR agonist RU26988 (Moguilevski and Raynaud 1980; Veldhuis et al. 1982) in a dose of 10 mg/kg s.c. to adrenalectomized male rats. Marked depletion of NA stores is observed after 4 h in large numbers of preoptic and hypothalamic NA nerve terminal systems but not in the median eminence CA nerve terminals (Figs. 25–27). These changes are associated with a lowering of NA utilization in several regions outside the median eminence using the TH inhibition method. It seems possible that these marked and widespread actions of the selective GR agonist on NA neurons are mediated via activation of GR in the NA cell bodies in the medulla oblongata, giving rise to the hypothalamic and preoptic NA innervation. Thus, via genomic actions GR may control the activity of NA-synthesizing enzymes and the production of receptor proteins controlling the excitability and the firing rate of the NA neurons. Recent findings also indicate that glucocorticoids control coexistence in both monoamine and peptide neurons.

Thus, upon adrenalectomy there is an increase in the number of NPY IR nerve terminals in the hypothalamus, partly caused by an increased amount of NPY IR in the phenylethanolamine-*N*-methyltransferase (PNMT) IR terminals (Fuxe et al. 1985a). Also, a number of papers (see Swanson et al. 1986) have shown that upon adrenalectomy CRF IR neurons of the paraventricular hypothalamic nucleus begin to exhibit vasopressin and angiotensin immunoreactivity. These effects are also reversed by treatment with glucocorticoids. These results indicate that glucocorticoids normally act to suppress the gene expression of peptide comodulators, at least in some monoamine and peptide neurons. When discussing the effects of glucocorticoids on the CRF neurons it is of special interest to note that the acetylcholine-induced secretion of CRF from CRF IR neurons in vitro is reduced upon in vitro treatment with glucocorticoids. These results again underline that glucocorticoids, by regulating the gene expression of various types of receptor-linked proteins belonging to the cholinergic receptor mechanism, may appropriately regulate the sensitivity of the CRF IR neurons (see Vale et al. 1985). In line with this interpretation it has also been noted by Schonbrunn (1985) that glucocorticoids decrease the number of somatostatin receptors in GH4 C1 cells.

Other presynaptic transmitter mechanisms which are influenced by glucocorticoid receptors are 5-HT utilization in the hippocampal formation (de Kloet 1985). Furthermore, vasoactive intestinal peptide (VIP) immunoreactivity is increased in the hippocampus of adrenalectomized animals (see McEwen 1982). Adrenalectomy also increases GABA uptake into synaptosomes of the hippocampal formation. All these effects are probably mediated via actions by glucocorticoids on hippocampal and/or raphe glucocorticoid receptors.

4.4 Studies on Monoamine Receptor Mechanisms

Evidence has been obtained for a glucocorticoid receptor involvement in the regulation of β-adrenergic receptors in the hippocampal formation and in the neocortex (Roberts et al. 1981; Ögren et al. 1981). It was found that the 6-hydroxydopamine-induced increase in the density values of ^{3}H-dihydroalprenolol binding sites in the hippocampal formation was significantly greater upon adrenalectomy (Roberts et el. 1981). Adrenalectomy alone did not influence the binding characteristics of the β-adrenergic receptors. In view of the learning impairments noted upon lesion of the ascending NA bundle to the cerebral cortex in combination with adrenalectomy (see Ögren et al. 1981), the further increase of the biochemical supersensitivity in the β-adrenergic receptors observed may reflect a potentiation of a compensatory biochemical phenomenon in the hippocampal formation to restore NA receptor activity and hippocampal function. The results amplify the existence of interactions between central noradrenergic mechanisms and central glucocorticoid receptors in avoidance learning, an interaction which may take place predominantly within the hippocampal formation and the cerebral cortex. It must also be considered that the NA cells in the locus coeruleus all contain a strong GR immunoreactivity predominantly located in their nuclei (Härfstrand et al. 1986), which must contribute to the interactions between NA and glucocorticoids in the intact rat.

Glucocorticoid receptors have also been shown to affect transduction mechanisms within the hippocampal formation. Thus, glucocorticoids facilitated histamine-stimulated cyclic AMP formation but suppressed NA-stimulated cyclic AMP formation as well as VIP-dependent cyclic AMP stimulation (see McEwen 1982; McEwen et al. 1985a).

It has also been shown that the receptor-receptor interactions (see Fuxe and Agnati 1985) are modulated by glucocorticoids. Thus, glucocorticoids increase the ability of VIP to increase the density values of 5-HT_1 receptors within the dorsal subiculum (see McEwen et al. 1985a). Not only the transduction mechanisms but also the substrates for the protein kinases are affected by the glucocorticoid receptors within the hippocampal formation. Thus, the phosphoprotein synapsin I is found in increased concentrations upon treatment with glucocorticoid hormones (see Nestler et al. 1981). Synapsin I is regulated via activation of cyclic AMP-dependent and calcium-dependent protein kinases.

In Fig. 28 it is illustrated that steroid hormones via nuclear actions may control the synthesis of receptors for peptides and catecholamines. As examples, the

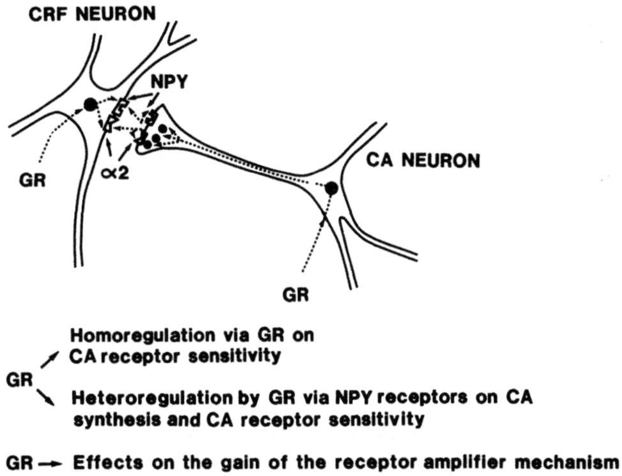

Fig. 28. Illustration of some principles for the steroid regulation of the pre- and postsynaptic features of catecholamine (*CA*) transmission and of corticotropin-releasing factor (*CRF*) neurons. Noradrenaline and adrenaline neurons both innervate the CRF immunoreactive neurons in the paraventricular hypothalamic neurons, and glucocorticoid receptor (*GR*) immunoreactivity has been demonstrated in both the nerve cell nuclei and the pericarya of the catecholamine as well as of the CRF neurons. Thus, the glucocorticoid receptors may regulate both the pre- and postsynaptic features of the classical transmission and the cotransmission lines in the catecholamine synapses. The direct regulation of the catecholamine transmission is termed "homoregulation" and the indirect regulation of the catecholamine synthesis and catecholamine receptor sensitivity, for example via modulation of neuropeptide Y (*NPY*) receptor mechanisms and NPY stores is called "heteroregulation". The available evidence indicates a profound effect of glucocorticoids on the regulation also of the synthesis of the cotransmitter, in this case NPY (see Fuxe et al. 1985a). It is also indicated that the glucocorticoid receptor, via the nuclear actions, can also regulate the receptor transducer mechanism (amplifier mechanism)

NPY receptors and the α_2-adrenergic receptors are shown. Thus, glucocorticoids may regulate the synthesis of the proteins containing the transmitter recognition sites leading to changes in the receptor density values and /or may regulate the coupling proteins such as the GTP-binding proteins which control the gain of the receptor-amplifying mechanism and/or may regulate the protein kinases involved in the phosphorylation of proteins, e.g. those in the membranes controlling transduction of signals. Another important aspect of Fig. 28 is the fact that glucocorticoids can regulate CRF synthesis and release not only via a direct action on the CRF IR neurons, but also via an influence on the afferent input. In this case the effects on the adrenergic input is illustrated. Thus, the glucocorticoids can influence the CRF neurons via both direct and indirect actions. The final outcome is the result of the integration of all these signals influencing the CRF neurons. Again, it should be emphasized that the glucocorticoids can modulate coexistence in neurons. In this case the effect on coexistence of NPY and A is exemplified.

It must be considered that glucocorticoids not only influence information handling, but also exert effects on the metabolic state of the neurons. These actions are probably of importance for the survival of the neurons. It has thus been seen that glucocorticoids can reduce the lesion-induced axon sprouting in the gyrus dentatus (see Scheff et al. 1980). Furthermore, McEwen and colleagues (1985a) have demonstrated that treatment with glucocorticoids produces not only a downregulation of the glucocorticoid receptors in the hippocampal formation, but also a nerve cell loss. Thus, glucocorticoids may, for example, inhibit the synthesis of a trophic factor in the pyramidal nerve cells. In this way the inhibitory effect of the hippocampal formation on the CRF neurons is lost (see Angelucci 1985), and the corticosterone serum levels become markedly increased, especially after stress.

Our observations that there are glucocorticoid receptors in glial cells, especially in the oligodendroglia cells, explain the ability of glucocorticoids to induce glycerol phosphate dehydrogenase in oligodendroglia cells (see McEwen 1982). WT in brain may be influenced by these effects of glucocorticoids on oligodendroglia cell, since these effects may lead to alteration in the myelination and nerve conduction velocities (Scheff et al. 1980; Friedrich and Bohn 1980; Henkin 1970). In addition, morphological changes have been noticed, such as hypertrophy of astroglia under the influence of glucocorticoids (see Scheff et al. 1980).

5 Actions of Thyroid Hormones on Wiring Transmission

5.1 General Aspects

In 1958 Ford and Gross provided evidence for the existence of a hypothalamic site of action of thyroid hormones by demonstrating the accumulation of radioactive T_4 and T_3 in the paraventricular region of the hypothalamus. Thyroid-concentrating neurons have also been demonstrated within other areas of the hypothalamus (Stumpf and Sar 1978). The thyroid hormone receptors have been characterized as nuclear receptors (Oppenheimer et al. 1974; Eberhart et al. 1976).

Light-microscopic autoradiograms of brain after intravenous administration of $^{125}I\text{-}T_3$ gives evidence for the localization of $^{125}I\text{-}T_3$ in discrete neuronal systems (Dratman et al. 1982). Furthermore, the circulating form of the hormone T_4 has been shown to be differentially taken up in various brain areas by a high-affinity transport mechanism. These results indicate that the central nervous system must be an important site of action for thyroid hormones. It has also been speculated that the amino acid hormones T_4 and T_3 may be substrates in metabolic pathways in the brain leading to the formation of adrenergically active neurotransmitters (see Dratman et al. 1984). It should be noticed, however, that T_4 does not produce any change in energy metabolism in the adult brain.

5.2 Studies on Presynaptic Features of Monoamine Neurons

Andersson and Eneroth have demonstrated that the long-term feedback action of thyroid hormones involves an action on hypothalamic catecholamine nerve terminal networks involved in the regulation of thyroid-stimulating hormone (TSH) secretion (Andersson et al. 1985; Andersson and Eneroth 1987). It was found that thyroidectomy leads to an activation of NA nerve terminal systems in the paraventricular hypothalamic nucleus and to an inactivation of the DA nerve terminal systems of the external layer of the median eminence. These effects were reversed by restitution therapy with T_3 or T_4. Chronic but not acute administration of T_3 and T_4 to the hypophysectomized rat was capable of increasing DA utilization within the external layer of the median eminence, as well as producing a reduction of NA utilization within the paraventricular hypothalamic nucleus. Thus, a long-term feedback action of thyroid hormones on TSH secretion involves inter alia the activation of DA nerve terminals in the median eminence, inhibiting the release of TSH-releasing hormone (TRH). It was suggested that these actions of thyroid hormones involve an activation of nuclear receptors in the medial basal hypothalamus, possibly located in some of the tuberoinfundibular DA nerve cell bodies. Furthermore, the feedback action of thyroid hormones on TSH secretion also involves the inactivation of a facilitory noradrenergic mechanism in the paraventricular hypothalamic nucleus. It seems possible that the thyroid hormone receptors in this case are located in the cell bodies of the TRH IR neurons, which via recurrent collaterals interact with the paraventricular NA nerve terminal systems, explaining the highly localized change in the paraventricular hypothalamic nucleus.

It is of substantial interest that the modulation of the pituitary thyroid activity could also produce highly discrete changes within the DA nerve terminal networks of the forebrain. Thus, thyroidectomy reduces DA utilization in the anterior part of the nucleus accumbens, while increasing DA utilization in the posterior part of the nucleus accumbens, where the costoring cholecystokinin (CCK)/DA nerve terminals exist. In agreement, chronic but not acute treatment with T_3 or T_4 of the hypophysectomized rat produces an increase in DA utilization within the anterior part of the nucleus accumbens. These results are of substantial interest, since DA nerve terminal systems within the nucleus accumbens participate in the control of locomotion and in the reward mechanisms (Fuxe et al. 1977b,

1986a; Ljungberg and Ungerstedt 1978). It may therefore be speculated that the hyperactivity found in patients with hyperthyroidism may be related at least partly to an activation of the dopaminergic mechanism within the anterior part of the nucleus accumbens. The results also imply the existence of nuclear receptors for thyroid hormones within the nucleus accumbens itself.

5.3 Studies on Monoamine Receptor Mechanisms

Many investigations have demonstrated that the number and functional activity of β-adrenergic receptors in peripheral tissues increase within increasing concentrations of thyroid hormones in the circulation (Williams and Lefkovits 1977).

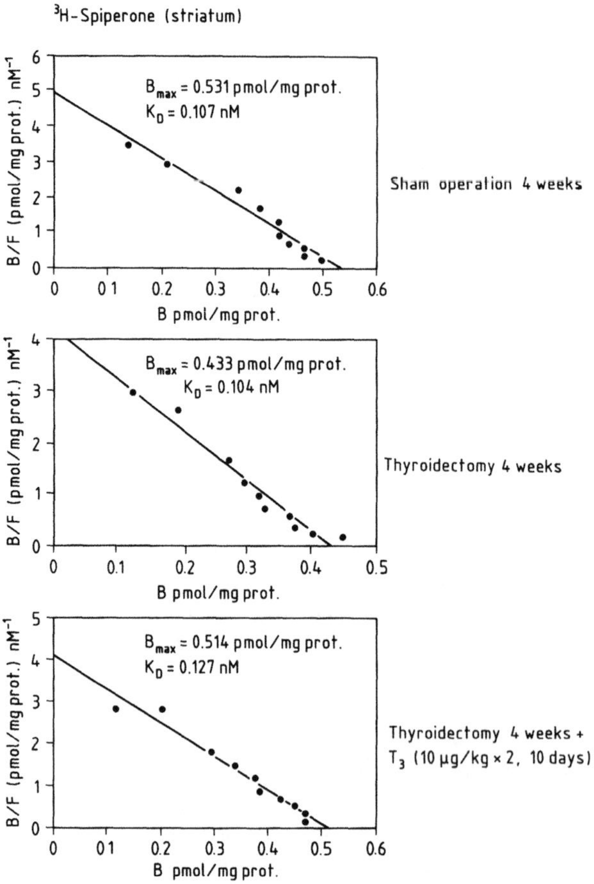

Fig. 29. Effects of thyroidectomy and restitution therapy with T_3 on the binding characteristics of ^{3}H-spiperone-labelled D_2 receptors in striatal membranes. The Scatchard plots are shown. A reduction of the B_{max} levels is shown following thyroidectomy, as well as prevention of such a fall by replacement therapy with T_3 (10 µg/kg, twice daily) (see Fuxe et al. 1984a)

Within the striatum of the rat brain we have made the observation that changes in the pituitary thyroid activity can produce substantial changes in the density of D2 receptors without changing DA utilization in various parts of the striatum (see Fuxe et al. 1984a). These results open up the possibility that thyroid hormones may accomplish a heterostatic regulation of the striatal DA synapses, i.e. they may change the sensitivity of the postsynaptic DA receptor mechanism by increasing the number of D2 receptors without any associated changes in the presynaptic features of the DA transmission line (Fig. 29). This regulation has been defined as a heterostatic regulation of the synaptic transmission. As seen in Fig. 30, it allows a change of the set point of the synapse. Thus, hormones such as thyroid hormones may control synaptic heterostasis, which probably represents an important part of functional synaptic plasticity (see also Fuxe et al. 1984a, 1986a, b; Agnati et al 1986a). Again it is visualized that the thyroid hormones control the DA receptor mechanisms via actions on nuclear thyroid hormone receptors present in the striatal neurons, leading to changes in the expression of the gene for the proteins carrying the dopamine recognition sites.

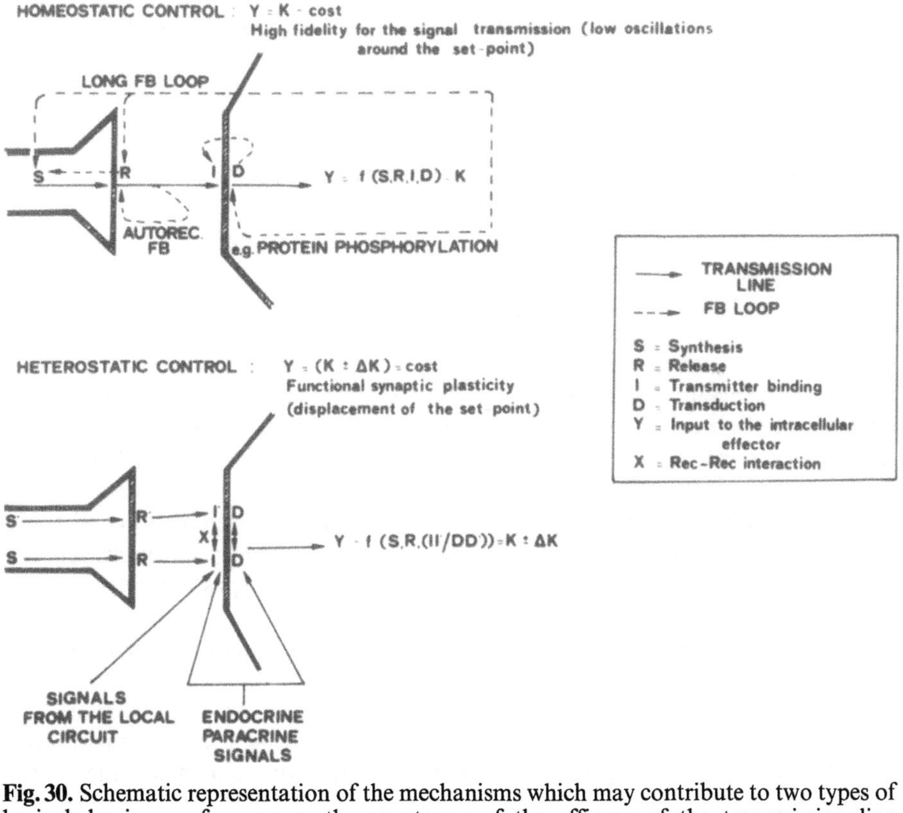

Fig. 30. Schematic representation of the mechanisms which may contribute to two types of basic behaviours of synapses: the constancy of the efficacy of the transmission line (synaptic homeostasis) and the change of the level at which this constancy is maintained (synaptic heterostasis)

6 The Humoral Modulation of Volume Transmission

It should be considered that steroid hormones, thyroid hormones and also peptide hormones can affect some of the circumventricular organs of the central nervous system and in this way influence the release of signals and the neuronal pathways from these organs. Some chemical signals may affect the neuronal networks via the ventricular system and the extracellular fluid. GR immunoreactivity exists within the nerve cells of the subfornical organ. The peptide hormones, such as angiotensin and atrial natriuretic peptides, may be active here, since there is no blood-brain barrier in the circumventricular organs and a high density of angiotensin II (Saavedra et al. 1986; Healey et al. 1986) and atrial natriuretic factor receptors (Saavedra et al. 1986; Bianchi et al. 1986) exists in this region.

As stated above, it seems likely that the steroid and thyroid hormones may influence the uptake of messengers and trophic factors in glial cells and neurons, as well as their release from glial cells and neurons. Thus, the release and the uptake of paracrine signals from neurons reaching a distant receptor population may be highly influenced by steroid and thyroid hormones, which thus can profoundly influence VT in the nervous system.

It is also conceivable that the extracellular fluid pathways which are of substantial importance in VT may be influenced by hormones in view of the ability of glucocorticoids to control astroglia functions.

7 Aspects on the Organization Principles of the CNS

It seems that the basic texture of some CNS structures is made up of elementary units. Thus, it can be surmised that in the CNS there are at least three types of basic organization: (a) Nuclei (which may be represented by subnuclei of classical anatomical nuclei, e.g. nucleus tractus solitarius), i.e. groups of neurons which operate to perform in an integrative fashion a certain elaboration of information, or to exert a certain trophic action. In particular, it is possible to distinguish the diffuse type of nuclear organization, such as the one of reticular formation, from the compact nuclear organization of the thalamus and the hypothalamus. (b) Elementary circuits, which show a highly repetitive geometry, without clear-cut morphological boundaries. This organization is represented by the cerebellum and hippocampus. (c) Modules, i.e. repetitive units made up by organized elements of neuronal structures. We can recognize different types of modules in the CNS.

1. "Structural" modules, i.e. repetitive clusters of cell bodies and/or terminals, characterized on the basis of transmitters and/or recognition sites for transducer mechanisms and/or metabolic and trophic features, as found by methods of chemical neuroanatomy. Modules may be heterogenous regarding size and shape. This organization is present in the striatum ("striosomes") (see Graybiel 1986).
2. Local circuit modules, i.e. repetitive aggregates of nerve terminals making synaptic contacts with one another controlling a neuronal input and/or a

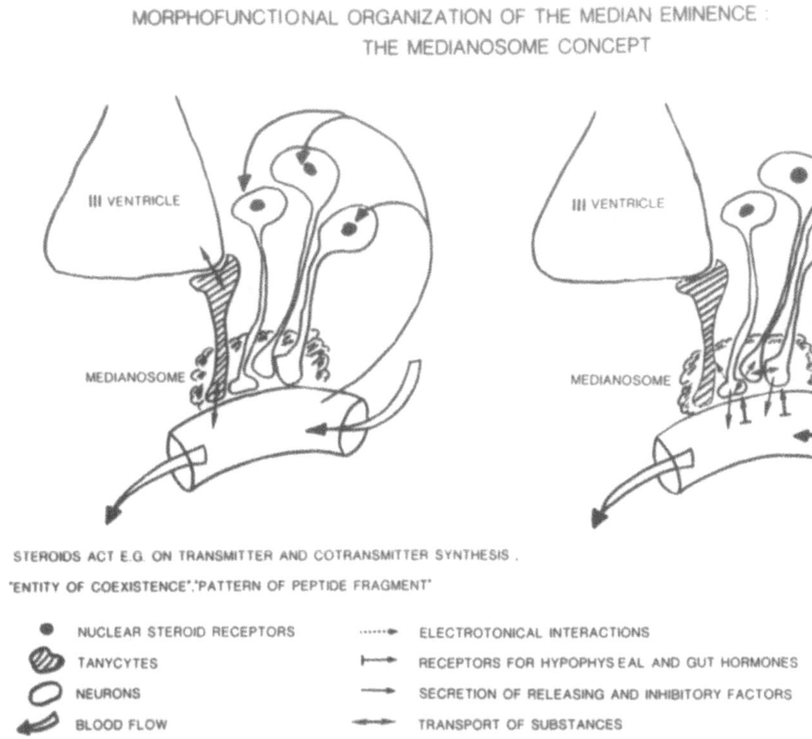

Fig. 31. Schematic illustration of the medianosome concept and of the sites of action of peptide, steroid and thyroid hormones

neuronal output. They can be recognized in the median eminence ("medianosomes") (Figs. 31 and 32) and in the olfactory bulb.

3. Columns, i.e. repetitive sets of cells which form a vertical structure throughout a brain region and are involved in the integration of specific inputs. They are homogenous regarding size and shape. This organization is present in the somatosensory, visual, motor and frontal association cortex.

Other principles of organization such as somatotopy (i.e. the segregation of a set of neurons or axons in the nuclei or pathways, respectively, according to the body surface area, to which they are connected) can be superimposed on the basic buildup of neuronal circuits. Genetic and epigenetic influences can exert their actions in the frame of these organizational principles. In the following paragraphs we will develop the concept of the modular organization of the CNS in the frame of the WT and VT.

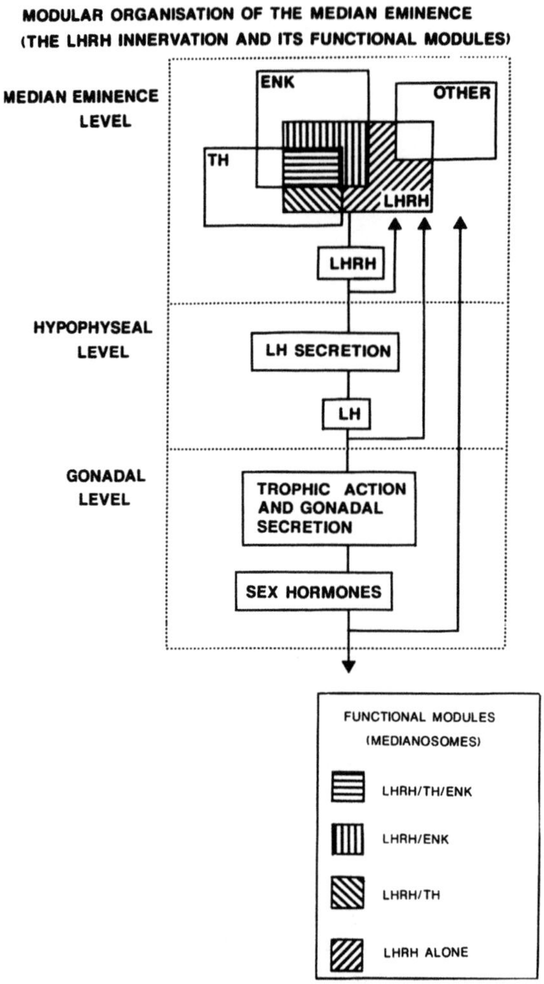

Fig. 32. Illustration of the structural luteinizing hormone-releasing hormone (LHRH) medianosome and its subdivision into functional or integrative medianosomes, represented by the various interaction zones (LHRH/TH, LHRH/ENK etc.). The modulation caused by LH and sex hormones is indicated (for further aspects see Fig. 34)

7.1 Modules of Wiring Transmission

The cortical columns were identified on the basis of functional criteria, e.g. the neurons of a column respond to the same sensory stimulus applied to a specific area of the skin. The columns have been shown to be modality specific and site specific and were subsequently demonstrated to have an anatomical correlate. Recently, studies in chemical neuroanatomy have supplied indications that the modular organization is not unique to the cerebral cortex but also exists in other brain areas, such as the striatum and the median eminence (Olson et al. 1972; Tenyson et al. 1972; Graybiel 1984; 1986; Fuxe et al. 1971, 1986c; Andersson et al. 1984). Structural modules, defined on the basis of markers of chemical transmission, have been recognized in the striatum. We have recently introduced the idea that there exist "integrative" modules in the brain, formed in the interaction zones between various structural modules (Agnati et al. 1986c; Fuxe et al. 1986c). The overlap zones between two, three or more structural modules correspond to integrative modules of higher and higher level (Fig. 32). Each integrative module of any level may have different spatial and temporal patterns of activity, resulting in a very ample spectrum of functional states. Such an integrative module may be considered as the functional counterpart of the module, which is basically defined by means of criteria borrowed from chemical neuroanatomy.

7.2 Modules of Volume Transmission

Recently, by means of automatic image analysis, we have also been able to establish the existence of islands of neurons which contain glucocorticoid receptor immunoreactivity within the striatum and the nucleus accumbens (Fig. 33). These results provide the first evidence for the existence, at least in parts of the CNS, of structural modules characterized on the basis of humoral inputs. GR IR islands have a rather uniform distribution and are different from the striatal modules of the WT mentioned above (see Zoli et al. 1988).

7.3 Functional Aspects on the Modular Organization

The best example of functional interactions between WT and VT modules is presented by the local circuits formed by various types of transmitter-identified nerve terminals at the median eminence level (Fuxe et al. 1986c). In fact, at median eminence level (an interface area between brain and the endocrine system), the various types of transmitter-identified nerve terminal networks form aggregates of nerve terminals, the medianosomes (see Fig. 31), located in distinct parts of the median eminence forming rostrocaudal strips. Such aggregates of nerve terminals regulate the secretion of one hypothalamic hormone (e.g. a releasing hormone) (see Fig. 34). The releasing factor secreted by the median eminence controls the release of one adenohypophyseal hormone. The differential activation of the various elements forming a medianosome results in a functional state which will provide the appropriate output to control anterior pituitary hormone secretion. Receptors for various hypothalamic and hypophyseal hormones, such as thyreotropic releasing hormone and corticotropin releasing hormone, probably exist in various portions of the cell membranes of the nerve terminals constituting the local circuits of the medianosomes (Taylor and Burt 1982; De Souza et al. 1985). Thus, one important role of the hypothalamic and of the hypophyseal hormones is to modulate the activity of the medianosomes, an action which probably underlies the ultrashort and short-loop feedback action of hypothalamic and of the hypophyseal hormones, respectively. From the above it becomes clear that one important site of action of hormones in the brain is a local circuit module, where the activated hormonal receptors of humoral modules interact with activated transmitter receptors of wiring modules to adjust the functional output.

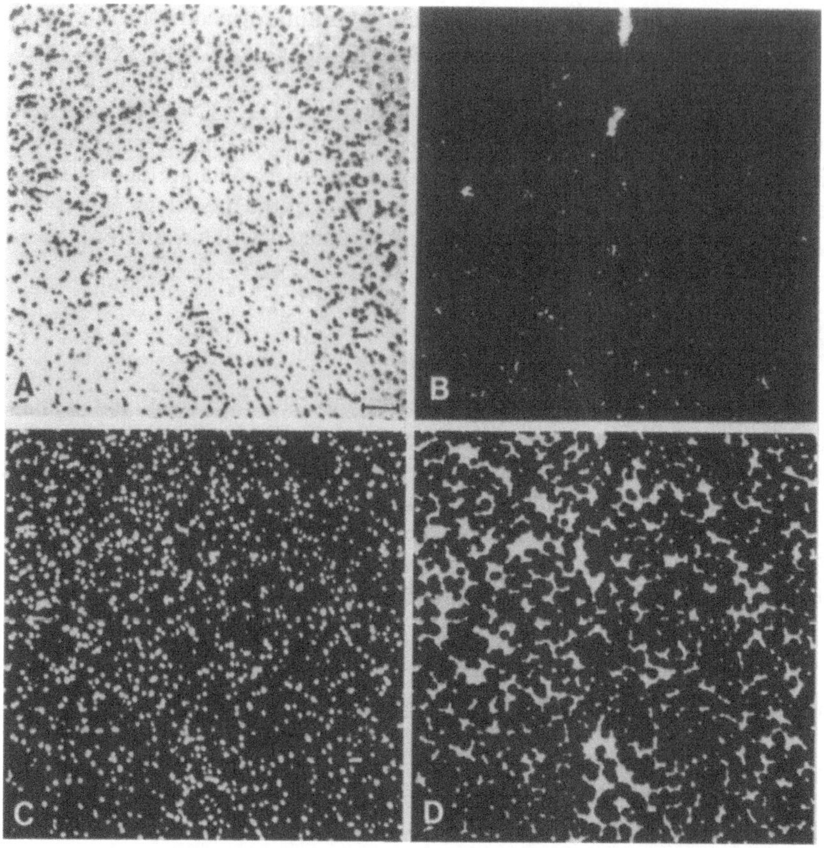

Fig. 33 A–D. Example of the elaboration of the sampled fields. **A** Original image as it appears on the IBAS screen. **B** The discrimination function has been performed on the original image to differentiate grey matter (*black* in the image) from white matter. This procedure allows the determination of the field area of the grey matter present in the sampled field. **C** The discrimination function has been performed on the original image to separate the specific staining (glucocorticoid receptor immunoreactive profiles: *white dots* in the discriminated image) from the background. This procedure allows the evaluation of the number of the discriminated profiles. **D** The close function has been performed on the discriminated image **C**. This procedure allows determination of the field area covered by the islands and the number and the size of the islands, defined as aggregates of three or more original profiles

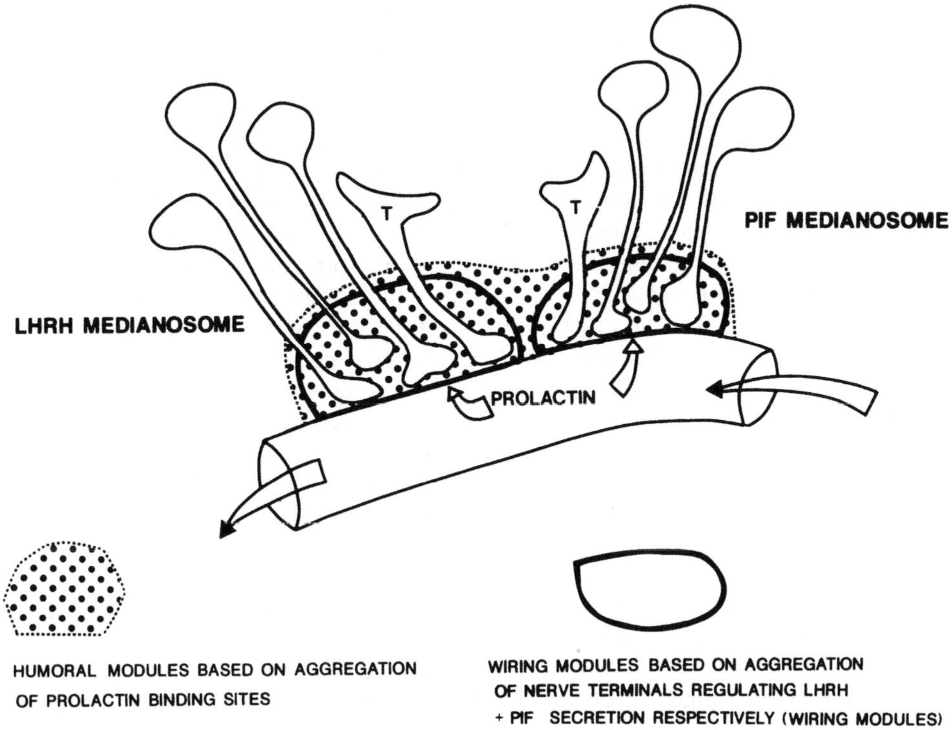

Fig. 34. Example of a possible interaction between volume transmission modules (humoral modules of prolactin binding sites) and of wiring transmission modules [wiring modules; structural medianosomes for luteinizing hormone-releasing hormone (*LHRH*) and prolactin release-inhibiting factor (*PIF*) secretion]

8 Summary

There are two types of chemical transmission in brain, namely wiring transmission (WT) and volume transmission (VT). WT is the classical type of transmission which is neuronally linked and operates with high speed, high safety and short term actions, their divergency and plasticity being low. VT is a type of transmission mainly operating via electronic and paracrine signals, diffusing in the extracellular fluid to reach the appropriate targets. To understand the actions of hormones and paracrine signals on WT, the complexity of the individual synapse must be understood. It consists of multiple transmission lines, which interact with one another at the pre- and postsynaptic membrane via intramembrane receptor-receptor interactions and via intracellular postreceptor signals.

There exists a blood-brain barrier, leading to the exclusion of a number of peripheral hormonal signals. However, there also exist chemical and physical "windows" through which the brain receives and delivers messages. Physical windows are represented by brain areas devoid of a blood-brain barrier, such as the area postrema, the median eminence and the subfornical organ. The chemical windows are represented by facilitated transport, active transports etc.

Steroid sex hormone-accumulating and estrogen IR nerve cells have been demonstrated in the central nervous system and are concentrated in the limbic forebrain, the medial preoptic area and the hypothalamus. The nuclear location of the estrogen immunoreactivity strongly supports the importance of the genomic actions of estrogens. Estradiol-17β, progesterone and androgens produce discrete changes in dopamine (DA), noradrenaline (NA) and adrenaline (A) levels and utilization within the hypothalamus and the preoptic area in male and female rats in various endocrine states. The monoamines participate in both the central inhibitory and facilitory feedback actions of estradiol-17β on LHRH secretion. The various catecholamine (CA) nerve terminal networks are probably influenced via local circuit interactions, in which the steroid target cells participate. However, an accumulation of steroid hormones has also been observed in DA and NA nerve cell bodies. Estrogens also influence striatal DA mechanisms, indicating that estrogens can also modulate motor functions and mental activities, such as mood. These effects may be produced via indirect actions involving the hypothalamus and/or the pituitary gland. However, the vast majority of the effects of estrogens and other gonadal steroids on the presynaptic properties of transmitter-identified neurons are probably secondary to primary actions on the genome of the estrogen immunoreactive neurons of the brain. This is also true for the ability of steroid hormones to influence the coexistence of peptides and CA in monoamine neurons, probably mainly representing an influence on the synthesis of the peptide comodulator.

Sex steroids also modulate the binding characteristics of central transmitter receptors, such as α- and β-adrenergic receptors. It seems likely that both genomic and nongenomic actions of sex steroids are involved in producing the changes in binding characteristics of monoamine and other transmitter receptors. Transduction mechanisms are also modulated by estrogens, as revealed by changes in adenylate cyclase activity. Effects on membrane fluidity should be considered when direct membrane actions are discussed. It must also be underlined that CA receptors can regulate steroid receptor numbers, so that the networks of the brain can regulate their own sensitivity to hormonal signals.

Corticosterone target neurons were first observed within the hippocampal formation. Recently by means of monoclonal antibodies against rat liver glucocorticoid receptor (GR) in combination with the indirect immunoperoxidase technique we have been able to demonstrate GR IR nerve and glial cells all over the brain and the spinal cord of the male rat. High densities of the moderately GR IR nerve cells were, inter alia, found all over the cortical hemispheres and in a vast majority of the thalamic and hypothalamic nuclei, illustrating the wide-spread distribution of GR IR nerve cells in the entire diencephalon and telencephalon. Two-color immunocytochemistry shows the vast majority of NA, A and 5-hydroxytryptamine (5-HT) of the lower brain stem to contain strong GR immunoreactivity. These results indicate that GR can in part control brain function via, for example, modulation of the synthesis and the release of NA, A and 5-HT and their respective peptide comodulators. In view of the known function of NA, A and 5-HT pathways it is now possible to better understand how glucocorticoids can regulate cardiovascular functions, the sleep-wakefulness cycle, attention and mood in man. Various DA cell groups in the brain showed variability with regard to the number of DA nerve cell bodies positive for GR. The arcuate DA nerve cell bodies all contained GR immunoreactivity. Furthermore, all the paraventricular CRF IR nerve cell bodies, and the arcuate GRF IR nerve cell bodies which project into the median eminence contain GR immunoreactivity. The results indicate that glucocorticoids can directly control the CRF and GRF neurons by an action at the nuclear level, probably representing the mechanism underlying the central feedback actions of glucocorticoids, leading to an inhibition of CRF and GRF synthesis (delayed feedback).

Studies on the effects of glucocorticoids on regional DA and NA levels and utilization have indicated the existence of an inhibitory NA mechanism in the median eminence, controlling CRF release via an axonic modulation. Selective GR agonists have been shown to deplete NA stores and to reduce NA utilization within the hypothalamus and the preoptic region. Thus, via nuclear actions in the NA cell bodies GR may control the synthesis of NA-synthesizing enzymes. Furthermore, they may control the production of several receptor proteins controlling the responses of the NA neurons to extracellular signals. Also, evidence has been obtained that glucocorticoids normally act to suppress expression of peptides in monoamine and peptide neurons.

Evidence has also been obtained for a GR involvement in the regulation of β-noradrenergic receptors in the hippocampal formation and in the neocortex. The results also amplify the interactions of central noradrenergic mechanisms and central GR in avoidance learning. GR also influence transduction mechanisms within the hippocampal formation and the receptor-receptor interactions between VIP and 5-HT_1 receptors within the dorsal subiculum. It must also be considered that glucocorticoids can influence not only the information handling but also the metabolic state of the neurons (Fig. 35). Such actions may influence the survival of neurons. Glucocorticoids have also been shown to induce morphological changes in oligodendroglia and astroglia cells.

Thyroidhormone-concentrating neurons have been demonstrated within the hypothalamus, and the thyroid hormone receptors have been characterized as nuclear receptors. Furthermore, the circulating form of the hormone T_4 has been shown to be preferentially taken up in various brain areas via a high-affinity transport mechanism. These results indicate that the central nervous system must be an important site of action for thyroid hormones. T_4 does not produce any change in energy metabolism in the adult brain. Evidence has been obtained that the long-term feedback action of thyroid hormones involves an action on hypothalamic DA and NA nerve terminal networks involved in the regulation of TRH secretion. We suggest that these actions of thyroid hormones involve an activation of nuclear receptors in the mediobasal hypothalamus, some of which may be located in the tuberoinfundibular DA nerve cell bodies. The feedback actions of thyroid hormones on TRH secretion also involve the inactivation of a facilitory noradrenergic mechanism in the paraventricular hypothalamic nucleus. Modulation of the pituitary-thyroid activity can also produce highly discrete changes within DA nerve terminal networks of the forebrain. Chronic treatment with T_3 or T_4 in the hypophysectomized rat can, for example, produce an increase of DA

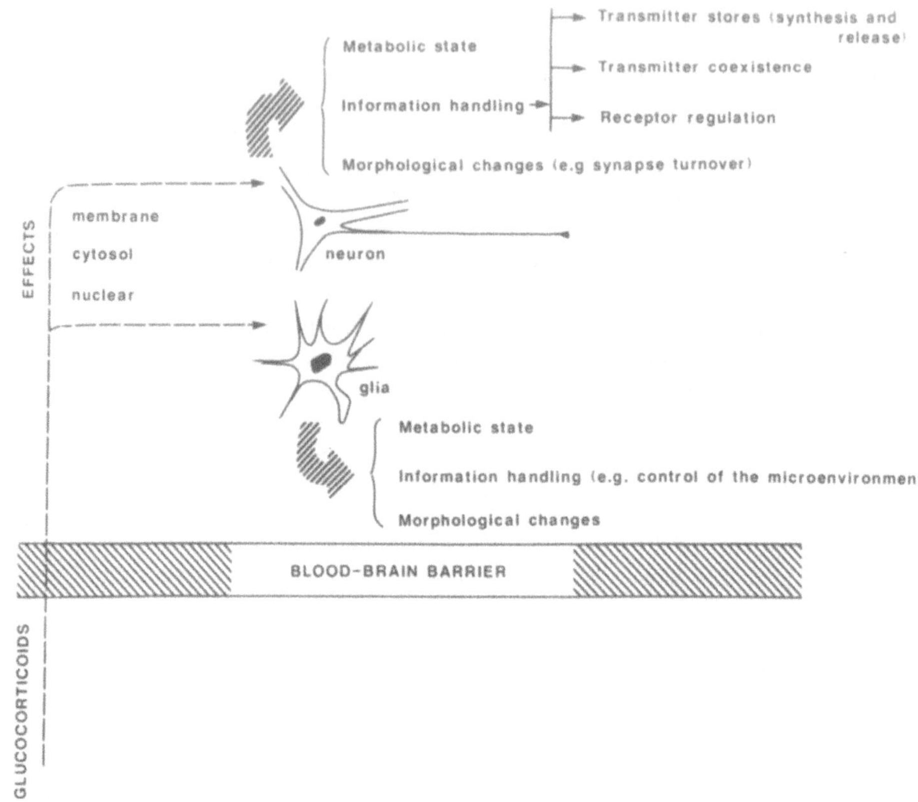

Fig. 35. Summary of the possible effects of glucocorticoids on the metabolic state, information handling and the morphological characteristics in central neurons and glia

utilization within the anterior part of the nucleus accumbens. These results amplify the existence of nuclear receptors for thyroid hormones within the nucleus accumbens itself and indicate that hyperactivity found in patients with hyperthyroidism may be related at least partly to an activation of the DA mechanism within the anterior part of the nucleus accumbens controlling locomotion. Within the striatum, changes in pituitary-thyroid activity can produce substantial changes in the density of D2 receptors without changing DA utilization in various parts of the striatum. These results open up the possibility that thyroid hormones may accomplish a heterostatic regulation of striatal DA synapses.

We have introduced the hypothesis that the activated overlap zones of structural modules of WT and VT represent the dynamic functional modules of the brain, where the appropriate integration of signals takes place for the performance of the task at hand. The best example of structural interaction zones between WT and VT modules is given by the overlap zones in the median eminence between various types of transmitter-identified nerve terminals. These zones have been called medianosomes. The medianosomes will, in a precise way, regulate the secretion of one or several hypothalamic hormones. Thus, they are thought of as integrative units in which various transmitter-identified nerve terminals containing various types of receptors for hypothalamic and hypophyseal hormones communicate with one another in well-defined local circuits formed in the interaction zones between the WT and VT modules. The ultrashort and the short-loop feedback actions of hypothalamic and hypophyseal hormones respectively probably involve modulation of activity in various types of medianosomes. Thus, one important site of action of hormones in the brain is the functional module where activated hormonal receptors interact with activated transmitter receptors to adjust the functional output.

In view of the very large number of neuronal and humoral signals involved in regulation of brain function, such as mood, it seems unlikely that an alteration in a single transmitter system is responsible for the development of mental illness. Instead, it seems likely that multiple errors in WT and VT must take place give rise to a disturbance sufficient to result, for instance, in a state of depression.

Acknowledgements. We are grateful to Anne Edggren and Monia Särd for excellent secreterial assistance.

References

Agnati LF, Fuxe K, Kuonen D, Blake CA, Andersson K, Eneroth P, Gustafsson J-Å, Battistini N, Calza L (1981) Effects of estrogen and progesterone on central α- and β-adrenergic receptors in ovariectomized rats: evidence for gonadal steroid receptor regulation of brain α- and β-adrenergic receptors. In: Fuxe K, Gustafsson J-Å, Wetterberg L (eds) Steroid hormone regulation of the brain. Pergamon, Oxford, pp 237–252

Agnati LF, Fuxe K, Benfenati F, Calza L, Battistini N, Ögren S-O (1983) Receptor-receptor interactions: possible new mechanisms for the action of some antidepressant drugs. In: Usdin E, Goldstein M, Friedhoff AJ, Gergotas A (eds) Frontiers in neuropsychiatric research. MacMillan, London, pp301–318

Agnati LF, Fuxe K, Benfenati F, Battistini N, Zini I, Camurri M, Hökfelt T (1984a) Postsynaptic effects of neuropeptide comodulators at central monoamine synapses. In: Usdin E, Carlsson A, Dahlström A, Engel J (eds) Catecholamines: neuropharmacology and central nervous system – theoretical aspects. Liss, New York, pp 191–198

Agnati LF, Fuxe K, Calza L, Giardino L, Zini I, Toffano G, Goldstein M, Marrama P, Gustafsson J-Å, Yu Z-Y, Cuello AC, Terenius L, Lang R, Ganten D, (1984b) Morphometrical and microdensitometrical studies on monoaminergic and peptidergic neurons in the aging brain. In: Agnati LF, Fuxe K (eds) Quantitative neuroanatomy in transmitter research. MacMillan, London, pp 91–112

Agnati LF, Fuxe K, Yu S-Y, Härfstrand A, Okret S, Wiksröm A-C, Goldstein M, Zoli M, Vale W, Gustafsson J-Å (1985) Morphometrical analysis of the distribution of corticotrophin releasing factor, glucocorticoid receptor and phenylethanolamine-*N*-methyltransferase immunoreactive structures in the paraventricular hyopthalamic nucleus of the rat. Neurosci Lett 54:147–153

Agnati LF, Fuxe K, Zoli M, Merlo Pich E, Benfenati F, Zini I, Goldstein M (1986a) Aspects on the information handling by the central nervous system: focus on cotransmission in the aged rat brain. In: Hokfelt T, Fuxe K, Pernow B (eds) Progress in brain research, vol 68. Elsevier, Amsterdam, pp 291–301

Agnati LF, Fuxe K, Zoli M, Zini I, Toffano G, Ferraguti F (1986 b) A correlation analysis of the regional distribution of central enkephalin and β-endorphin immunoreactive terminals and of opiate receptors in adult and old male rats. Evidence for the existence of two main types of communication in the central nervous system: the volume transmission and the wiring transmission. Acta Physiol Scand 128:201–207

Agnati LF, Fuxe K, Zoli M, Härfstrand A, Benfenati F, Ferraguti F, Cimino M, Goldstein M (1986c) Neurochemical aspects on the modular organization of the neostriatum of the male rat. ENA, Marseille

Agnati LF, Fuxe K, Zoli M, Ferraguti F, Benfenati F, Quimet C, Walaas I, Goldstein M, Greengard P (1987) Morphometrical analysis of tyrosine-hydroxylase enkephalin and DARPP-32 like immunoreactive profiles and their codistribution at three rostrocaudal levels of the rat neostriatum. Evidence for a complex islandic and striae organization. Neuroscience (to be published)

Andersson K, Eneroth P, (1987) Thyroidectomy and central catecholamine neurons of the male rat. Neuroendocrinology 45:14–27

Andersson K, Fuxe K, Eneroth P, Blake CA, Agnati LF, Gustafsson J-Å (1981) Effects of androgenic and adrenocortical steroids on hypothalamic and preoptic catecholamine nerve terminals and on the secretion of anterior pituitary hormones. In: Fuxe K, Gustafsson J-Å, Wetterberg L (eds) Steroid hormone regulation of the brain. Pergamon, Oxford, pp 117–134

Andersson K, Fuxe K, Agnati LF, Eneroth P, Camurri M (1984) Luteinizing hormone-releasing hormone increases dopamine turnover in the lateral palisade zone of the median eminenence and reduces noradrenaline turnover in the nuc. preopticus medialis of the hypophysectomized male rat. Neurosci Lett 45:253–258

Andersson K, Eneroth P, Roos P (1985) Effects of TRH and a rat TSH preparation on discrete hypothalamic and forebrain catecholamine nerve terminal networks in the hypophysectomized male rat. Eur J Pharmacol 111:293–307

Angelucci L (1985) A function for the hippocampal glucocorticoid receptor. In: Baulieu EE, Fuxe K, McEwen BS (eds) Table ronde roussel uclaf no 53. Steroids and the brain. Paris, pp 52–53

Baulieu EE (1981) Steroid hormones in the brain: several mechanisms? In: Fuxe K, Gustafsson J-Å, Wetterberg L (eds) Steroid hormone regulation of the brain. Pergamon, Oxford, pp 3–14

Bedard PJ, Di Paolo T, Langelier P, Poyet P, Labrie F (1981) Behavioural and biochemical evidence of an effect of estradiol on striatal dopamine receptors. In: Fuxe K, Gustafsson J-Å, Wetterberg L (eds) Steroid hormone regulation of the brain. Pergamon, Oxford, pp 331–340

Blum M, McEwen, Roberts JL (1985) Estrogen regulation of tyrosine hydroxylase gene transcription in the arcuate nucleus of the rat hypothalamus. In: Baulieu EE, Fuxe K, McEwen BS (eds) Table ronde Roussel Uclaf no 53. Steroids and the brain. Paris, pp 57–58

Bianchi C, Gutkowska J, Ballak M, Thibault G, Garcia R, Genest J, Cantin M (1986) Radioautographic localization of ^{125}I-atrial natriuretic factor binding sites in the brain. Neuroendocrinology 44:365–372

Carette B, Barry J, Linkie D, Ferin M, Mester J, Baulieu EE (1979) Effets de 1'„oestradiol-7 α-acide butyrique" au niveau de cellules hypothalamique. CR Acad Sci 288:631–634

Cintra A, Fuxe K, Härfstrand A, Agnati LF, Miller LS, Greene JL, Gustafsson J-Å (1986) On the cellular localization and distribution of estrogen receptors in the rat tel- and diencephalon using monoclonal antibodies to human estrogen receptor. Neurochem Int 8(4):587–595

Cintra A, Fuxe K, Härfstrand A, Agnati LF, Wikström A-C, Okret S, Vale W, Gustafsson J-Å (1987) Presence of glucocorticoid receptor immunoreactivity in corticotrophin releasing factor and in growth hormone releasing factor immunoreactive neurons of the rat di- and telencephalon. Neurosci Lett 77:25–30

Cottingham SL, Pfaff D (1986) Interconnectedness of steroid hormone-binding neurons: existence and implications. In: Ganten D, Pfaff D (eds) Morphology of hypothalamus and its connections. Springer, Berlin Heidelberg New york Tokyo, pp 223–250

Dahlström A, Fuxe K (1964) Evidence for the existence of monoamine containing neurons in the central nervous system. I. Demonstration of monoamines in the cell bodies of brainstem neurons. Acta Physiol Scand 63 [Suppl 232]:1–55
De Kloet ER (1985) Heterogeneity, function and control of adrenal receptor systems. In: Baulieu EE, Fuxe K, McEwen BS (eds) Table ronde roussel uclaf no 53. Steroids and the brain. Paris, pp 30–31
De Souza EB, Insel TR, Perrin MH, Rivier J, Vale WS, Kuhar MJ (1985) Corticotrophin-releasing factor receptors are widely distributed within the rat central nervous system: an autoradiographic study. J Neurosci 5(12):3189–3203
Dratman MB, Futaesaku Y, Crutchfield FL, Berman N, Payne B, Sar M, Stumpf WE (1982) Iodine-125-labeled triiodothyronine in rat brain: evidence for localization in discrete neural systems. Science 215:309–312
Dratman MB Floy L, Crutchfield L, Gordon JT (1984) Thyroid hormones and adrenergic neurotransmitters. In: Catecholamines: neuropharmacology and central nervous system – theoretical aspects. Liss, New York, pp 425–439
Dupont A, Di Paolo T, Merand Y, Gagne B, Lepine J, Barden N, Labrie (1981) Biochemical effects of estrogens and thyroid hormones in the CNS. In: Fuxe K, Gustafsson J-Å, Wetterberg L (eds) Steroid hormone regulation of the brain. Pergamon, Oxford, pp 217–236
Eberhardt N, Valcana T, Timiras PS (1976) Hormone-receptor interactions in brain: uptake and binding of thyroid hormones. Psychoneuroendocrinol 1:399–409
Euvrard C, Boissier JR (1981) Estrogens and striatal cholinergic function. In: Fuxe K, Gustafsson J-Å, Wetterberg L (eds) Steroid hormone regulation of the brain. Pergamon, Oxford, pp 341–354
Ford DH, Gross J (1985) The metabolism of I^{131}-labeled thyroid hormones in the hypophysis and brain of the rabbit. Endocrinology 62:416–436
Friedrich VL, Bohn MC (1980) Glucocorticoids inhibit myelination in developing rat. Soc Neurosci Abstr 6:380
Fuxe K, Agnati LF (1985) Receptor-receptor interactions in the central nervous system. A new integrative mechanism in synapses. Med Res Rev 5(4):441–482
Fuxe K, Hökfelt T, Ungerstedt U (1971) Localization of monoamines in the central nervous system. In: de Ajuriaguerra J, Gauthier G (eds) Monoamines noyaux gris centraux et syndrome de Parkinson. Georg, Geneva, pp 23–60
Fuxe K, Löfström A, Eneroth P, Gustafsson J-Å, Skett P, Hökfelt T, Wiesel F-A, Agnati LF (1977a) Involvement of central catecholamines in the feedback actions of 17β-estradiolbenzoate on luteinizing hormone secretion in the ovariectomized female rat. Psychoneuroendocrinology 2:203–225
Fuxe K, Ögren S-O, Fredholm B, Agnati LF, Hökfelt T, Perez de la Mora M (1977b) Possibilities of a differential blockade of central monoamine receptors. In: de Ajuriaguerra J, Tissot R (eds) Rhinencéphalon neurotransmetteurs et psychoses. Symposium Bel-Air V, Geneva, pp 253–289
Fuxe K, Andersson K, Ögren S-O, Perez de la Mora M, Schwarcz R, Hökfelt T, Eneroth P, Gustafsson J-Å, Skett P (1979a) GABA neurons and their interaction with monoamine neurons: an anatomical, pharmacological and functional analysis. In: Krogsgaard-Larsen P, Scheel-Krüger J, Kofod H (eds) GABA neurotransmitters. Pharmacochemical, biochemical and pharmacological aspects. Munksgaard, Copenhagen, pp 79–94
Fuxe K, Andersson K, Löfström A, Hökfelt T, Ferland L, Agnati LF, Pérez de la Mora M, Schwarcz R, Eneroth P, Gustafsson J-Å, Skett P (1979b) Neurotransmitter mechanisms in the control of the secretion of hormones from the anterior pituitary. In: Fuxe K, Hökfelt T, Luft R (eds) Central regulation of the endocrine system. Plenum, New York, pp 349–380
Fuxe K, Andersson K, Schwarcz R, Agnati LF, Pérez de la Mora M, Hökfelt T, Goldstein M, Ferland L, Possani L, Tapia R (1979c) Studies on different types of dopamine nerve terminals in the forebrain and their possible interactions with hormones and with neurons containing GABA, glutamate and opioid peptides. In: Poirier LF, Sourkes TL, Bédard PJ (eds) Advances in neurology, vol 24. Raven, New York, pp 199–214

Fuxe K, Agnati LF, Andersson K, Locatelli V, Eneroth P, Hökfelt T, Mutt V, McDonald T, El Etreby MF, Zini I, Calza L (1980) Concepts in neuroendocrinology with emphasis on neuropeptide-monoamine interactions in neuroendocrine regulation. In: Brambilla F, Racagni G, de Wied D (eds) Progress in psychoneuroendrocrinology. Elsevier, Amsterdam, pp 47–61

Fuxe K, Andersson K, Blake CA, Eneroth P, Gustafsson J-Å, Agnati LF (1981 a) Effects of estrogen and combined treatment with estrogen and progesterone on central dopamine, noradrenaline and adrenaline nerve terminal systems of the ovariectomized rat. Relationship of changes in amine turnover to changes in LH and prolactin secretion and in sexual behaviour. In: Fuxe K, Gustafsson J-Å, Wetterberg L (eds) Steroid hormone regulation of the brain. Pergamon, Oxford, pp 73–92

Fuxe K, Agnati LF, Andersson K, Martire M, Ögren S-O, Giardino L, Battistini N, Grimaldi R, Farabegoli C, Härfstrand A, Toffano G (1984 a) Receptor-receptor interactions in the central nervous system. Evidence for the existence of heterostatic synaptic mechanisms. In: Vizi ES, Magyar K (eds) Regulation of transmitter function: basic and clinical aspects. Akadémia Kiadó, Budapest, pp 129–140

Fuxe K, Agnati LF, Calza L, Andersson K, Giardino L, Benfenati F, Camurri M, Goldstein M (1984 b) Quantitative chemical neuroanatomy gives new insights into the catecholamine regulation of the peptidergic neurons projecting to the median eminence. In: Usdin E, Carlsson A, Dahlström A, Engel J (eds) Catecholamines: neuropharmacology and central nervous system – theoretical aspects. Liss, New York, pp 441–449

Fuxe K, Agnati LF, Andersson K, Zoli M, Benfenati F, Eneroth P, Cuello C (1985 a) Development of quantitative methods for the evaluation of the entity of coexistence of neuroactive substances in nerve terminal populations in discrete areas of the central nervous system: evidence for hormonal regulation of cotransmission. In: Agnati LF, Fuxe K (eds) Quantitative neuroanatomy in transmitter research. MacMillan, London, pp 331–348

Fuxe K, Agnati LF, Härfstrand A, Wikström A-C, Okret S, Yu Z-Y, Goldstein M, Cintra A, Jansson AM, Zoli M, Miller LS, Greene JL, Gustafsson J-Å (1985 b) Morphofunctional studies on central glucocorticoid and estrogen receptors: new openings for the understanding of the hormone regulation of brain function. In: Baulieu EE, Fuxe K, McEwen BS (eds) Table ronde roussel uclaf no 53. Steroids and the brain. Paris, pp 17–21

Fuxe K, Härfstrand A, Agnati LF, Yu Z-Y, Cintra A, Wikström A-C, Okret S, Cantoni E, Gustafsson J-Å(1985 c) Immunocytochemical studies on the localization of glucocorticoid receptor immunoreactive nerve cells in the lower brain stem and spinal cord of the male rat using a monoclonal antibody against rat liver glucocorticoid receptor. Neurosci Lett 60:1–6

Fuxe K, Wikström A-C, Okret S, Agnati LF, Härfstrand A, Yu Z-Y, Granholm L, Zoli M, Vale W, Gustafsson J-Å (1985 d) Mapping of glucocorticoid receptor immunoreactive neurons in the rat tel- and diencephalon using a monoclonal antibody against rat liver glucocorticoid receptor. Endocrinology 117:1803–1812

Fuxe K, Andersson K, Härfstrand A, Agnati LF (1986 a) Increases in dopamine utilization in certain limbic dopamine terminal populations after a short period of intermittent exposure of male rats to cigarette smoke. J Neural Transm 67:15–29

Fuxe K, Agnati LF, Härfstrand A, Ögren S-O, Goldstein M (1986 b) Heterogeneities on the dopamine neuron systems and dopamine cotransmission in the basal ganglia and the relevance of receptor-receptor interactions. In: Fahn S, Marsden CD, Jenner P, Teychenne P (eds) Recent developments in Parkinson's disease. Raven, New York, pp 17–32

Fuxe K, Andersson K, Härfstrand A, Agnati LF, Eneroth P, Janson AM, Vale W, Thorner M, Goldstein M (1986 c) Medianosomes as integrative units in the external layer of the median eminence. Studies on GRF/catecholamine interactions in the hypothalamus of the male rat. Neurochem Int 9, (1):155–170

Fuxe K, Cintra A, Agnati LF, Härfstrand A, Wikström A-C, Zoli M, Miller LS, Greene GL, Gustafsson J-Å (1987) Studies on the cellular localization and distribution of glucocorticoid receptor and estrogen receptor immunoreactivity in the central nervous sys-

tem of the rat and their relationship to the monoaminergic and petidergic neurons of the brain. J Steroid Biochem (to be published)
Ganten D, Pfaff D (eds) (1982) Adrenal actions on brain. Springer, Berlin Heidelberg New York (Current topics in neuroendocrinology, vol 2)
Gordon JH (1985) Estrogenic modulation of striatal dopamine receptor sensitivity. In: Baulieu EE, Fuxe K, McEwen BS (eds) Table ronde Roussel Uclaf no 53. Steroids and the brain. Paris, p 51
Ginsburg M, MacLusky NJ, Morris ID, Thomas PJ (1977) The specifity of oestrogen receptor in brain, pituitary and uterus. Br J Pharmacol 59:397–402
Graybiel AM (1984) Neurochemically specified subsystems in the basal ganglia. In: Evered D, O'Connor M (eds) Functions of the basal ganglia. Ciba Foundation symposium 107. Pitman, London, pp 114–149
Graybiel AM (1986) Dopamine-containing innervation of the stratium: subsystems and their striatal correspondents. In: Fahn S, Marsden CD, Jenner P, Teychenne P (eds) Recent developements in Parkinson's disease. Raven, New York, pp 1–32
Härfstrand A, Fuxe K, Cintra A, Agnati LF, Zini I, Wikström A-C, Okret S, Yu Z-Y, Goldstein M, Steinbusch H, Verhofstad A, Gustafsson J-Å (1986) Glucocorticoid receptor immunoreactivity in monoaminergic neurons of rat brain. Proc Natl Acad Sci USA 83:9779–9783
Healey DP, Maciejewski AR, Printz MP (1986) Localization of central angiotensin II receptors with (^{125}I)-sar$_1$, ile^8-angiotensin II: periventricular sites of the anterior third ventricle. Neuroendocrinology 44:15–21
Heritage AS, Grant LD, Stumpf WE (1977) ^{3}H-Estradiol in catecholamine neurons of rat brain stem: combined localization by autoradiography and formaldehyde-induced fluorescence. J Comp Neurol 176:607–630
Hökfelt T, Rehfeld JF, Skirboll L, Ivemark B, Goldstein M, Markey K (1980) Evidence for the coexistence of dopamine and CCK in meso-limbic neurones. Nature, 285 (5765):476-478
Kato J (1985) Thyroid hormone and arachidonic acid as possible regulatory and/or modulatory factors of brain sex steroid receptors. In: Baulieu EE, Fuxe K, McEwen BS (eds) Table ronde roussel uclaf no 53. Steroids and the brain. Paris, p 35
Ljungberg T, Ungerstedt U (1978) Classification of neuroleptic drugs according to their ability to inhibit apomorphine-induced locomotion and gnawing: evidence for two different mechanisms of action. Psychopharmacology 56:239–247
Löfström A, Bäckström T (1981) Plasma steroid–tissue catecholamine relationships in limbic and related areas of the brain. In: Fuxe K, Gustafsson J-Å, Wetterberg L (eds) Steroid hormone regulation of the brain. Pergamon, Oxford, pp 147–160
McEwen BS (1982) Glucocorticoids and hippocampus: receptors in search of a function. In: Ganten D, Pfaff D (eds) Adrenal actions on brain. Springer, Berlin Heidelberg New York (Current topics in neuroendocrinology, vol 2, pp 1–22)
McEwen BS, Weiss JM, Schwartz LS (1969) Uptake of corticosterone by rat brain and its concentration by certain limbic structures. Brain Res 16:227–241
McEwen BS, Zigmond RE, Azmitia EC Jr, Weiss JM (1970) Steroid hormone interaction with specific brain regions. In: Bowman RE, Datta SP (eds) Biochemistry of brain and behaviour. Plenum, New York, pp 123–167
McEwen BS, Biegon A, Rainbow T, Paden C, Snyder L, DeGroff V (1981) The interaction of estrogens with intracellular receptors and with putative neurotransmitter receptors: implications for the mechanism of activation of sexual behaviour and ovulation. In: Fuxe K, Gustafsson J-Å, Wetterberg L (eds) Steroid hormone regulation of the brain. Pergamon, Oxford, pp 15–30
McEwen BS, Sapolsky RM, Harrelson A, Krey LC, Rostene W (1985 a) Stress, adrenal steroids and the hippocampus. In: Baulieu EE, Fuxe K, McEwen BS (eds) Table ronde Roussel Uclaf no 53. Steroids and the brain. Paris, p 54
McEwen BS, Weiss JM, Schwartz L (1985 b) Selective retention of corticosterone by limbic structures in rat brain. Nature 220:911–912
Meyer JS (1985) Biochemical effects of corticosteroids on neural tissues. Physiol Rev 4:946–1020

Miselis RR, Shapiro RE, Hand PJ (1979) Subfornical organ efferents to neural systems for control of body water. Science 205:1022–1025
Moguilevski M, Raynaud JP (1980) Evidence for a specific mineralocorticoid receptor in rat pituitary and brain. J Steroid Biochem 2:309–314
Moss RL, Dudley CA (1985) Electrophysiologic actions of estrogen and antiestrogen on hypothalamic nerve cell activity. In: Baulieu EE, Fuxe K, McEwen BS (eds) Table ronde Roussel Uclaf no 53. Steroids and the brain. Paris, pp 46–48
Nestler EJ, Rainbow TC, McEwen BS, Greengard P (1981) Effect of steroid hormones on the level of protein I in rat brain. In: Fuxe K, Gustafsson J-Å, Wetterberg L (eds) Steroid hormone regulation of the brain. Pergamon, Oxford, pp 205–216
Nicholson C (1980) Measurements of extracellular ions in the brain. TINS:215–218
Oegren S-O, Archer T, Fuxe K, Eneroth P (1981) Glucocorticoids, catecholamines and avoidance learning. In: Fuxe K, Gustafsson J-Å, Wetterberg L (eds) Steroid hormone regulation of the brain. Pergamon, Oxford, pp 355–379
Olson L, Seiger Å, Fuxe K (1972) Heterogeneity of striatal and limbic dopamine innervation: highly fluorescent islands in developing and adult rats. Brain Res 44:283–288
Oppenheimer JH, Schwartz HL, Surks MI (1974) Tissue differences in the concentration of triiodothyroine nuclear binding sites in the rat: liver, kidney, pituitary, brain, spleen and testis. Endocrinology 95:897–903
Quimet CC, Miller P, Hemmings HC Jr, Walaas I (1984) DARPP-32, a dopamine- and adenosine 3′:5′-monophosphate-regulated phosphoprotein enriched in dopamine-innervated brain regions. III. Immunocytochemical localization. J Neurosci 4(1):111–124
Ramirez VD (1985) In vitro progesterone (P_4) activates LHRH and striatal dopamine nerve terminals by a non-genomic mechanism. In: Baulieu EE, Fuxe K, McEwen BS (eds) Table ronde roussel uclaf no 53. Steroids and the brain. Paris, pp 38–39
Roberts DCS, Shoemaker WJ, Back S, Bloom FE (1981) Modulation of ^{3}H-dihydroalprenolol binding in the hippocampus by corticosterone. In: Fuxe K, Gustafsson J-Å, Wetterberg L (eds) Steroid hormone regulation of the brain. Pergamon, Oxford, pp 379–387
Roberts JL, Wilcox JN, Blum M (1985) The regulation of proopiomelanocortin gene expression by estrogen in the rat hypothalamus. In: Baulieu EE, Fuxe K, McEwen BS (eds) Table ronde roussel uclaf no 53. Steroids and the brain. Paris, pp 59–60
Rostene WH, Sarrieu A, Dussaillant M, Moguilewsky M, McEwen B, Philibert D (1985) In: Baulieu EE, Fuxe K, McEwen BS (eds) Table ronde roussel uclaf no 53. Steroids and the brain. Paris, pp 26–27
Saavedra JM, Correa FMA, Plunkett LM, Israel A, Kurihara M, Shigematsu K (1986) Binding of angiotensin and atrial natriuretic peptide in brain of hypertensive rats. Nature 320:758–760
Sar M, Stumpf WE (1973) Neurons of the hypothalamus concentrate ^{3}H-progesterone or metabolites of it. Science 182:1266–1268
Sar M, Stumpf WE (1977) Distribution of androgen target cells in rat forebrain and pituitary after ^{3}H-dihydrotestosterone administration.J Steroid Biochem 8:1131–1135
Scheff SW, Bernado LS, Cotman CW (1980) Hydrocortisone administration retards axon sprouting in the rat dentate gyrus. Exp Neurol 68:195–201
Schonbrunn A (1985) Mechanisms by which somatostatin inhibits hormone secretion in a rat pituitary cell line. In: Baulieu EE, Fuxe K, McEwen BS (eds) Table ronde roussel uclaf no 53. Steroids and the brain. Paris, p 61
Stumpf WE (1968) Estradiol concentrating neurons: topography in the hypothalamus by dry-mount autoradiography. Science 162:1001–1003
Stumpf WE, Sar M (1975a) Anatomical distribution of corticosterone concentrating neurons in rat brain. In: Stumpf WE, Grant LD (eds) Anatomical neuroendocrinology. Karger, Basel pp 254–261
Stumpf WE, Sar M (1975b) Autoradiographic techniques for localizing steroid hormones. In: O'Malley BW, Hardman JG (eds) Hormone action: steroid hormones. Academic, New York (Methods in enzymology, vol XXXVI, pp 135–156)

Stumpf WE, Sar M (1975c) Hormone architecture of mouse brain with ^{3}H-estradiol. In: Stumpf WE, Grant LD (eds) Anatomical neuroendocrinology. Karger, Basel, pp 254–261
Stumpf WE, Sar M (1978) Anatomical distribution of estrogen, androgen, progestin, corticosteroid and thyroid hormone target sites in the brain of mammals: phylogeny and ontogeny. Am Zool 18:435–445
Stumpf WE, Sar M (1981) Steroid hormone sites of action in the brain. In: Fuxe K, Gustafsson J-Å, Wetterberg L (eds) Steroid hormone regulation of the brain. Pergamon, Oxford, pp 41–50
Swansson LW, Sawchenko PE, Lind RW (1986) Regulation of multiple peptides in CRF parvocellular neurosecretory neurons: implications for the stress response. In: Hökfeld T, Fuxe K, Pernow B (eds) Progress in brain research. Amsterdam, pp 169–190
Taylor RL, Burt DR (1982) Species differences in the brain regional distribution of receptor binding of thyrotropin-releasing hormone. J Neurochem 38:1649–1656
Tennyson VM, Barret RE, Cohen G, Cote L, Heikkila R, Mytilineou C (1972) The developing neurostriatum of the rabbit: correlation of fluorescence histochemistry, electron microscopy, endogenous dopamine levels, and ^{3}H-dopamine uptake. Brain Res 46:251–285
Vale WW, Rivier C, Plotzky P, Brown, Lim A, Bilezikjian L, Perrin M, Chen M, Sawchenko P, Wanson L, Fischer L, Spiess J, Rivier J (1985) Glucocorticosteroid and corticotropin releasing factor. In: Baulieu EE, Fuxe K, McEwen BS (eds) Table ronde roussel uclaf no 53. Steroids and the brain. Paris, pp 55–56
Veldhuis HD, Van Koppen C, Van Ittersum M, De Kloet ER (1982) Specificity of the adrenal steroid receptor system in rat hippocampus. Endocrinology 110:2044–2051
Warwich R, Williams P (1973) Gray's anatomy, 35th ed. Longman, Norwich
Wilkinson M (1978) Pharmacological and physiological correlates of variable receptor sensitivity. Biochem Soc Trans 6:853–858
Wilkinson M, Herdon H, Pearce M, Wilson CA (1979a) Precocious puberty and changes in α- and β-adrenergic receptors in the hypothalamus and cerebral cortex of immature female rats. Brain Res 167:195–199
Wilkinson M, Herdon H, Pearce M, Wilson CA (1979b) Radioligand binding studies on hypothalamic noradrenergic receptors during the oestrous cycle or after steroid injection in ovariectomized rats. Brain Res 168:652–655
Wilkinson M, Herdon H, Wilson CA (1981) Gonadal steroid modification of adrenergic and opiate receptor binding in the central nervous system. In: Fuxe K, Gustafsson J-Å, Wetterberg L (eds) Steroid hormone regulation of the brain. Pergamon, Oxford, pp 253–264
Williams LT, Lefkowitz RJ (1977) Thyroid hormone regulation of β-adrenergic receptor number. J Biol Chem 252:2787–2789
Wuttke W, Mansky T, Stock KW, Sandmann R (1981) Modulatory actions of estradiol on catecholamine and GABA turnover and effects on serum prolactin and LH release. In: Fuxe K, Gustafsson J-Å, Wetterberg L (eds) Steroid hormone regulation of the brain. Pergamon, Oxford, pp 135–146
Zoli M, Agnati LF, Fuxe K, Tesini M, Cantoni E, Gustafsson J-Å (1986) Morphofunctional studies on glucocorticoid receptor immunoreactivity in central nervous system. In: Monoclonal antibodies: basic principles, experimental and clinical application in endocrinology. Sereno symposia publications, vol 30. Raven, New York, pp 159–170
Zoli M, Agnati LF, Fuxe K, Zini I, Cintra A, Aronsson M, Wikström A-C, Okret S, Gustafsson J-Å (1988) Evidence for the existence of a humoral mosaic organization in the dorsal striatum and nucleus accumbens of the male rat: studies on glucocorticoid receptor immunoreactive nerve cells. Brain Res (submitted)

Clinical Studies with Corticotropin Releasing Hormone: Implications for Hypothalamic-Pituitary-Adrenal Dysfunction in Depression and Related Disorders

P. W. Gold[1], M. A. Kling[1], M. A. Demitrack[1], H. Whitfield[1], K. Kalogeras[1], D. L. Loriaux[2] and G. P. Chrousos[3]

Contents

1 Introduction

We now appreciate that the brain is the most prolific of all endocrine organs, producing scores of neurohormones within and beyond the boundaries of the endocrine hypothalamus. The idea that the brain functions as a gland, however, is not new. Indeed, the evolution of thought leading to the identification of corticotropin releasing hormone (CRH) began around 400 B.C. (Zuingerus 1669; Iason 1946). At that time, Hippocrates, in his work entitled *De Glandulus*, stated explicitly "The flesh of the glands is different from the rest of the body, being spongy and full of veins; they are found in the moist part of the body where they receive humidity ... and the brain is a gland as well as the mammae."

Following this farsighted postulate of Hippocrates, elucidation of the brain's endocrine functions proceeded at an extraordinarily slow pace. For instance, 500 years passed before one of the principal target tissues of the neurohormones was inadvertently identified by Galen, which he mistook for a mucous (*pituita*) secretion of the brain (Medvei 1984). Before the hypothalamic-pituitary system of portal vessels was described, another 1600 years passed (Lieutaud 1792), and another 200 years elapsed before its functional significance was understood. It was then,

1 Section of Clinical Neuroendocrinology, Biological Psychiatry Branch, National Institute of Mental Health, Bethesda, MD 20892, USA
2 Developmental Endocrinology Branch, Intramural Research Program, National Institute of Child Health and Human Developement
3 Section of Developmental Endocrinology, Developmental Endocrinology Branch, National Institute of Child Health and Human Development

Current Topics in Neuroendocrinology, Vol. 8

in 1948, that Harris advanced the hypothesis that the hypothalamus produced humoral factors which traversed the hypophyseal portal system to regulate the secretion of anterior pituitary hormones into the systemic circulation (Harris 1948).

By 1955, stimulated by Harris's postulate, the major groundwork for the identification of CRH was laid when Saffran et al. (1955) showed that hypothalamic fragments possessed remarkable corticotropin (ACTH) releasing properties when incubated with pituicytes in vitro. Within 30 years, Vale et al. (1981) announced the sequence of a 41-amino-acid peptide isolated from ovine hypothalami, which showed greater in vivo and in vitro corticotropin releasing potency than any previously identified synthetic or endogenous peptide. Shortly therafter, Schally et al. (1981) described the composition of porcine CRH, and Rivier et al. (1983) the sequence of rat CRH (rCRH). Finally, Numa's group sequenced the genes of both ovine CRH (oCRH) and human CRH (hCRH) and deduced the amino acid sequence of the corresponding peptides (Furatani et al. 1983; Shibahara et al. 1983). Surprisingly, both the rCRH and the hCRH appeared to be chemically identical. Moreover, oCRH and hCRH are also structurally similar, each containig 41 amino acids and showing 83% homology (Fig. 1).

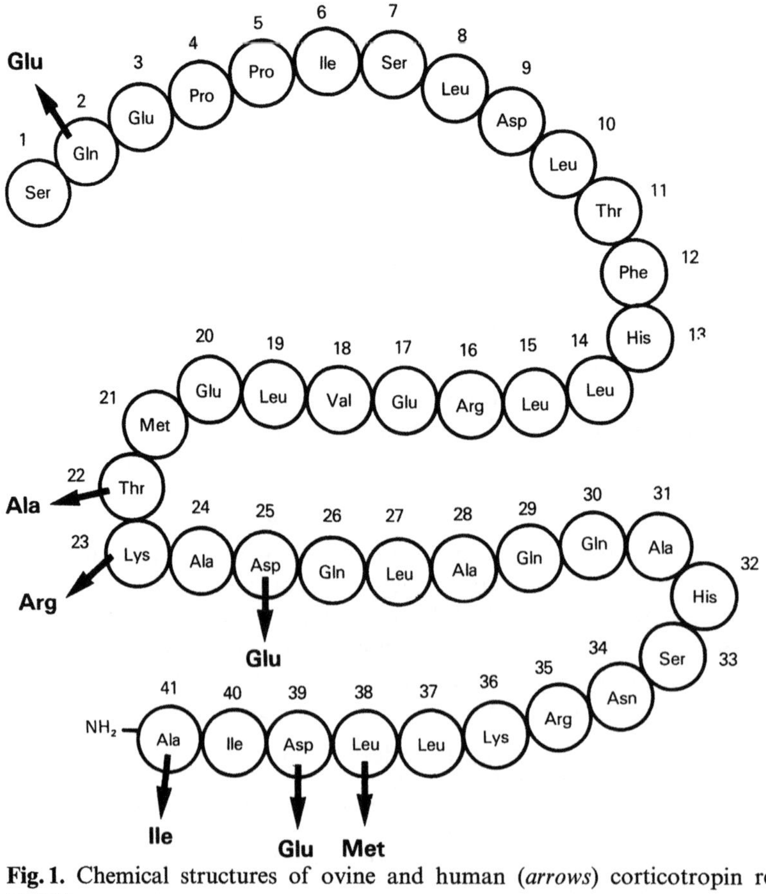

Fig. 1. Chemical structures of ovine and human (*arrows*) corticotropin releasing hormones

The sequencing and subsequent synthesis of CRH has greatly enhanced the capacity for clinical neuroendocrinologists to explore the hypothalamic-pituitary components of Cushing's disease and adrenal insufficiency. CRH is also of importance to clinical psychoneuroendocrinologists since its discovery occured in the context of two other discoveries which have heightened their interest in the hypothalamic-pituitary- adrenal (HPA) axis and its regulation. First, many patients with depression (Sacher et al. 1970; Carrol et al.), anorexia nervosa (Gerner and Gwirtsman 1981), alcoholism (Stokes 1973), and obsessive-compulsive neurosis (Insel et al. 1982) manifest a hyperactive pituitary-adrenal axis. In fact, the hypercortisolism seen in depressive illness and alcoholism can be so severe that it is difficult to distinguish them from Cushing's disease; hence each of these psychiatric entities has been referred to colloquially by some endocrinologists as a pseudo-Cushing's state. Secondly, it has been shown that ACTH is secreted synchronously with beta-endorphin, one of the principal endogenous opioid peptides (Guillemin 1977), and that both hormones are contained within the sequence of a common precursor molecule, proopiomelanocortin (POMC; Mains et al. 1977). CRH is the principal central signal for the cleavage of pituitary POMC into biologically active peptides.

CRH is also of interest to psychiatrists and neurobiologists for reasons other than its putative role in regulating the pituitary-adrenal axis. CRH is not only synthesized by the hypothalamus for transport by hypophyseal portal blood, rather like other hypothalamic hormones, but is widely distributed and/or synthesized beyond the boundaries of the hypothalamus to play an important role in coordinating complex behavioral and physiological processes of survival value. Neuroanatomically, it has been shown that there are extensive hypothalamic aggregations of CRH cell bodies and terminal fields in the limbic system, cortex, and in close association with the central autonomic system and the locus ceruleus (Bloom et al. 1982; De Souza et al 1984; Olschowka et al. 1982). This distribution of CRH within and beyond the hypothalamus provides an anatomical context for the observation that CRH can simultaneously activate and coordinate metabolic (Brown et al. 1982), circulatory (Brown et al. 1982), and behavioral responses that are adaptive in stressful situations (Britton et al. 1982; Sirinathsinghji et al. 1983; Sutton et al. 1982). Hence, in the rat, intracerebroventricular (ICV) administration of CRH leads not only to activation of the HPA axis but also to activation of the sympathetic nervous system (Brown et al. 1982) and to several behavioral changes characteristic of the stress response, including decreased feeding (Britton et al. 1982) and sexual behavior (Sirinathsinghji et al. 1983), assumption of a freeze posture in a foreign environment (Sutton et al. 1982), and decreased exploration in familiar surroundings (Sutton et al. 1982). In addition, in collaboration with Weiss and Post, our group has shown that CRH given ICV to the rat causes a marked increase in hostility and induces limbic seizures which show cross-sensitization with electrically kindled seizures (Weiss et al. 1987).

Given CRH's significant role in HPA regulation and its intriguing effects on CNS function, we embarked on a series of clinical studies with CRH both in normal volunteers and in pschiatric patients with major psychiatric disorders whose illnesses were at times characterized by hypercortisolism (e.g., primary affective disorder, anorexia nervosa, panic anxiety disorder, and schizophrenia). In addition, we studied patients with endocrine disturbances characterized by abnormal HPA function, including subjects with Cushing's disease, ectopic ACTH secretion, primary and secondary adrenal insufficiency, and Nelson's syndrome. In volunteers, we hoped to examine the physiological relevance of CRH to pituitary-adrenal function in man as well as to explore the differential biological effects and pharmacokinetics of oCRH and hCRH under varying conditions. Using our patient populations, we hoped to answer the following questions. (1) Can CRH help determine whether the hypocortisolism in depression reflects an alteration in the set point for feedback inhibition of cortisol on ACTH secretion of endogenous CRH? (2) Can CRH help in the differential diagnosis of the various hypercortisolemic psychiatric syndromes? (3) Can CRH help determine whether depression and Cushing's disease lie on a common pathophysiological continuum or represent distinct abnormalities of the HPA axis? (4) Can CRH help establish the differential diagnosis of disturbances of HPA function which can be difficult to distinguish from one another, such as depression from Cushing's disease, ectopic ACTH production from Cushing's disease, or hypothalamic from pituitary-adrenal insufficiency? (5) Is CRH of possible relevance to the overall symptom complex of major psychiatric illnesses such as depression and anorexia nervosa?

2 Relevance of CRH to Pituitary Adrenal Regulation in Man

Even before the sequencing of CRH, this peptide was postulated to play a major role in the regulation of pituitary-adrenal function. This role ascribed to CRH is schematically illustrated in Fig. 2, which depicts the classic formulation of HPA regulation that has held sway for at least the past two decades. According to this formulation, CRH is secreted in seven to ten discrete episodes per day (under unstressed conditions), which in turn produce a similar number of ACTH and cortisol secretory episodes. The majority of these pulses occur in the early morning hours. Since cortisol has a longer half-life than ACTH (and presumably CRH), the frequent cortisol pulses occuring in the early morning produce a situation in which a new cortisol secretory episode may be initiated before the last episode is terminated, producing a rise in cortisol with each successive ACTH pulse. Hence, with an increasing frequency of CRH and ACTH pulses, there emerges an increase in the level or amplitude of plasma cortisol secretion which we recognize as the early morning plasma corticol surge. This phenomenon of the early morning rise in amplitude of cortisol secretion is an example of frequency modulation (i.e., timing of CRH, ACTH, and cortisol pulses) being translated into amplitude modulation (i.e., amplitude of cortisol secretion).

Over the years (and particularly since the sequencing of CRH), this classic formulation has been modified by additional information. For instance, it has been known for some time that vasopressin is capable of stimulating the pituitary corticotroph cell, and it has even been postulated to be the major hypothalamic CRH peptide. It is now known that arginine vasopressin (AVP) itself is not as potent as hCRH(1–41) in stimulating ACTH secretion, but that AVP markedly potentiates CRH's corticotropin releasing properties (Gilles et al. 1982). Moreover,

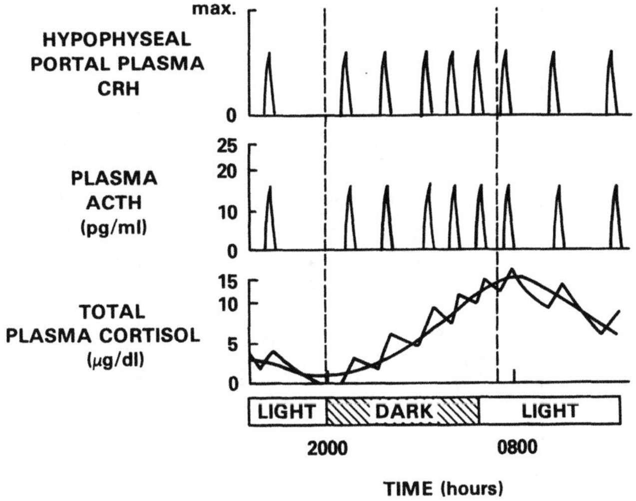

Fig. 2. Daily plasma ACTH and cortisol rhythm in a normal, unstressed person. Secretory episodes of ACTII (7–10/24 h) occur more frequently in early morning hours. The corresponding cortisol secretory episodes at that time "accumulate", a result of the relatively long plasma half-life of cortisol, and produce the circadian morning cortisol surge. Thus, modulation of frequency of ACTH secretory episodes results finally in modulation of the amplitude in daily cortisol rhythm. Presumably, secretory episodes of cortisol could result from an equal number of secretory episodes of CRH secreted into the hypophyseal portal blood

other substances such as the catecholamines also potentiate CRH-induced ACTH secretion (Vale et al. 1983). In light of these data, and because of a paucity of clinical studies with hCRH, one of the first studies our group undertook with this peptide was to explore its possible relevance to the physiologic regulation of the pituitary-adrenal axis in man. To address this question, we first studied normal controls to compare naturally occurring ACTH pulses to those induced by hCRH administration. Figure 3 shows that a comparison of spontaneous endogenous ACTH and cortisol surges (isolated during normative circadian studies) is identical to those induced by an intravenous bolus of hCRH of 1 μg/kg (Avgerinos et al. 1987 a). This finding, that hCRH-induced plasma ACTH and cortisol pulses mimicked the spontaneously occuring secretory episodes of these hormones, provided the first compelling evidence that CRH may be of physiological relevance to the pituitary-adrenal axis in man.

The finding that hCRH produces ACTH pulses which closely resemble endogenous ACTH secretory episodes led to an additional experiment designed further to explore the relationship between pulsatile CRH secretion and the pattern of ACTH and cortisol secretion in the basal state. In this study by Avgerinos et al. (1987 b) we asked the following question: could hCRH, given in pulses that simulate the hypothetical pattern shown in Fig. 2, restore the function of the pituitary-adrenal axis in patients with CRH deficiency (i.e., secondary adrenal insufficiency due to corticotroph-sparing [superpituitary] lesions)? The patients included in this study had secondary adrenal insufficiency determined by a standard

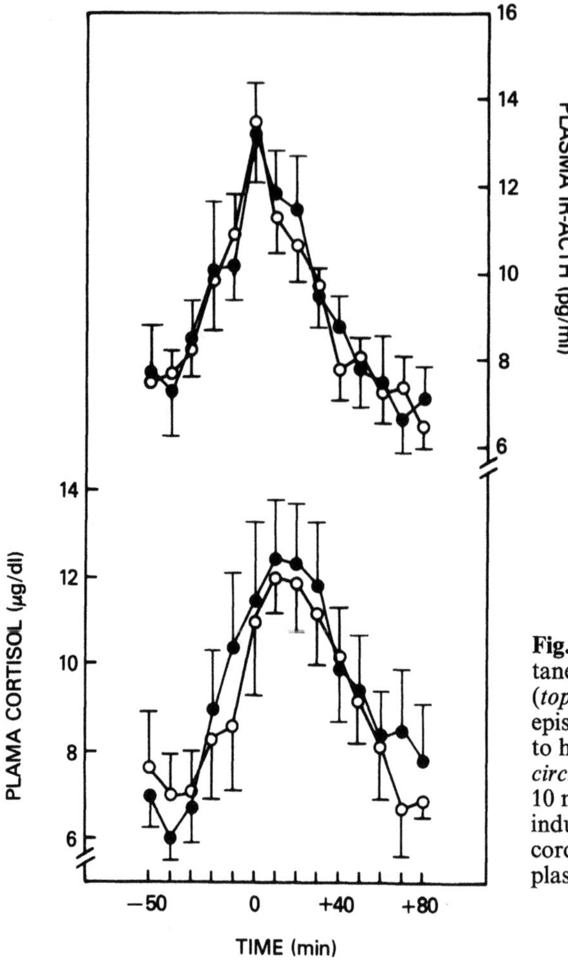

Fig. 3. Comparison of natural, spontaneously occuring plasma ACTH (*top*) and cortisol (*bottom*) secretory episodes (*open circles*) and responses to human CRH (hCRH; *filled circles*). Samples were drawn every 10 min. All spontaneous or hCRH-induced episodes were centered according to their highest (peak) plasma ACTH value at time 0

48-h ACTH stimulation test. Thus, during a standard 3-h evening oCRH test, the plasma cortisol response of these patients was diminished while the plasma ACTH response was normal or exaggerated and showed a delayed pattern, suggestive of a hypothalamic lesion. hCRH was given as a bolus of 1 µg/kg eight times during the 24-h period. The timing of each pulse of hCRH was chosen to correspond to the expected times of ACTH pulsation under natural conditions. Hence, the majority of the pulses were given in the early morning hours to correspond with the morning control surge. To a degree which was unexpected, such a paradigm of pulsatile hCRH administration reproduced the normal amplitude and circadian variation of cortisol secretion in patients with hypothalamic CRH deficiency (Fig. 4). Moreover, the administration of eight pulses of synthetic hCRH, as described above, normalized 24-h urinary 17-hydroxysteroid and urinary free cortisol secretion in these patients (Avgerinos et al. 1987b). Parenthetically, we have previously noted that a continuous infusion of oCRH in normal volunteers for 24 h produces a pattern of cortisol secretion which includes pre-

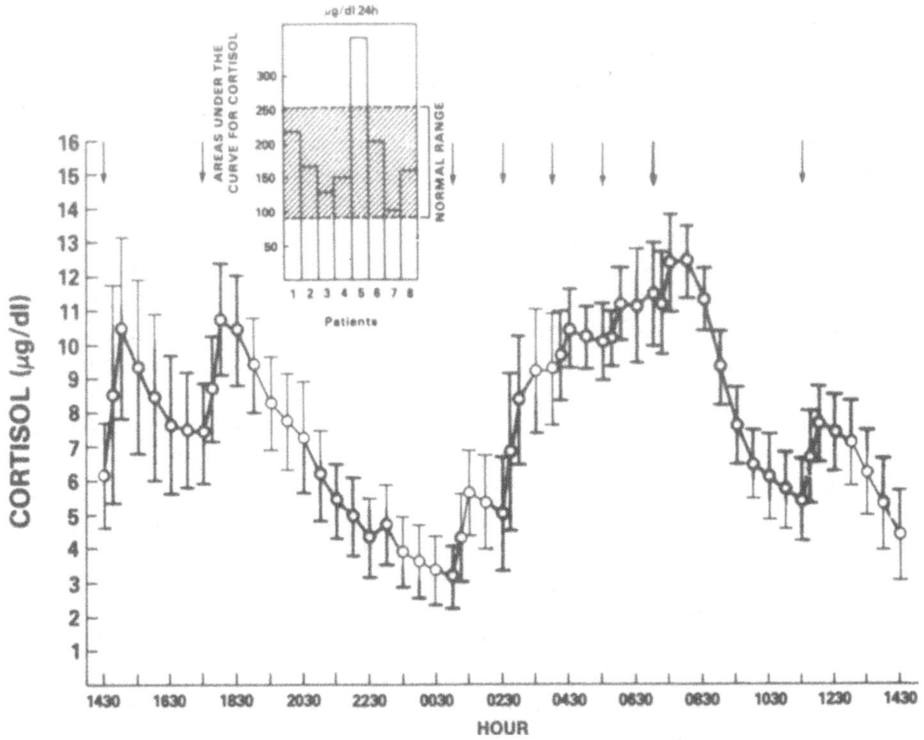

Fig. 4. The mean plasma cortisol responses to pulses of hCRH at the times indicated by the *arrows*. The shaded areas represent the range in eight normal subjects studied in the basal state. *Inset*, individual areas under the plasma cortisol curve during the 24 h of pulsatile hCRH administration

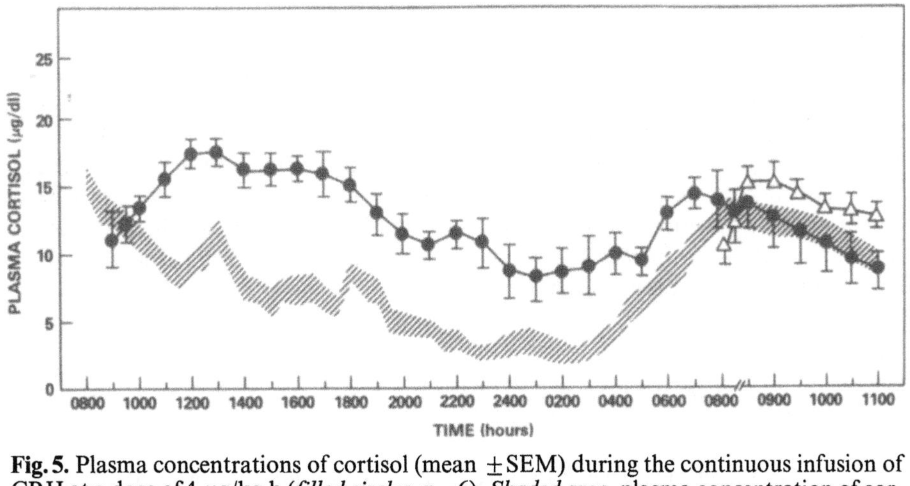

Fig. 5. Plasma concentrations of cortisol (mean ±SEM) during the continuous infusion of CRH at a dose of 1 μg/kg h (*filled circles*, $n=6$). *Shaded area*, plasma concentration of cortisol (mean ±SEM) during normal saline infusion ($n=11$). *Open triangles*, plasma cortisol concentration after a bolus of CRF 1 μg/kg given at *arrow*

servation of a circadian rhythm (Fig. 5) though the amplitude is blunted compared to the naturalistic rhythm of that induced by the hCRH pulses (Schulte et al. 1985 a). Thus, although the basal circadian cortisol secretion may depend on endogenous CRH secretion, this circadian pattern of pituitary-adrenal function may also involve a component of a circadian rhythm in the responsiveness to CRH itself (Schulte et al. 1985 a). This is further suggested by our data on pulsatile hCRH administration in patients with hypothalamic CRH deficiency (in which the responses to the early morning pulses are greater than the responses to the daytime pulses; Avgerinos et al. 1987 b) and by our own delineation of the naturalistic pattern of circadian ACTH secretion in normal volunteers. This latter study showed that the early morning cortisol surge is associated with an increase in both the pulse frequency and amplitude of ACTH secretion (Schuermeyer et al., unpublished observations).

In addition to studies of the relevance of CRH to basal and circadian ACTH and cortisol secretion, we have also attempted to assess the relationship of CRH to pituitary-adrenal function during stress. To accomplish this task, we have attempted to see whether the ACTH responses to frequent pulses of hCRH given at 30 to 90 min intervals from 1800 to 2000 would produce an ACTH secretory pattern resembling that seen during the standard insulin tolerance test. We noted that ACTH and cortisol responses to repeated pulses of hCRH were much less than those seen during the insulin tolerance test (Avgerinos et al., unpublished observations). Although neutralization of endogenous CRH in rats by administration of CRH-specific antibodies abolishes more than 75% of the ACTH response to insulin-induced hypoglycemic stress (Rivier and Vale 1983), our data is compatible with previous suggestions that factors other than CRH play an important role in producing the ACTH responses seen during stress. Putative factors which may underlie these extra-CRH influences on stress-induced ACTH secretion include the catecholamines and AVP (Gilles et al. 1982), both of which are known to increase during hypoglycemia and other forms of stress (Rivier and Vale 1983; Plotsky et al. 1985; Axelrod and Reisine 1984). In support of the possible role of vasopressin is our data obtained in normal volunteers which shows that osmotically induced vasopressin secretion markedly potentiates CRH-induced ACTH secretion (Rittmaster et al., unpublished observations), compatible with other studies showing synergism between exogenous vasopressin administration and CRH in human volunteers. We doubt that oxytocin participates in CRH-induced ACTH secretion, despite a suggestion from in vitro studies, where it has been shown to have synergy with CRH in causing ACTH secretion. We have observed no synergy of oxytocin with hCRH in eliciting ACTH secretion in man (Nieman et al., unpublished observations).

3 Development of a Clinically Applicable CRH Stimulation Paradigm

To ascertain the clinical applications of CRH, we initiated a series of studies in volunteers to assess the following questions. (1) Which peptide (oCRH or hCRH) might it be best to use in acute challenges of the pituitary-adrenal axis? (2) What dose should be administered in these studies and for how long should hormonal responses be sampled? (3) What time of day would be best suited for the performance of dynamic stimulation of the human pituitary-adrenal axis? We conducted pharmacokinetic and dose-response studies with both oCRH and hCRH. Some of these studies were performed at different times of day to correspond to periods when the HPA axis is normally quiescent or most active.

The first dose–response study with oCRH in primates was performed by our group in cynomolgus macaques (Schulte et al. 1982a). Corresponding studies in man yielding similar results were performed by Grossman et al. (1982) and Orth et al. (1983). These studies show that the lowest maximal stimulatory dose for cortisol secretion was 1 µg/kg; moreover, this dose produced clear-cut plasma cortisol and ACTH secretion in all volunteers and experimental animals without de-

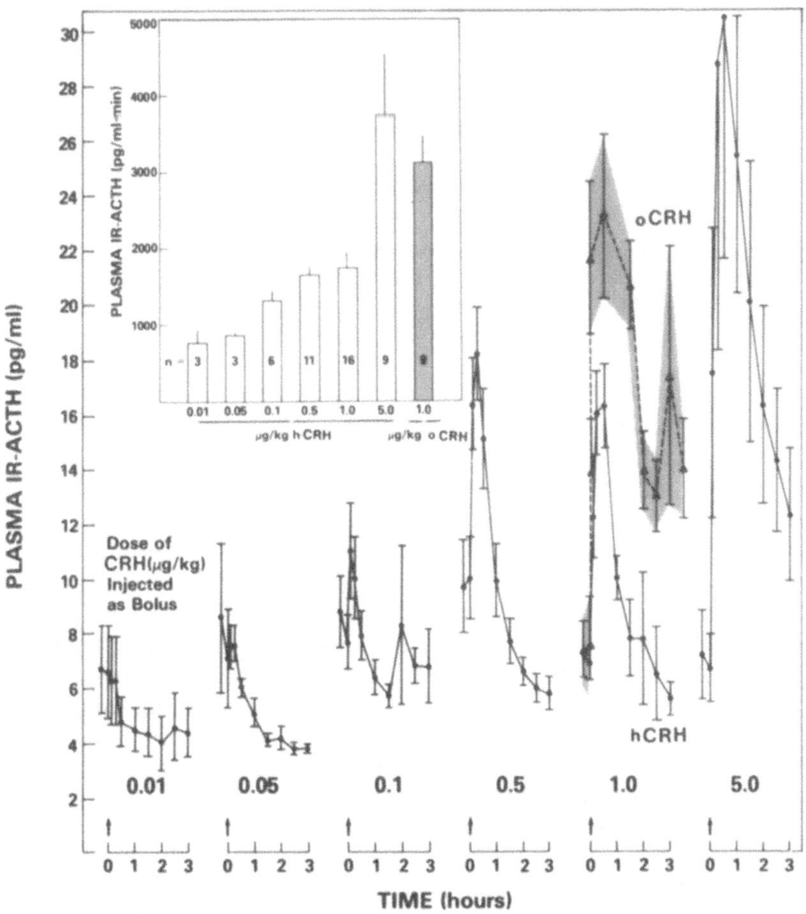

Fig. 6. Dose-response relation of the effects of intravenous hCRH on plasma ACTH. *Shaded area*, the response observed after oCRH 1 µg/kg for comparison. *Inset*, time-integrated (over 3 h) levels of plasma ACTH in response to graded doses of hCRH. *Shaded bar*, the response to oCRH 1 µg/kg

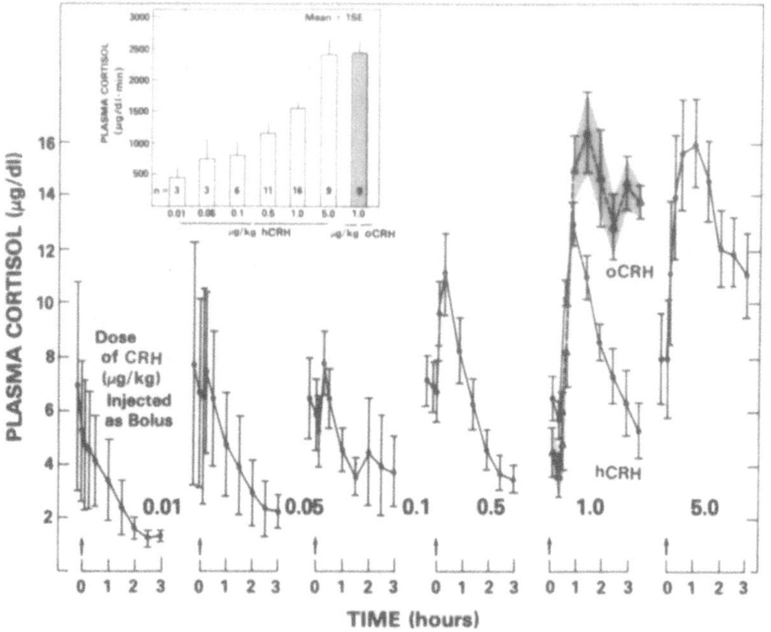

Fig. 7. Dose-response relation of the effects of intravenous hCRH on plasma cortisol. *Shaded area*, the response observed after oCRH 1 µg/kg for comparison. *Inset*, time-integrated (over 3 h) levels of plasma cortisol in response to graded doses of hCRH. *Shaded bar*, the response to oCRH 1 µg/kg

tectable adverse effects. Of particular interest was the fact that the ACTH and cortisol responses to oCRH were prolonged, remaining clearly elevated at the end of the 3-h sampling period (Schulte 1982a).

In our similar dose–response studies with hCRH in nonhuman primates (Schuermeyer et al. 1985) and man (Schuermeyer et al. 1984) we noted a dose-dependent increase of plasma ACTH and cortisol concentrations with graded doses of hCRH (Fig. 6 and 7). Peak plasma ACTH and cortisol responses to hCRH were significantly lower than those achieved by oCRH (Schuermeyer et al. 1984, 1985). Moreover, the ACTH and cortisol responses to hCRH were of much shorter duration than those to oCRH. Accordingly, comparisons of the integrated secretory responses of both ACTH and cortisol following hCRH administration indicate that oCRH is at least five times more potent than hCRH (Schuermeyer et al. 1984, 1985). This difference is mainly due to the longer-lasting effect of oCRH upon ACTH and cortisol secretion.

These longer-lasting effects of oCRH on ACTH and cortisol secretion can presumably be accounted for on the basis of the different pharmacokinetic properties of oCRH (Schulte et al. 1982b) and hCRH (Schuermeyer et al. 1984, 1985). Hence, in our study directly comparing the metabolic clearance rate of these two peptides in human volunteers, noted that hCRH is cleared from plasma much more rapidly than oCRH (Schuermeyer et al. 1984, 1985) (Fig. 8). On the basis of the relatively slower clearance and more prolonged biological effects of oCRH,

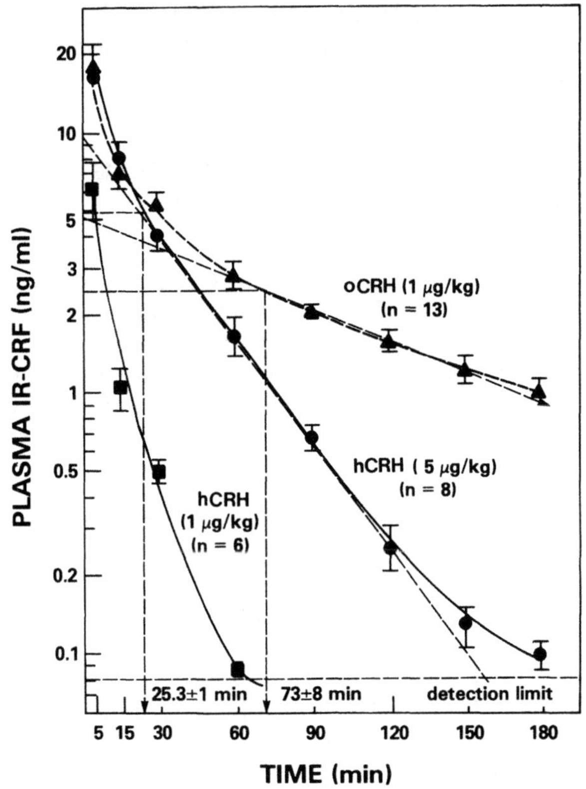

Fig. 8. Disappearance curves of immunoreactive (IR) ovine CRH and hCRH from plasma after administration of a bolus injection. The calculated metabolic clearance rate (MCR = dose/area under curve) of hCRH was three to four times that of oCRF

we elected to use this peptide to characterize the functional integrity of the pituitary corticotroph in our clinical populations. We reasoned that the extra information provided by a pulse of oCRH might provide the kind of additional information that could be helpful in exploring the pathophysiology of HPA activity in different patient subgroups. Moreover, we felt that this additional information could also be helpful in determinig subtle differences in responses between patient subgroups and hence more helpful in establishing differential diagnoses between clinical entities. On the other hand, as illustrated by the studies in patients with hypothalamic CRH deficiency (Avgerinos et al. 1987b), the much more rapidly cleared hCRH seems far more suitable than oCRH for studies of pulsatile ACTH secretion.

An additional factor of relevance to the establishment of a clinically applicable CRH stimulation test is the determination of an optimal time of day for administration of the peptide to patients. To explore this question, we administered a bolus of oCRH 1 μg/kg at 0900, near the time of day when the axis is most active, and at 2000 when the axis is normally dormant. We found that owing to the lower baseline cortisol levels seen in the evening, the net integrated cortisol responses to oCRH are greater at this time (Schulte et al. 1985 b). Hence, we decided that our CRH stimulation test would consist of a bolus of oCRH 1 μg/kg given at the 2000 time.

4 Clinical Studies with oCRH Patients with Major Psychiatric Disorders

The first major finding utilizing CRH in psychiatry was made by our group when we noted that 12 of our drug-free depressed patients showed a significantly blunted ACTH response to oCRH (Gold et al. 1984). This finding was subsequently reproduced by Holsboer et al. (1984) in 12 subjects. We have continued to see this pattern in a larger series of depressed patients, and most recently have replicated our finding in a large series of 32 patients (Fig. 9) (Gold et al. 1986 a).

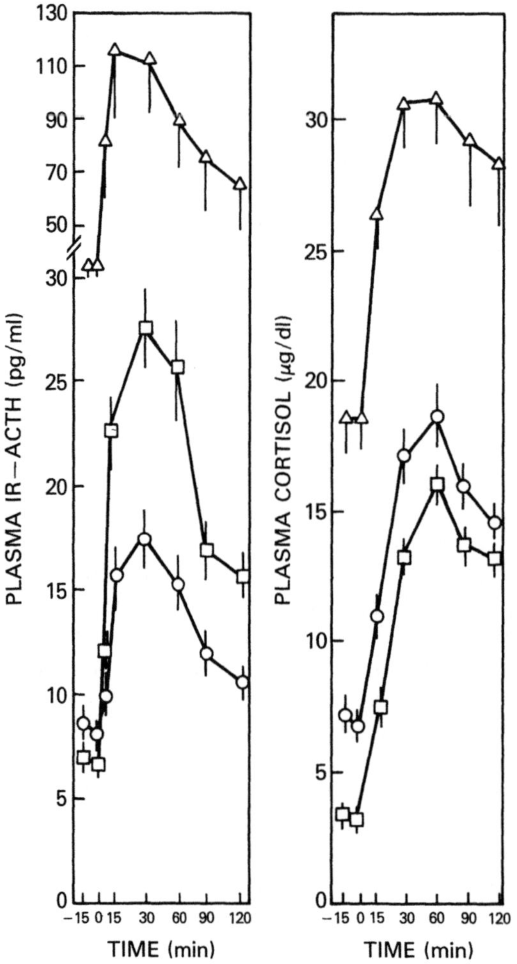

Fig. 9. Comparison of plasma (*left*) and cortisol (*right*) responses to oCRH in controls (*squares*, $n=34$), patients with depression (*circles*, $n=30$), and patients with Cushing's disease (*triangles*, $n=29$). All data are expressed as mean ±SEM

The finding of an attenuated response to oCRH in depression suggested that the pituitary corticotroph cell in depressed patients was appropriately restrained by the negative feedback effects of elevated cortisol levels (Gold et al. 1984, 1986a; Gold and Chrousos 1985). This hypothesis was supported by finding of significant negative correlation between basal cortisol levels and the ACTH response to CRH in depression (Gold et al. 1984, 1986a; Gold and Chrousos 1985).

In light of the apparently normal corticotroph cell function in depressed patients, we first advanced the hypothesis that hypercortisolism in depression represents a defect at or above the hypothalamus which results in the hypersecretion of endogenous CRH. To test this hypothesis, we attempted to replicate in normal controls a situation in which the pituitary corticotroph cell is exposed to excess CRH. In order to accomplish this, we administered a continous infusion of oCRH for 24 h and evaluated the ACTH and cortisol responses (Schulte et al 1985a). As noted earlier, Fig. 5 shows the cortisol responses to this continous infusion of oCRH. Of interest is the fact that the circadian rhythm of cortisol is preserved despite the continuous administration of oCRH, suggesting that the pituitary corticotroph cell shows a diurnal sensitivity to exogenous CRH. This is of interest in light of the fact that the circadian rhythm of cortisol is generally preserved in depression (Halbreich et al. 1985). Of additional interest is that the mean amplitude of cortisol secretion during continous CRH infusion is elevated by about 40% to 50%, and that the mean urinary free cortisol secretion during CRH infusion was 150–200 μg/day (Schulte et al. 1985a). Hence, the amplitude of plasma cortisol during the 24-h period and the magnitude of urinary free cortisol hypersecretion are very similar during continuous administration of oCRH to controls and in the endogenously depressed state. We concluded, therefore, that a continous infusion of CRH to normal volunteers reproduces the pattern and magnitude of hypercortisolism typically associated with depression. Additional data compatible with the idea that CRH is hypersecreted in depressed patients derives from the data of Nemeroff et al. (1984) who showed that the level of CSF CRH is elevated in depression. Although we could not demonstrate a significant elevation of CSF CRH in depressed patients (Roy and Gold, unpublished observations), these subjects did manifest a significant positive correlation between postdexamethasone cortisol levels and the CSF level of CRH (Roy et al., unpublished observations). Moreover, we noted that CSF CRH is significantly higher in depressed patients who are dexamethasone nonsupressors than in dexamethasone supressors.

Inspection of our ACTH and cortisol responses to CRH in depression revealed other salient features of HPA dysfunction in depressed patients. For instance, we noted that depressed patients showed a robust total and free cortisol response despite the very small amount of ACTH released during CRH stimulation (Gold et al. 1984, 1986a; Gold and Chrousos 1985). In fact, the free cortisol response to CRH was even greater in depressed patients than in controls (Gold et al. 1986a). We surmise from this data that the adrenal cortex in depression has grown hyperresponsive to ACTH, compatible with the well-described phenomenon of progressive functional and anatomical hypertrophy of the adrenal cortex seen either during experimentally induced stress (Selye 1936) or during the course

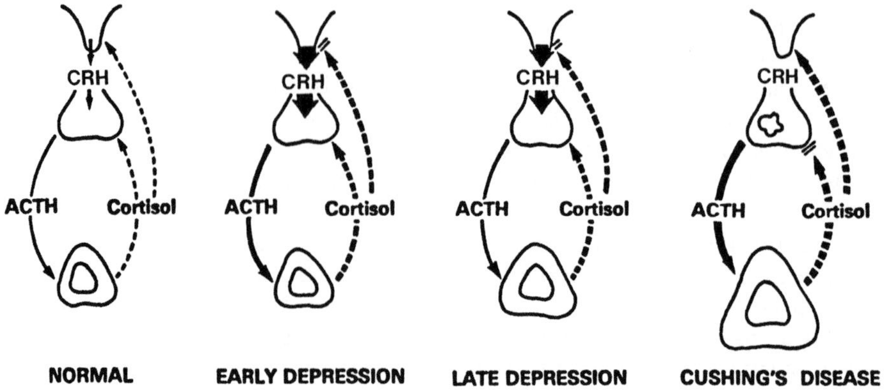

Fig. 10. Schematic representation of a hypothetical model depicting the functional status of the HPA axis in depression and Cushing's disease

of chronic and repeated hyperstimulation of the adrenal cortex by ACTH in man (Renold et al. 1952). This suggestion of adrenal hyperresponsiveness to ACTH in depression is compatible with the data of Amsterdam et al. (1938), which showed that chronically depressed patients manifest greater cortisol responses to a bolus of exogenous ACTH than normal subjects.

Although our depressed patients were hypercortisolemic, it is noteworthy that basal ACTH levels remained in the normal range (Gold et al. 1984, 1986a; Gold and Chrousos 1985). This "normal" plasma ACTH level in depression most likely reflects a normal corticotroph cell caught in the balance between forces (i.e., negative feedback exerted by a hyperactive adrenal cortex from below and a predominating excess of CRH drive from above). Hence, the corticotroph cell, though restrained by the negative feedback to secrete at a rate that produces ACTH levels in the normal range, is nevertheless sufficiently driven by CRH to promote excessive cortisol secretion by hyperplastic adrenals. Presumably, depressed patients would have shown elevated levels of ACTH in the beginning of their depressive illness. A schematic diagram of these proposed relationships is shown in the first three panels from the left in Fig. 10 (Gold et al. 1986a).

We have also explored the pathophysiology of hypercortisolism in other major psychiatric disorders including anorexia nervosa. We note that underweight patients with anorexia nervosa show hypercortisolism which is even more severe than that seen in depressed patients. However, like the depressed patients, these underweight anorexics manifest a markedly attenuated ACTH response to exogenous CRH (Gold et al. 1986b). This finding strongly suggests that, as in the depressed patients, the pituitary corticotroph cell in anorexia nervosa is appropriately restrained by the negative feedback effects of hypercortisolism, and also suggests that anorexics, like depressed patients, show a defect in the secretion of CRH. In support of this hypothesis is our finding that underweight patients with anorexia nervosa show significantly higher CSF CRH levels than controls (Kaye and Gold, unpublished observations). When these underweight anorexic subjects were restudied after their weight had stabilized at 100% of normal body weight, their basal hypercortisolism had resolved, suggesting normalization of the central defect which resulted in the hypersecretion of endogenous CRH (Grossman et al. 1982). Moreover, the level of CRH in the CSF also returned to normal at this time (Kay and Gold, unpublished observations). However, despite normalization of the apparent central defect in CRH secretion which had produced their hypercortisolism, these normal-weight anorexics continued to show a markedly attentuated ACTH response to CRH (Gold et al. 1986 b). Although we cannot definitely account for this finding, it may represent the persistence of a functionally hypertrophied adrenal cortex into this phase of short-term recovery. On the other hand, normal-weight bulimic subjects studied for 1 month after a voluntary absence from bingeing and vomiting and a group of subjects with anorexia nervosa who had maintained normal body weight for at least 6 months showed normal basal ACTH and cortisol values in their responses to exogenous CRH (Gold et al. 1986 b). From these studies, we conclude that the basic pathophysiology of hypercortisolism in anorexia nervosa is similar to that seen in the depressed phase of primary affective disorder and that subtle defects in HPA function persist in patients with anorexia nervosa despite the return to eucortisolism after short-term correction of the weight loss.

The apparently similar pathophysiology of hypercortisolism in depression and anorexia nervosa supports the notion that these two disorders lie on a pathophysiologic continuum. Clinical data also support this idea. For instance, both patients with depression and those with anorexia nervosa manifest not only hypercortisolism but also hypothalamic hypogonadism and anorexia. Moreover, many anorexia nervosa patients are depressed and show strong family histories for depression (Cantwell 1977). Parenthetically, we observed a significant positive correlation between depression ratings and CSF CRH levels in anorexic patients (Kay and Gold, unpublished observations), suggesting that CRH may be an important link between primary affective disorder and anorexia nervosa. We have also noted attenuated ACTH responses to CRH in hypercortisolemic patients with anxiety disorder (Roy-Byrne et al. 1987), suggesting that these patients show a similar pathophysiology of HPA function to that seen in subjects with depression and anorexia nervosa.

In contrast to the three hypercortisolemic psychiatric subgroups who showed blunted ACTH responses to CRH, a group of severely psychotic schizophrenic patients showed normal basal cortisol values and normal ACTH responses to exogenous CRH (Roy et al. 1986). These data are compatible with the previously published data of Nemeroff et al. (1984) showing normal CSF CRH in schizophrenics. Moreover, these data are compatible with previous studies which show that a smaller percentage of patients with schizophrenia than of those with affective disturbance fail to suppress their cortisol levels after dexamethasone. Parenthetically, when eight of these drug-free schizophrenic patients were studied after treatment with fluphenazine, the ACTH and cortisol responses to CRH were similar to those seen during the drug-free state. In light of this eucortisolism and the normal ACTH responses to CRH in these schizophrenic patients, it is intriguing to speculate that perhaps they showed a defect in the translation of their enormous distress into activation of the CRH neuron. Additional work on exploring the perturbability of the CRH neuron in schizophrenic subjects following experimentally induced stress (i.e., hypoglycemic or exercise) would seem warranted to further test this hypothesis.

Whether CRH plays a role in any human disease apart from the rare cases of Addison's disease secondary to CRH deficiency or Cushing's syndrome secondary to ectopic CRH secretion remains to be established. Its possible involvement in depression is intriguing in light of the following four sets of findings taken from the disciplines of developmental psychology, clinical psychiatry, and neurophysiology (Gold et al. 1984).

1. Laboratory animals subjected to maternal deprivation during the neonatal period show significant hyperactivity of the hypothalamic-pituitary-adrenal axis during stress throughout adult life (Thomas et al. 1968). Hence, such animals presumably show a permanent change in the responsivity of their CRH neuron.

2. Individuals who are depression-prone are thought to show greater than usual incidence of early noxious stress or maternal deprivation. Clinical experience shows that such a history seems to produce a tendency to repetitively relive the intense anxiety and dysphoria associated with this early deprivation throughout adult life whenever a significant frustration or important loss occurs. Thus, such individuals also seem prone to a hyperresponsivity of their CRH neuron intermittently throughout life.

3. CRH given ICV to primates not only stimulates the HPA axis (Rock et al. 1984) but also activates the locus ceruleus (Valentino et al. 1983), produces decreased eating (Britton et al. 1983) and sexual behavior (Sirinathsinghji et al. 1983), and causes significant changes in activity (Sutton et al. 1982)

4. CRH has been reported to induce limbic seizures which cross-sensitize with electrically kindled seizures (Weiss et al. 1987). These findings, taken together, suggest that a CRH model of depression could help integrate dynamic formulations which take into account early losses and subsequent internal and external stress as factors which can predispose to or precipitate major depression, and the

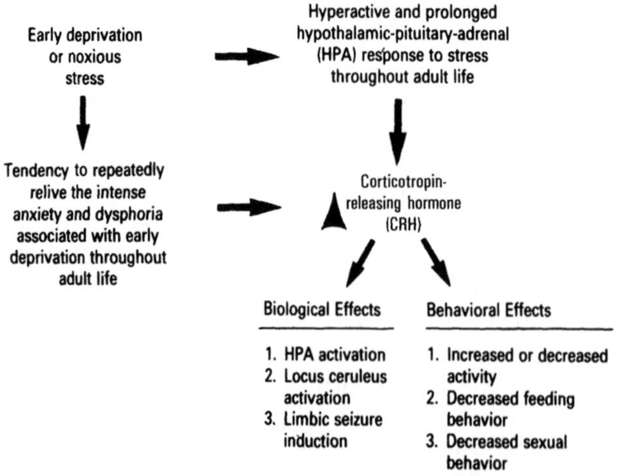

Fig. 11. Schematic representation of a hypothetical model of depression. This model suggests that the putative effect of early deprivation in increasing the vulnerability to depression may be translated biologically via regulatory peptides such as CRH

observations that depressed subjects often show hypercortisolism, significant anxiety, anorexia, diminished libido, hypo- or hyperactivity, and respond at times to limbic anticonvulsants (Post 1982). A schematic representation of this model is shown in Fig. 11. That changes in CRH may be related to depressive symptomatology is also supported by empirical observations that depression is perhaps the only major symptom represented in a substantial number of patients with each of the various psychiatric disorders characterized during their course by sustained or episodic hypercortisolism.

5 The CRH Stimulation Test: Implications for Hypercortisolism in Depression and Cushing's Disease

The hypercortisolism of depression can be of sufficient magnitude that it has been termed a pseudo-Cushing's state. Conversely, patients with Cushing's disease often show signs of clinical depression. Although there has been controvery over the years concerning the etiology of the hypercortisolism associated with affective illness and Cushing's disease, the overlap in the clinical and biochemical manifestations of these illnesses has prompted some to suggest that they share common pathophysiological features. Of clinical significance is the fact that patients with primary depression who may be hirsute or obese and who manifest high plasma and urinary free cortisol levels can be impossible to distinguish from patients with mild or early Cushing's disease. Indeed, depression can often be the first manifestation of Cushing's disease preceding the physical stigmata such as the buffalo hump or purple striae by months or even years.

Data from our group and others show that despite profound basal hypercortisolism, patients with Cushing's disease show a marked hyperresponsiveness of the pituitary corticotroph cell to exogenous CRH (Gold et al. 1986a; Chrousos et al. 1984a, b, 1987; Lytras et al. 1984; Muller et al. 1983; Nakahara et al. 1983; Orth 1984; Orth et al. 1982; Peters et al. 1983) (Fig. 9). Thus, in contrast to patients with depression , who show a pituitary corticotroph cell normally responsive to the negative feedback effects of glucocorticoids, patients with Cushing's disease manifest a pituitary corticotroph cell which is grossly unresponsive to cortisol negative feedback. Our data also suggest that the differences in pituitary corticotroph cell function between depressed patients and those with Cushing's disease seem to be accompanied by differences in hypothalamic CRH neuron function. Specifically, we have shown that many of our patients with Cushing's disease whom we studied 1 week after selective transsphenoidal adenectomy (at a time when basal ACTH and cortisol were uniformly undetectable) showed clear but subnormal or nearly normal plasma ACTH responses to exogenous CRH (Oldfield et al., unpublished observations). We surmise that the adrenal insufficiency in each of these postoperative patients reflects hypofunction of CRH neurons which had been physiologically suppressed by exposure to the negative feedback of the long-standing hypercortisolism. This formulation is supported by our recent finding that CSF CRH is significantly lower in Cushing's disease patients than in depressed patients and controls (Kling and Gold, unpublished observations). A schematic diagram of the proposed differences in pituitary and hypothalamic function between patients with depression and Cushing's disease is shown in Fig. 10

The differential pathophysiology of hypercortisolism which we propose for Cushing's disease and depression is manifested by the fact that responses to CRH in these disorders are in the opposite direction (e.g., an exaggerated ACTH response in Cushing's disease and a blunted one in depression) (Fig. 9). In all other diagnostic tests which have been utilized to differentiate depression from Cushing's disease such as the dexamethasone suppression test and serial urinary free cortisol determinations, responses and/or levels in depression and Cushing's disease were in the same direction. Thus, the CRH stimulation test is in a unique position to assist in the differential diagnosis of depression and early Cushing's disease. Indeed, our data do show that the CRH stimulation test is helpful in the differential diagnosis of these disorders, with only a small percentage of depressed patients showing ACTH responses to CRH in the Cushing's disease range and vice versa (Gold et al. 1986a).

6 Differential Diagnosis of Cushing's Disease from Other Types of Cushing's Syndrome

Cushing's syndrome, as a spontaneous pathophysiological entity, can be divided into three types (for a review, see Gold 1979): Cushing's disease due to pituitary hypersecretion of ACTH (as discussed previously), hypercortisolism secondary to ectopic secretion of ACTH, and hypercortisolism caused by autonomous secretion of cortisol by an adrenal adenoma or carcinoma. Thus, Cushing's syndrome can be divided into ACTH-dependent (caused by pituitary and ectopic ACTH secretion) and ACTH-independent (the cortisol-producing adrenal neoplasms) subsets. The differential diagnosis of the two types of ACTH-dependent Cushing's syndrome is often difficult. In contrast, adrenal tumors are usually diagnosed radiologically or by ultrasound. The most sensitive procedure for this diagnosis is high-resolution computerized axial tomography of the adrenal glands.

We and others have shown that the CRH stimulation test appears to differentiate between Cushing's disease and the ectopic ACTH syndrome (Chrousos et al. 1984b, 1987; Lytras et al. 1984; Muller et al. 1983; Pieters et al. 1983). Thus, whereas almost all patients with Cushing's disease show exaggerated or robust ACTH responses to CRH, patients with the ectopic ACTH syndrome generally fail to respond to CRH. Nine such patients were examined by us (Chrousos et al. 1984b, 1987) and four by other groups (Lytras et al. 1984; Muller et al. 1983; Pieters et al. 1983). Only one patient with ectopic ACTH secretion was responsive to CRH (Muller et al. 1983). This response was not found on repetition of the test, suggesting that randomly occuring secretory episodes of ACTH may provide a spurious diagnosis during a CRH test. Since medical or surgical correction of hypercortisolism in these patients is followed by a rapid return (within 3 days) of pituitary-adrenal axis responsiveness to CRH hypercortisolism during testing is a prerequisite for assessing the response to CRH in these conditions.

Patients with ACTH-independent Cushing's syndrome had undetectable levels of plasma ACTH throughout the test and their plasma cortisol concentrations remained unaltered, like the patients with ectopic ACTH secretion. Medical or surgical correction of the hypercortisolism was followed quickly by normalization of the CRH response (Chrousos et al. 1984b, 1987).

We have concluded that CRH testing assists in the differential diagnosis of Cushing's disease, the ectopic ACTH syndrome, and adrenal causes of Cushing's syndrome. The available data, cited above, indicate about two out of 44 (4.5%) false negatives and one out of 13 (7.6%) false positives in differentiating pituitary from ectopic causes of Cushing's syndrome.

7 Summary and Conclusions

CRH is a 41-amino-acid peptide first isolated from ovine and subsequently from porcine, rat, and human hypothalami. We have conducted a series of clinical studies with oCRH and hCRH in volunteers and patients with various disorders of HPA function. In volunteers, we demonstrated that hCRH administration produces ACTH and cortisol responses which closely mimic naturally occuring secretory episodes. These data, as well as our demonstration that pulsatile hCRH can reestablish normal ACTH and cortisol secretion in patients with hypothalamic CRH deficiency, strongly argue that CRH is of physiological relevance to the human pituitary-adrenal axis. However, since the ACTH response to an insulin tolerance test is greater than the maximal ACTH response to CRH, other factors such as vasopressin may be relevant to stress-induced ACTH secretion in man.

Following the demonstration that CRH seems to be of physiological relevance to human subjects, we developed a CRH stimulation test based on pharmacokinetic and dose-response studies with oCRH and hCRH. Based on these data, which revealed that oCRH functions as a long-acting analogue of hCRH, and the demonstration that hormonal responses to CRH are greatest in the evening, patient groups with abnormalities of the HPA axis were tested with intravenous oCRH at a dose of 1 μg/kg given at 2000.

This CRH stimulation test has proved helpful in clarifying the pathophysiology of hypercortisolism in a variety of psychiatric disorders characterized by this endocrine abnormality. Thus, blunted ACTH responses in hypercortisolemic patients with depression, anorexia nervosa, and panic anxiety disorder indicate normality of the pituitary corticotroph in these patient subgroups. These data, along with the finding that a continuous infusion of CRH to normal volunteers reproduces the pattern and magnitude of hypercortisolism in depression and anorexia nervosa, suggest that the hypercortisolism in these disorders represents a defect at or above the hypothalamus resulting in the hypersecretion of CRH. This hypothesis is particularly intriguing in light of the demonstration that CRH administration to experimental animals produces many of the physiologic and behavioral responses classically associated with depression and anorexia nervosa, including hypercortisolism, hypothalamic hypogonadism, and decreases in libido and appetite.

The CRH stimulation test has also helped to resolve one of the oldest endocrinologic dilemmas, namely whether the hypercortisolism of depression and Cushing's disease share a common or dissimilar pathophysiological basis. As noted, our data indicate a hypothalamic locus for depression and a pituitary locus for Cushing's disease. Moreover, the CRH stimulation test has also proven helpful in the often diffcult differential diagnosis of depression and early Cushing's disease. Hence, the wide divergence of ACTH responses to CRH in depression and Cushing's disease constitutes the basis for distinguishing most cases of Cushing's disease from depression.

The CRH test has also proven helpful in the differential diagnosis of Cushing's syndrome. Hence, in contrast to patients with Cushing's disease, patients with ectopic ACTH secretion show no response of either ACTH or cortisol to CRH, whereas patients with the hypercortisolism of adrenal origin have low or undetectable plasma ACTH, high plasma cortisol, and no response to CRH. Patients with primary adrenal insufficiency had increased plasma ACTH and diminished cortisol responses to CRH.

We conclude that CRH is an important modulator of the human pituitary-adrenal axis which has helped to elucidate the pathophysiology of abnormalities in HPA function in patients with depression, anorexia nervosa, panic anxiety disorder, Cushing's disease, and secondary adrenal insufficiency. Moreover, CRH has proven helpful in the differential diagnosis of pseudo-Cushing's and Cushing's disease, Cushing's disease and ectopic ACTH secretion, and pituitary and hypothalamic adrenal insufficiency. Although no therapeutic applications of CRH have yet been implemented, administration of CRH by implantable infusion pumps should constitute a more physiologic treatment of the hypothalamic CRH deficiency than glucocorticoid replacement.

References

Amsterdam JD, Winokur A, Abelman E, Lucki I, Rickels K (1938) Co-syntropin (ACTH 1–24) test in depressed patients and healthy subjects. Am J Psychiatry 140:907–909

Avgerinos PC, Schuermeyer TH, Nieman L, Udelsman R, Gold PW, Loriaux DL, Chrousos GP (1987a) Synthetic human CRH as a probe of pulsatile ACTH and cortisol secretion. In: Crowley W Hormones and pulsatility. Plenum, New York (press)

Avgerinos PC, Schuermeyer TH, Gold PW, Tomai TP, Loriaux DL, Sherins RJ, Cutler GB Jr, Chrousos GP (1987b) Pulsatile administration of human corticotropin releasing factor (hCRF) in patients with secondary adrenal insufficiency in restoration of normal cortisol secretory pattern. J Clin Endocrinol Metab (press)

Axelrod J, Reisine TD (1984) Stress hormones: their interaction and regulation. Science 224:452–459

Bloom FE, Battenberg ELF, Rivier J, Vale W (1982) Corticotropin releasing factor (CRF): immunoreactive neurons and fiber in rat hypothalamus. Regul Pept 4:43–48

Brown MR, Fisher LA, Siess J, Rivier C, Rivier J, Vale W (1982) Corticotropin releasing factor: actions on the sympathetic nervous system and metabolism. Endocrinology 111:928–931

Britton DR, Koob GF, Rivier J (1981) Intraventricular corticotropin releasing factor enhances behavioral effects of novelty. Life Sci 31:363–367

Cantwell DP, Stuzenberger S, Burnough J et al. (1977) Anorexia nervosa. An affective disorder? Arch Gen Psychiatry 34:1087–1094

Carroll BJ, Curtis GC, Mendels J (1900) Neuroendocrine regulation in depression, vol I, Limbic system – adrenocortical dysfunction.
Chrousos G, Schulte HM, Oldfield EH, Gold PW, Cutler GB Jr, Loriaux DL (1984 a) The corticotropin releasing factor stimulation test: an aid in the differential diagnosis of patients with Cushing's syndrome. N Engl J Med 310:622–627
Chrousos GP, Oldfield EH, Nieman L, Nisula B, Cutler GB Jr, Schulte HM, Gold PW, Loriaux DL (1987) The corticotropin releasing factor test in Cushing's syndrome. N Engl J Med (Letter) (press)
De Souza EB, Perrin HM, Insel T, Rivier J, Vale WW, Kuhar MJ (1984) Corticotropin-releasing factor receptors in rat forebrain: autoradiographic identification. Science 224:1449–1450
Furatini Y, Morimoto Y, Shubahara S, Noda M, Takahashi H, Hirose T, Asai M, Inayama S, Hayashida H, Miyata T, Numa S (1983) Cloning and sequence analysis of cDNA for ovine corticotropin-releasing factor precursor. Nature 301:537–540
Gerner GH, Gwirtsman HE (1981) Abnormalities of dexamethasone suppression test and urinary MHPG on anorexia nervosa. Am J Psychiat 138:650–653
Gilles GE, Lingo EA, Lowry PJ (1982) Corticotropin releasing activity of the new CRF is potentiated several times by vasopressin. Nature 299:355–357
Gold PW, Chrousos GP (1985) Clinical studies with corticotropin releasing factor: implications for the diagnosis and pathophysiology of depression, Cushing's disease, and adrenal insufficiency. Psychoneuroendocrinology (press)
Gold PW, Chrousos GP, Kellner CH, Post RM, Roy A, Avgerinos P, Schulte HM, Oldfield EH, Loriaux DL (1984) Psychiatric implications of basic and clinical studies with corticotropin releasing factor. Am J Psychiatry 141:619–627
Gold PW, Loriaux DL, Roy A, Kellner CH, Post RM, Kling MA, Calabrese JR, Oldfield EH, Pickar D, Avgerinos PC, Paul SM, Cutler GB Jr, Chrousos GP (1986 a) Physiology of hypercortisolism in depression and Cushing's disease. N Engl J Med (press)
Gold PW, Gwirtsman H, Avgerinos PC,Nieman L, Jimerson D, Kaye W, Loriaux DL, Chrousos GP (1986 b) Abnormal hypothalamic-pituitary-adrenal function in anorexia nervosa: pathophysiological mechanisms in underweight and weight-recovered patients. N Engl J Med (press)
Grossman A, Perry L, Schally AV, Rees LH, Tomlin D, Coy D, Comary-Schull AM, Besser GM (1982) The new hypothalamic hormone corticotropin releasing factor specifically stimulates the release of adrenocorticotropic hormone and cortisol in man. Lancet 921–922
Guillemin R, Vargo T, Rossier J, Minick S, Ling N, Rivier C, Vale W, Bloom F (1977) Beta-endorphin and adrenocorticotropin are secreted concomitantly by the pituitary gland. Science 197:1367–1368
Halbreich V, Asnis GM, Slindledecker R, Zumoff B, Nathan RS (1985) Cortisol secretion in endogenous depression II time-related functions. Arch Gen Psychiatry 42:909–915
Harris GW (1948) Neural control of the pituitary gland. Physiol Rev 28:134–179
Holsboer F, Genken H, Stalla GK, Muller GH (1984) Blunted ACTH responses to human CRH in depression. N Engl J Med (Letter) 311:1127
Iason AH (1946) The thyroid gland in medical history. Froben Press, New York
Insel TR, Kalin WH, Guttmacher LB, Cohen RM, Murphy DL (1982) The dexamethasone suppression test in obsessive-compulsive disorder. Psychiatry Res 6:153–160
Lieutaud J (1742) Essais anatomiques, contenant l'histoire exact de toutes les parties qui composent le corps de l'homme, avec la manier de dissequer. Huart, Paris
Lytras N, Grossman A, Tomlin PS, Waas JAH, Coy DH, Schally AV, Rees LH, Besser GM (1984) Corticotropin releasing factor responses in normal subjects and patients with disorders of the hypothalamus and pituitary. Clin Endocrinol 20:71–84
Mains R, Eipper E, Ling N, (1977) Common precursor to corticotropin and endorphins. Proc Natl Acad Sci USA 74:3014–3018
Medvei VC (1984) A history of endocrinology. MTP Press, Boston
Muller OH, Stalla GK, Werder K (1983) Corticotropin releasing factor: a new tool for the differential diagnosis of Cushing's syndrome. J Clin Endocrinol Metab 56:227–229

Nakahara M, Shibasaki T, Shizume K, Kiyosawa Y, Odagiri E, Toshihiro S, Yamaguchi H, Tsushima T, Demura H, Maeda T, Wakabayashi I, Ling N (1983) Corticotropin-releasing factor test in normal subjects and patients with hypothalamic-pituitary-adrenal disorders. J Clin Endocrinol Metab 57:963–968

Nemeroff CB, Widerlov E, Bissette G, Wallens H, Karllson I, Eklund K, Kilts G, Loosen P (1984) Elevated immunoreactive corticotropin releasing hormone in depressed patients. Science 224:1342–1344

Olschowka JA, O'Donohue TL, Mueller GP, Jacobowitz DM (1982) The distribution of corticotropin releasing factor-like immunoreactive neurons in rat brain. Peptides 3:995–1015

Orth D (1984) The old and the new in Cushing's syndrome. N Engl J Med 310:649–651 (Editorial)

Orth DN, Debold CR, Decherney GS, Jackson RV, Alexander AN, Rivier J, Rivier C, Spiess J, Vale W (1982) Pituitary microademonas causing Cushing's disease respond to corticotropin-releasing factor. J Clin Endocrinol Metab 55:1017–1019

Orth DN, Jackson RV, DeCherney GS, Debold CR, Alexander AN, Island DP, Rivier J, Rivier C, Spiess J, Vale W (1983) Effect of synthetic ovine corticotropin-releasing factor: dose response of plasma adrenocorticotropin and cortisol. J Clin Invest 71:587–595

Pieters GFFM, Hermus ARMM, Smals AGH, Bartelink AKM, Benraad THJ, Kloppenborg PWG (1983) Responsiveness of the hypophyseal adrenocortical axis to corticotropin-releasing factor in pituitary-dependent Cushing's disease. J Clin Endocrinol Metab 57:513–515

Plotsky PM, Rubin TO, Vale W (1985) Hypophysiotropic regulation of adrenocorticotropin secretion in response to insulin-induced hypoglycemia. Endocrinology 117:323–329

Post RM (1982) Use of the anticonvulsant carbamazepine in primary and secondary affective illness: clinical and theoretical implications. Psychol Med 12:70–104

Renold AE, Jenkins D, Thorn GW (1952) The use of intravenous ACTH. A study in quantitative adrenocortical stimulation. J Clin Endocrinol Metab 12:763–797

Rivier C, Vale W (1983) Modulation of stress-induced ACTH release by corticotropin releasing factor, catecholamines and vasopressin. Nature 305:325–327

Rivier J, Spiess J, Vale W (1983) Characterization of rat hypothalamic corticotropin releasing factor. Proc Natl Acad Sci USA 80:4851–4855

Rock JP, Oldfield EH, Schulte HM, Chrousos GP, Gold PW, Cutler GB Jr, Kornblith PL, Loriaux DL (1984) Corticotropin releasing factor administered into ventricular CSF stimulates pituitary ACTH release. Brain Res 323:365–368

Roy A, Pickar D, Doran A, Paul SM, Chrousos GP, Gold PW (1986) Normal ACTH and cortisol responses to CRH in schizophrenia. Am J Psychiatry (press)

Roy-Byrne PP, Uhde T, Post RM, Galluci WT, Chrousos GP, Gold PW (1987) Blunted ACTh responses to ovine CRH in panic-anxiety disorder. Am J Psychiatry (press)

Sachar EJ, Hellman L, Fukushima DK, Gallagher TF (1970) Cortisol production in depressive illness: a clinical and biochemical classification. Arch Gen Psychiatry 23:289–298

Saffran M, Schally AV, Bentey BG (1955) Stimulation of the release of corticotropin from the adenohypophysis by a neurohypophyseal factor. Endocrinology 57:439–444

Schally AV, Chang RC, Arimura A, Redding TW, Fishback JB, Vigh S (1981) High molecular weight peptide with corticotropin-releasing factor activity from porcine hypothalami. Proc Natl Acad Sci USA 78:5197–5201

Schuermeyer TH, Avgerinos PC, Gold PW, Tomai TP, Galluci WT, Cutler GB Jr, Loriaux DL, Chrousos GP (1984) Human corticotropin releasing factor: dose response and time course of ACTH and cortisol secretion in man. J Clin Endocrinol Metab 59:1103–1108

Schuermeyer TH, Gold PW, Gallucci WT, Tomai TP, Cutler GB Jr, Loriaux DL, Chrousos GP (1985) Effect and pharmacokinetic properties of the rat/human corticotropin releasing factor (r/h CRH) in rhesus monkeys. Endocrinology 117:300–306

Schulte HM, Chrousos GP, Oldfield EH, Gold PW, Cutler GB Jr, Loriaux DL (1982a) The effects of corticotropin releasing factor on the anterior pituitary function of stalk-

sectioned cynomologus macaques: dose response of cortisol secretion. J Clin Endocrinol Metab 55:810–812
Schulte HM, Chrousos GP, Gold PW, Oldfield EH, Phillips JM, Munson PJ, Cutler GB Jr, Loriaux DL (1982b) Metabolic clearance rate and plasma half-life of radioiodinated ovine corticotropin releasing factor in a primate. J Clin Endocrinol Metab 55:1023–1027
Schulte HM, Chrousos GP, Gold PW, Booth JP, Oldfield TH, Cutler GB Jr, Loriaux DL (1985a) Continuous infusion of CRF in normal volunteers. Psysiological and pathophysiological implications. J Clin Invest 75:1781–1785
Schulte HM, Chrousos GP, Oldfield EH, Gold PW, Cutler GB Jr, Loriaux DL (1985b) Corticotropin releasing factor administration in normal men: pituitary and adrenal responses in the morning and evening. Horm Metab Res (press)
Selye H (1936) Stress syndrome. A syndrome produced by diverse noxious agents. Nature 132–138
Shibahara S, Morimoto Y, Furatani Y, Notake M, Takahashi H, Shimizu S, Horikawa S, Numa S (1983) Isolation and sequence analysis of the human corticotropin-releasing factor precursor gene. EMBO J 2:775–779
Sirinathsinghji DJS, Rees LH, Rivier J, Vale W (1983) corticotropin releasing factor is a potent inhibitor of sexual receptivity in the female rat. Nature 305:232–235
Stokes PE (1973) Adrenocorticol activation in alcoholics during chronic drinking. Ann NY Acad Sci 215:77–81
Sutton RE, Koob GF, Le Moal M, Rivier J, Vale W (1982) Corticotropin releasing factor produces behavioral activation in rats. Nature 297:331–333
Thomas EB, Levine ES, Arnold WJ (1968) Effects of maternal deprivation and incubation rearing upon adrenocortical activity in the adult rat. Dev Psychobiol 1:21–23
Vale W, Spiess J, Rivier C, Rivier J (1981) Characterization of a 41-residue ovine hypothalamic peptide that stimulates secretion of carticotropin and beta-endaphin. Science 213:1394–1397
Vale W, Vaughn J, Smith M, Yamamoto G, Rivier J, Rivier C (1983) Effects of synthetic ovine corticotropin releasing factor, glucocorticoids, catecholamines, neurohypophyseal peptides and other substances on culture corticotropin cells. Endocrinology 113:1121–1131
Valentino R, Foote SL, Aston-Jones G (1983) Corticotropin releasing factor activates noradrenergic neurons of the locus coeruleus. Brain Res 220:363–367
Weiss SRB, Post RM, Gold PW, Chrousos GP, Pert A (1987) Behavioral activation and amygdala seizures following central CRH administration. Brain Res (press)
Zuingerus T (1669) Commentarii Hippocratis. Basel Episcoporium Opera, pp 324–385

Biological Rhythms and Mood Disorders

J. Mendlewicz and P. Linkowski

Contents

1 Introduction

Biological rhythms have always modulated living systems. The 24-h cycle is present in almost all biological systems, remaining constant in various environments. This observation has led to the concept of endogenous biological clocks. The exact synchronization of the 24-h period also involves external synchronizers (Zeitgeber). The precise nature and control of biological clocks remains unclear, but the hypothalamo-pituitary axis seems to play a major role in the maintenance of circadian rhythms in both animals and man (Richter 1958). It is now clear that the pituitary gland constitutes a major link in the neuroendocrine axis in man and that central neurotransmitters regulate the secretion of the hypothalamic neurohormones, which in turn may affect brain monoamine metabolism. Most neuroendocrine parameters are subject to circadian variations both in animals and in man and may be implicated in the pathogenesis of periodic psychoses. A new approach, chronobiology, permits the study of temporal changes in clinical and physiological symptoms in diseases such as cancer, endocrine disturbances, and cardiovascular and psychiatric conditions. Furthermore, treatment response may also show important variations, probably due to circadian variations in drug metabolism and receptor sensivity (chronopharmacology). Some affective disorders are characterized by alternation of depressive and manic episodes and by periodic as well as diurnal disturbances in mood and in physiological functions such as sleep, energy, appetite and sex (evening improvement and morning worsening). These diurnal changes may be related to alterations in day-night variation of the mood and drive systems. According to this concept, desynchronization may be an important pathophysiological aspect of depression, with some biological

University Clinics of Brussels, Department of Psychiatry University of Brussels, Ersme Hospital, 808 Route de Lennik, Brussels 1070, Belgium

Current Topics in Neuroendocrinology, Vol. 8

rhythms following a nonendogenous, or free-running, circadian cycle deviated from the normal 24-h period. Furthermore, experimental studies have shown that jet day and night shift modifies energy level and concentration abilities, while sleep deprivation has been reported to temporarily alleviate depression.

The hypothesis of a desynchronization of biological rhythms was first tested by Gjessing et al. (1932) in their classic reports on the periodical psychoses and the periodical catatonias. In 1947, Georgi suggested that circadian rhythms in depressed patients might be phase-advanced compared to clock time. In 1975, Papousek hypothesized that the early occurrence of REM sleep in depressed patients was consecutive to a phase advance of REM sleep in depression. This hypothesis was further expanded by the contributions of Wehr and Wirz-Justice (1981). Several clinical and biological observations corroborate the hypothesis of an anomaly of circadian rhythms in depression. From the biological point of view, the study of temperature rhythms has shown a phase advance (Avery et al. 1982; Schulz and Lund 1983), but some negative findings have also been reported (Beersma et al. 1983).

On the other hand, preliminary data from Rubin et al. (1980) and Branchey et al. (1982) have suggested an advance of the cortisol rhythm, which is synchronized by a common oscillator with the REM and temperature rhythms. The phase advance hypothesis is also supported by two other groups of observations. An advance of sleep onset of 6 h has a clear-cut antidepressant effect (Wehr et al. 1979) which is also accompanied by normalization of the specific abnormalities of paradoxical sleep. Similar results may be obtained with antidepressants (imipramine and/or lithium) (Kripke et al. 1979). Lithium also delays circadian rhythms in normal men (Kripke et al. 1979) and lengthens the period of circadian rhythms in free-running rodents (MacEachron et al. 1982). More recently, a new affective disorder has been described, based on observations by Rosenthal et al. (1984), and named "seasonal affective disoreder" (SAD), and chronobiological abnormalities have been suggested in this condition.

SAD is a cyclic illness characterized by recurrent episodes of fall/winter depression alternating with periods of summer normal mood or hypomania. In the winter, the patients present with some atypical depressive symptoms such as hypersomnia, increased appetite, weight gain, fatigue, daytime drowsiness and craving for carbohydrates.

The symptomatology during the summer months is quite opposite: elevation of mood, increased activity and libido and decreased appetite and sleep are observed. Women are 3–4 times more frequently affected than men (Rosenthal et al. 1984). It has been shown that success in treatment of SAD can be achieved using bright arificial light of an intensity of at least 2500 lux. This light intensity is known to suppress human nocturnal melatonin secretion and play an effect in the entrainment of human circadian rhythms; ordinary indoor light does not have these effects. Successful treatment of seasonal depression was reported in one study to have been obtained (Lewy et al. 1987) by exposure to bright light in the morning (but not in the evening), suggesting that timing is critical for the antidepressant effects of bright light. This antidepressant response to morning light was also accompanied by an advance (shift to an earlier time) in the onset of nighttime melatonin production, leading to the hypothesis that in patients with SAD, circadian rhythms are abnormally phase-delayed and that light in the morning offers a corrective phase advance. In SAD patients before light treatment, prebaseline times of melatonin onset were significantly delayed compared to those in the group of normal control subjects (Lewy et al. 1987). Various aminergic systems are involved in the regulation of biological rhythms, and recently the GABA pathway from the suprachiasmatic area has been implicated in experiments showing phase advance and delay of motor activity in hamsters after the administration of a benzodiazepine (Turek and Losee-Olson 1986). The study of circadian and ultradian rhythms of biological functions is thus of great importance in psychopathology, in particular in depressive disorders. According to the desynchronization hypothesis, some cases of depression may be conceptualized as "biological clock" disorders. Phase shifts in biological circadian rhythms in depressive patients have also been suggested for steroids, electrolyte rhythms, neurochemical metabolities and body temperature. While these studies based on brief observations suggested there are circadian disturbances in some depressives, longitudinal observations are essential to demonstrate consistent circadian patterns of affectively ill patients. The existence of circadian variations in the release of several pituitary hormones, such as adrenocorticotropic hormone (ACTH), thyroid-stimulating hormone (TSH), and prolactin (PRL), and adrenal hormones, such as cortisol, in man has been well documented. Moreover, for several of these hormones, such as ACTH, TSH, cortisol and melatonin, it has been shown that higher frequency nonperiodic oscillations, corresponding to secretory episodes, were superimposed on the basal circadian rhythm.

As a consequence of this hormonal variability, consistent studies of hormonal secretion imply repeated blood sampling over long periods of time (i.e., 24-h periods). This approach was used by Sachar et al. (1970), who found that the normal 24-h pattern of cortisol secretion was disrupted in some depressed patients. There was an increase in the number of secretory episodes with active secretion of all peaks of plasma cortisol throughout the 24 h. This abnormality was not confined to the daytime waking hours but also involved nocturnal hours, and was not related to the clinical severity of depressive illness (Sachar et al. 1973). The patterns almost returned to normal after the patients recovered. More recently, Jarrett et al. (1983) reported a reduced latency between sleep onset and the nocturnal increase in cortisol secretion in depressive patients (Jarrett et al. 1983). Other studies suggest that the hypercortisolism in depression cannot be explained entirely as a simple stress response, since these abnormalities are present in unanxious patients during sleep and are not corrected after the administration of large doses of sedative medications (Stockes 1972). The cortisol hypersecretion in depression may be transiently corrected by pretreatment with dextroamphetamine, indicating that a noradrenergic deficit may be present (Sachar et al. 1980). In light of the biogenic amine hypothesis of affective illness, recent studies have also shown alterations in circadian and seasonal rhythms of various neurotransmitter substances in the plasma of depressed patients (Riederer et al. 1974). Previous studies in our laboratory (Golstein et al. 1980; Mendlewicz et al. 1979, 1980) and from other groups (Halbreich et al. 1979; Weeke and Weeke 1978) have shown alterations in the circadian secretion of PRL, TSH and melatonin, a pineal hormone in depressive illness. The above arguments led us to examine circadian variations of pituitary and adrenal hormone levels in bipolar and unipolar depressive disorders in comparison to normal controls. This chapter provides new data on the 24-h profiles of ACTH, cortisol and growth hormone (GH) in depression.

2 Methods

All patients studied were free of medication for at least 2 weeks prior to the investigation and were inpatients in a sleep unit with a primary depressive episode severe enough to warrant hospitalization. The severity of the depressive illnesss was assessed by the Hamilton Rating Scale and diagnoses were made using the Research Diagnostic Criteria of Spitzer et al. (1978).

Age- and sex-matched controls were also investigated. Blood samples were drawn for 25 h through a plastic indwelling catheter. All patients were confined to bed, had normal breakfast, lunch and supper and their nocturnal sleep was not interrupted. Daytime sleep was prevented and sleep times were recorded by trained nurses. After one night of habituation, polygraphic recordings of sleep were obtained for four consecutive nights. On the last day, blood was sampled every 15 min for 25 h. ACTH, cortisol and GH were measured by radioimmunoassay. The diagnostic workup also included a dexamethasone suppression test (DST) performed after the 24 h investigation. Proportions of DST suppressors and non-suppressors were not statistically different in uni- and bipolar patients.

Each individual hormonal profile was submitted to a detailed statistical analysis, including seperate evaluation of circadian variation and of the episodic fluctuations. The circadian variation was evaluated by building a best-fit curve, obtained by repeated periodogram calculations using a method described in detail elsewhere (Van Cauter 1978). The analysis of the episodic fluctuations consisted, for each hormonal profile, of the calculation of the number and characteristics of the significant episodic spikes using a computer program previously described (Van Cauter 1981). The overall similarity between ACTH and cortisol diurnal variations was further estimated by calculating the coefficient of cross-correlation for each individual. Statistics were analyzed using nonparametric one-way analysis of variance. All patients and controls were investigated throughout the calendar year.

3 Results

We report our results on 24-h GH, ACTH and cortisol secretion in relation to sleep-wakefulness cycles. Pathologically short REM latencies (lower than 40 min) were observed in nine of 18 depressive patients. Sleep time was increased, with a resulting decrease in stages I and II. Reductions in slow wave (SW) sleep were also observed in bipolar patients and, to a lesser degree, in unipolars.

The GH profiles are illustrated in Fig. 1.

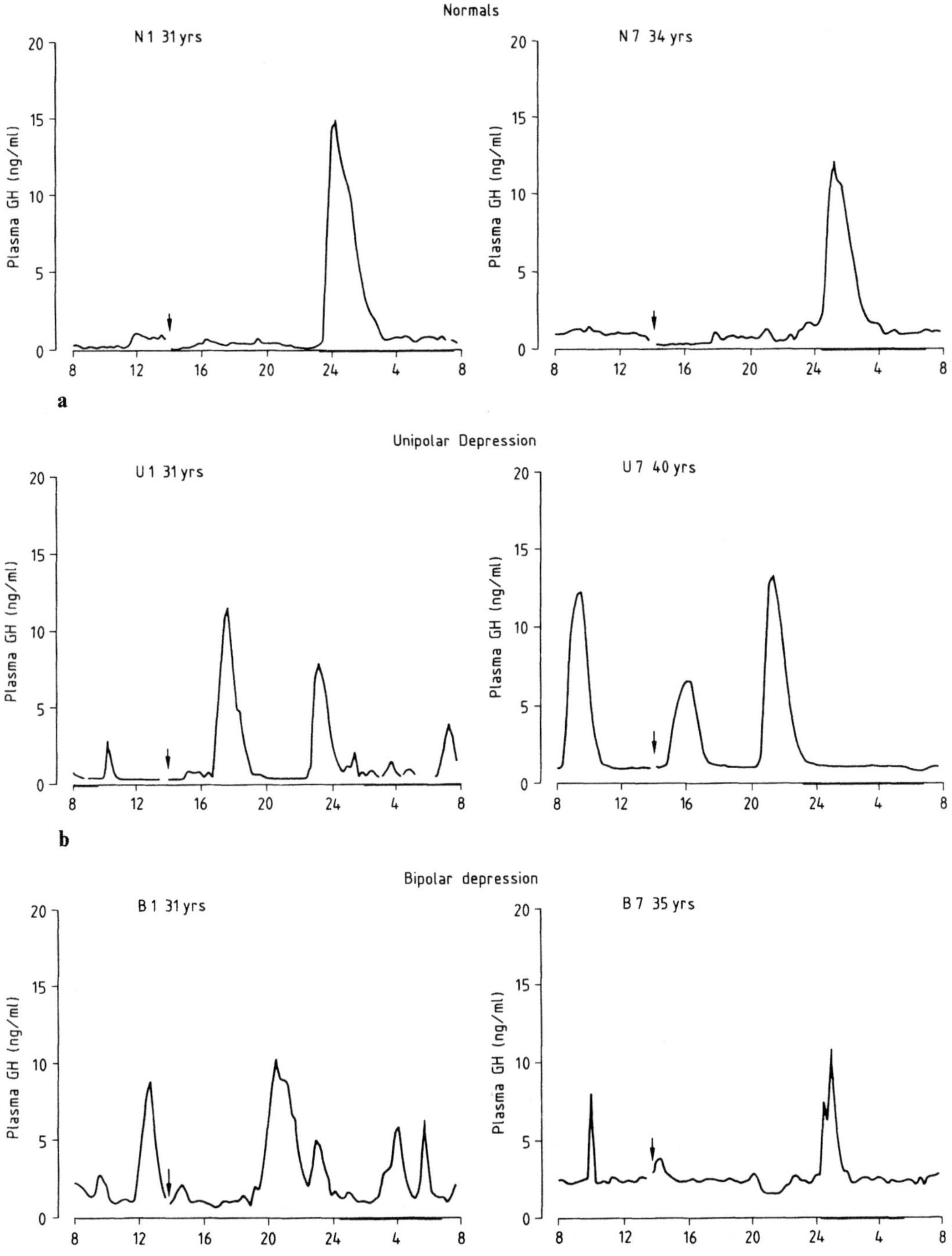

Fig. 1 a–c. GH secretion. **a** Normals; **b** unipolar depressives; **c** bipolar depressives. *Black bars* indicate the sleep period. (From Mendlewicz et al. 1985)

Table 1. Slow wave (SW) sleep and growth hormone (GH) spikes in normal controls and in unipolar and bipolar depressives

Time of GH spike		Proportion of SW during ascending portion (%)
Young normals		
B	02.15	96
	04.15	0
C	01.00	91
	03.30	0
D	00.45	89
	04.00	44
E	00.45	50
	01.45	41
F	01.15	86
Older normals		
VII	01.15	0
VG	–	–
VE	00.30	0
	05.30	1
D	00.30	48
Pi	01.45	40
Unipolars		
Ra	02.30	17
	05.00	0
Ma	03.30	2
F	01.30	0
	07.15	2
Gi	–	–
Go	–	–
La	–	–
Bipolars		
Po	–	–
J	02.00	79
Du	02.00	1
	04.30	0
Me	23.45	0
	01.15	0
	04.30	0
V	00.45	0
	02.00	7
	04.15	28
G	01.00	30

Both unipolar and bipolar patients secreted more GH than control subjects during the 24-h span. This hypersecretion occurred during wakefulness rather than during sleep and corresponded, on average, to an increase in daytime secretion about sixfold in unipolars and fourfold in bipolars.

The amount of GH secreted during sleep did not differ significantly between depressive patients and normal subjects. The increased daytime secretion of GH appeared to be due to an increase in the number of spikes (1.1 ± 1.4 in controls vs 3.4 ± 1.7 in depressed patients), since the magnitudes of the daytime spikes detected in controls and depressed subjects did not differ significantly. Seven of the eight normal volunteers had an early sleep GH spike, culminating within 90 min after sleep onset. None of the unipolar patients had a spike in that time interval, and half of them had no sleep spike at all. However, seven of the eight unipolar patients exhibited instead a GH peak occurring shortly before sleep onset, in the time interval 21:00–00:00. Such a presleep GH spike was also present in five of the eight bipolar profiles, as well as in three of the control subjects. These figures are, however, not sufficiently high to exclude the possibility of a fortuitous event. No consistent disturbance of the temporal association between sleep onset and GH secretion was found in bipolar depressed patients. The amount of GH secreted during sleep is greater in bipolars, but there is more waking time in unipolars. There were no differences in REM, SW sleep and stages I and II. Table 1 illustrates the association between SW sleep and GH spikes in normal controls and in unipolar and bipolar patients.

All young controls have a GH spike shortly after sleep onset. An average of $82\% \pm 18\%$ of the ascending portion of these spikes was spent in SW sleep. In three of five older controls, an early GH spike was also observed. Only two of these spikes were associated with SW sleep. In the unipolar depressives, no GH peak occured in the 3-h period following sleep onset and only one of the five nocturnal GH spikes was associated with concomitant SW sleep. Three unipolars did not show any GH peak during the night. In the bipolars, five of six patients had an early sleep GH spike but only one of them was concomitant with SW sleep. Six of a total of 10 nocturnal GH spikes occurred without any concomitant stage III or IV. Thus the classical association between nocturnal GH secretion and SW sleep was absent in older male controls, as well as in depressed patients.

Both unipolar and bipolar patients had elevated 24-h mean cortisol levels compared to control subjects. No significant differences were observed between unipolars and bipolars. The 24-h mean cortisol was higher in DST nonsuppressors than in suppressors. Hypercortisol secretion as evidenced by DST nonsuppression was also associated with a significantly greater shortening of sleep REM latency in depression (Mendlewicz et al. 1984). There were no significant differences in mean 24-h ACTH levels among unipolars, bipolars and controls.

Figure 2 illustrates the profiles of ACTH and cortisol in a bipolar depressed patient.

The timing of the early morning increase in cortisol secretion appeared to be a function of age in normals, occurring earlier in older subjects. An advance of the early morning increase of cortisol secretion was observed in unipolar patients irrespective of age. In bipolar patients, a similar trend was observed but did not reach significance. DST suppressors and nonsuppressors did not differ in terms of advance of early morning increase of cortisol. Hypersecretion of cortisol appeared to be due to an increase in the magnitude but not in the number of spikes. The magnitudes of the ACTH spikes did not differ among the three groups of subjects. Overall, episodic cortisol fluctuations were less well correlated with concomitant ACTH spikes than in normal subjects. We also observed a phase advance for the nadir and the quiescent period of ACTH cortisol secretion in depressive patients (more pronounced in unipolars) compared to normal controls.

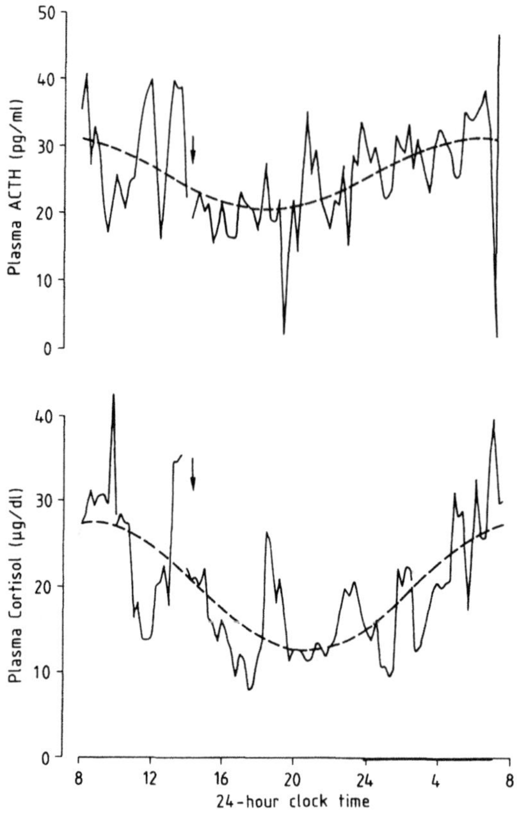

Fig. 2. Circadian rhythm of ACTH and cortisol secretion in a bipolar patient. *Black bars* indicate the sleep period; *dotted lines* represent the best-fit patterns. (From Linkowski et al. 1985). A more precise characterization of the curve is obtained after linear transformation by considering the slope values (*lower panels*), which are given with the respective 95% confidence interval

On average, depressed patients secreted more GH and cortisol than normal subjects. GH hypersecretion occurred during the daytime and was due to an increase in the number of secretory spikes. In contrast, cortisol hypersecretion occurred throughout the 24-h span and was due to an increase in the magnitude, but not in the number, of secretory spikes. ACTH levels remained in the normal range. These observations suggest the existence of different controls for daytime and for sleep-related GH release. The finding of hypercortisolism associated with normal ACTH levels could be explained by (1) elevations in ACTH secretion, resulting in plasma levels remaining in the normal range; (2) increased sensitivity of the adrenals to ACTH stimulation; (3) enhanced contribution of extrapituitary neural inputs to the adrenals. These results are consistent with the hypothesis of Gold et al. (1986) of an enhanced endogenous corticotropin release factor activation from the hypothalamus and of the persistence of a negative feedback of high cortisol levels on the corticotroph pituitary cells in depression. In addition to these alterations in amount of secretion, disturbances in the temporal organization of secretion were observed and appeared more pronounced and more consistent in unipolar than in bipolar patients. The advance of the quiescent period of the ACTH cortisol rhythms, as well as the appearance of presleep peak of GH

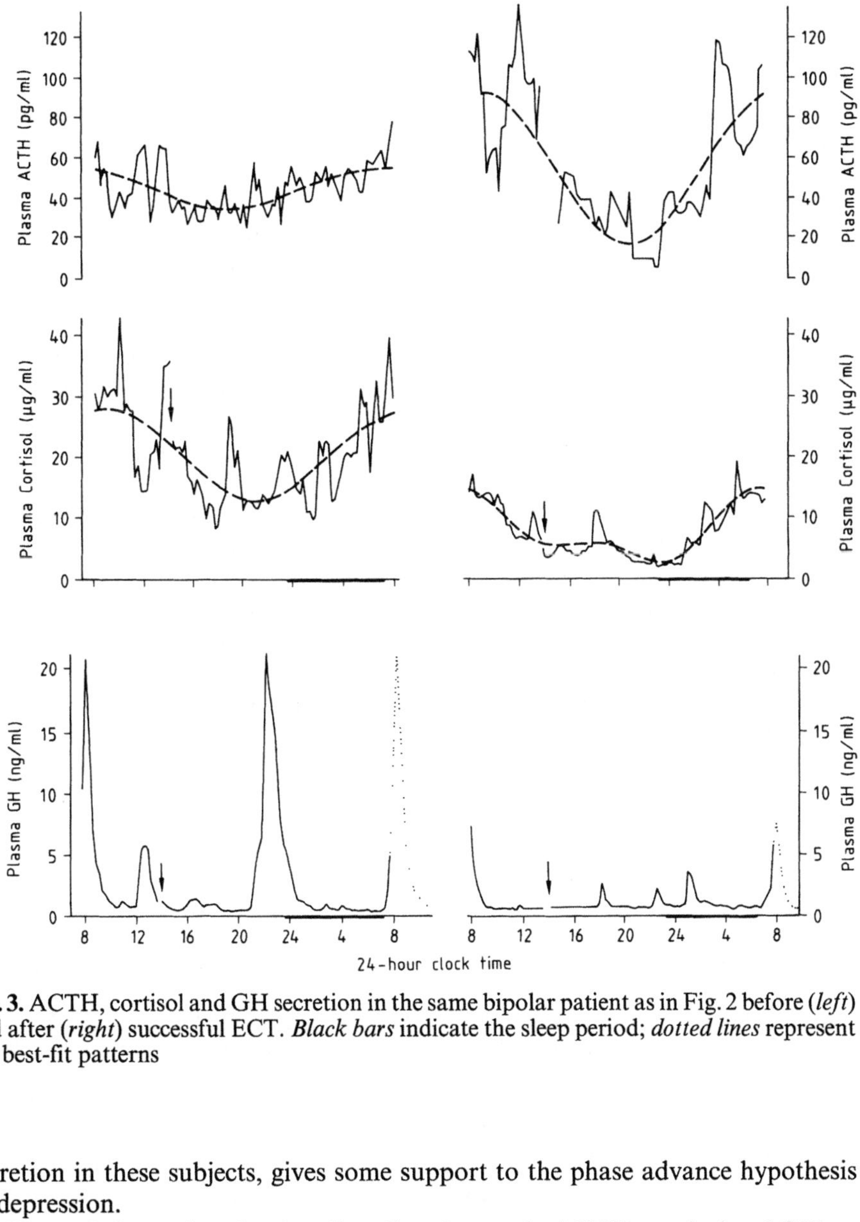

Fig. 3. ACTH, cortisol and GH secretion in the same bipolar patient as in Fig. 2 before (*left*) and after (*right*) successful ECT. *Black bars* indicate the sleep period; *dotted lines* represent the best-fit patterns

secretion in these subjects, gives some support to the phase advance hypothesis of depression.

Figure 3 shows that the circadian disturbances in ACTH, cortisol and GH secretion observed during the depressive phase completely normalize during clinical remission following successful ECT. This indicates that these disturbances have to be considered as state rather than trait marker anomalies in depressive illness.

4 Discussion

Alterations in circadian rhythms for plasma pituitary, pineal and adrenal hormones such as PRL, GH, TSH, ACTH, cortisol and melatonin are thus present in some depressed patients as state-dependent variables and may be related to basic disturbances in circadian oscillators of central catecholaminergic and serotonergic activity in affective illness. Moreover, it is possible that cholinergic-adrenergic interactions are also of significance. The rapid changes in mood and the abruptness of the desynchronization observed in manic-depressive patients make it unlikely that these phenomena are related to primary oscillations in central neurotransmitter synthesis or turnover rate. More fruitful hypotheses may be formulated in terms of infradian, ultradian or circadian variation in specific brain receptor sensitivity or more complex behavioral modulation through endogenous neuropeptide substances. The brain distribution of other releasing factors and peptides has not yet been reported, and when the brain effects of hypothalamic hormones and peptides are further elucidated, the clinical conditions in which their actions are investigated may be better understood. Nevertheless, the chronobiological studies combining sleep and neuroendocrine evaluation over long periods of time in the study of abnormal behavior are most promising and may enable us better to understand cyclical alterations of hypothalamic and pituitary functions in man, although it is still premature to draw firm conclusions from the neuroendocrine abnormalites as to the specific nature of the underlying neurotransmitter or neuropeptide disturbances in depressive disorders.

References

Avery D, Wildschiodtz G, Rafaelsen O (1982) Nocturnal temperature in affective disorder. J Affective Disord 4:61–71

Beersma D, Van Den Hoofdakker RH, Van Berkenstijn H (1983) Circadian rhythms in affective disorders. Body temperature and sleep physiology in endogenous depressives. Adv Bio Psychiatry 11:114–127

Branchey L, Weinberg V, Branchey M, Linkowski P, Mendlewicz J (1982) Simultaneous study of 24-h patterns of melatonin and cortisol secretion in depressed patients. Neuropsychobiology 8:225–229

Georgi F (1947) Psychophysiologische Korrelationen. Psychiatrische Probleme im Licht der Rhythmusforschung. Schweiz Med Wochenschr 49:1267–1288

Gjessing R (1932) Beiträge zur Kenntnis der Pathophysiologie des katatonen Stupors. 1. Über periodisch rezidivierend Katatonen-Stupor, mit lytischem Beginn und Abschluß. Arch Psychatr Nervenkr 96:319–392

Gold PW, Loriaux L, Roy A, Kling MA, Calabrese JR, Kellner CH, Nieman LK, Post RM, Pickar D, Calluci W, Avgerinos BS, Paul S, Oldfield EH, Cutler GB, Chrousos G (1986) Responses to corticotropin-releasing hormone in the hypercortisolism of depression and Cushing's disease. N Engl J Med 314(21):1329–1335

Golstein J, van Cauter E, Linkowski P, Vanhaelst L, Mendlewicz J (1980) Thyrotropin nyctohemeral pattern in primary depression: differences between unipolar and bipolar women. Life Sci 27:1695

Halbreich U, Grunhaus L, Ben-David M (1979) Twenty-four hour rhythm of prolactin in depressive patients. Arch Gen Psychiatry 36:1183
Jarrett DV, Coble PA, Kupfer DJ (1983) Reduced cortisol latency in depressive illness. Arch Gen Psychiatry 40:506–511
Kripke DF, Mullaney DF, Atkinson ML, Huey LY, Hubbard B (1979) Circadian rhythm phases in affective illness. Chronobiologia 6:365–371
Lewy AJ, Sack RL, Miller S, Slaban TM (1987) Antidepressant and circadian phase shifting effects of light. Science 235:352–354
Linkowski P, Mendlewicz J, Leclerqo R, Brasseur M, Hubain P, Golstein J, Copinschi G, van Cauter E (1985) The 24-h profile of ACTH and cortisol in major depressive illness. J Clin Endocrinol Metab 61:429–438
MacEachron DL, Krijpke DF, Hawkins R, Shaw E, Pavlinac D, Deftos L (1982) Lithium delays biochemical circadian rhythms in rats. Neuropsychobiology 8:12–19
Mendlewicz J, Linkowski P, Branchey L, Weinberg U, Weitzman E, Branchey M (1979) Abnormal 24-h pattern of melatonin secretion in depression. Lancet II:1362
Mendlewicz J, van Cauter E, Linkowski P, l'Hermite M, Robyn c (1980) The 24-h profile of prolactin in depression. Life Sci 29:2015
Mendlewicz J, P, Kerkhofs M, Hoffmann G, Linkowski P (1984) Dexamethasone suppression test and REM sleep in patients with major depressive disorder. Br J Psychiatry 145:383–388
Mendlewicz J, Linkowski P, Kerkhofs M, Desmedt D, Golstein J, Copinschi G, van Cauter E (1985) Diurnal hypersecretion of growth hormone in depression. J Clin Endocrinol Metab 60:505–512
Papousek M (1975) Chronobiology aspects of cyclothymia. Fortschritte der Neurologie, Psychatrie und ihrer Grenzgebiete 43:381–423
Richter CP (1958) Abnormal but regular cycles in behaviour and metabolism in rats and catatonic-schizophrenics. In: Reiss M (ed) Psychoneuroendocrinology. Grune and Stratton, New York, pp 161–181
Riederer P, Birkmayer W, Neumayer E, Ambrozi L, Linauer W (1974) The daily rhythm of HVA, VMA, (VA), and 5HIAA in depression syndrome. J Neural Transm 35:23–45
Rosenthal NE, Sack DA, Gillin JC, Lewi AJ, Goodwin FK, Davenport Y, Mueller PS, Newsome DA, Wehr TA (1984) Seasonal affective disorder: a description of the syndrome and preliminary findings with light therapy. Arch Gen Psychiatry 41:42–80
Rubin RT, Poland RE, Blodgett ALN, Winston RA, Forster B, Carroll BJ (1980) Cortisol dynamics and dexamethasone pharmacokinetics in primary endogenous depression: preliminary findings. In: Brambilla F, Racagni G, de Wied D (eds) Progress in psychoneuroendocrinology. Elsevier North-Holland, Amsterdam, pp 223–234
Sachar EJ, Hellman L, Fukushima DK (1970) Cortisol production in depressive illness. Arch Gen Psychiatry 23:289–298
Sachar EJ, Hellman L, Roffwarg H, Halpern F, Fukushima D, Gallacher T (1973) Disrupted 24-h patterns of cortisol secretion in psychotic depression. Arch Gen Psychiatry 28:19
Sachar EJ, Asnis G, Nathan S, Halbreich U, Tabrizi MA, Halpern FS (1980) Dextroamphetamine and cortisol in depression. Arch Gen Psychiatry 37:755–757
Schulz H, Lund R (1983) Sleep onset Rem episodes are associated with circadian parameters of body temperature. A study in depressed patients and normal controls. Biol Psychiatry 18(12):1411–1426
Spitzer RL, Endicott J, Robbins E (1978) Research diagnostic criteria: rationale and reliability. Arch Gen Psychiatry 35:773–782
Stockes PE (1972) Studies on the control of adrenocortical function in depression. In: Williams T, Katz S, Shiels J (eds) Recent advances in the psychobiology of depressive illnesses. US DEHW, Washington DC, pp 199–200
Turek F, Losee-Olson S (1986) A benzodiazepine used in the treatment of insomnia phase-shifts the mammalian circadian clock. Nature 321:167–168
Van Cauter E (1978) Method for the characterization of 24-h temporal variations of blood components. Am J Physiol 237:255–264

Van Cauter E (1981) Quantitative methods for the analysis of circadian and episodic hormone fluctuations. In: van Cauter E, Copinschi G (eds) Human pituitary hormones: circadian and episodic variations. Nyhoff, The Hague, pp 1–25
Weeke A, Weeke J (1978) Disturbed circadian variation of serum thyrotropin in patients with endogenous depression. Acta Psychiatr Scand 57:281
Wehr TA, Wirz-Justice A (1981) Internal coincidence model for sleep deprivation and depression. In: Koella WP (ed) Sleep 1980. Karger, Basel
Wehr TA, Wirz-Justice A, Goodwin FK, Duncan W, Christian-Gillin JC (1979) Phase advance of the circadian sleep wake cycle as an antidepressant. Science 206:710–713

Recurrent Affective Disorders: Lessons from Limbic Kindling

R. M. Post, S. R. B. Weiss, and D. R. Rubinow

Contents

Various endocrinological and biochemical conceptualizations of affective illness relate to different aspects of its phenomenology and course. Most endocrinological and animal modeling investigatory approaches focus on acute state changes of depressive behavior or its opposing phase of mania. In this chapter, we focus on the longitudinal course of affective illness, which is often characterized by a tendency for recurrence of episodes and a speeding up over time. This aspect of affective illness will be considered from the perspective of two different phenomena that can be studied in the laboratory – behavioral sensitization to psychomotor stimulants and limbic kindling.

As discussed extensively elsewhere, neither sensitization nor kindling is a strictly homologous animal model of affective disorders (Post 1988). There is a lack of commonality in etiology, pathophysiology, behavioral phenomenology, and pharmacological responsitivity between the experimental procedures and affective illness. Nevertheless, elucidation of sensitization and kindling may help us formulate theoretical or conceptual mechanisms helpful in understanding alterations in the longitudinal course of these illnesses in particular, and in long-term changes in behavioral responsivity in general.

Behavioral sensitization and kindling are linked and discussed together in this chapter because of their common ability to induce increasing behavioral change in response to the same stimulation over time. The mechanisms that underlie the two phenomena are apparently quite different, and we presume that these mech-

Biological Psychiatry Branch, National Institute of Mental Health, Bethesda, MD 20892, USA

Current Topics in Neuroendocrinology, Vol. 8

anisms are also quite different from those observed in affective disorders; nonetheless, there may be common threads in the increasing developement of behavioral pathology and in some of the principles that are applicable.

1 Sensitization Phenomena in Affective Disorders

The affective disorders tend to be recurrent. As illustrated in Fig. 1, the median course of manic and depressive episodes in 82 consecutively admitted bipolar patients tends to be progressive. This is illustrated in several measures: the first episode is often milder than subsequent episodes, and the well interval between episodes tends to shorten as a function of interval number. Figure 2 and Table 1 illustrate this phenomenon as a general characteristic of the course of affective illness in studies from Kraepelin's time to the present. With repetition, the onsets of episodes may become more rapid and precipitous, the intervals may shorten, and, in some cases, rapid or continuous cycling may develop (Table 2). While there is ample evidence that early episodes may be associated with psychosocial stresses and definable precipitants (Brown et al. 1975; Amelas 1979), this is increasingly unlikely to be the case with episodes late in the course of illness which are, at times, occurring with such rapidity and rhythmicity that specific precipitants are difficult to postulate or rationalize (Fig. 3). The tendency for opposite mood phases to occur as the illness progresses is also evident.

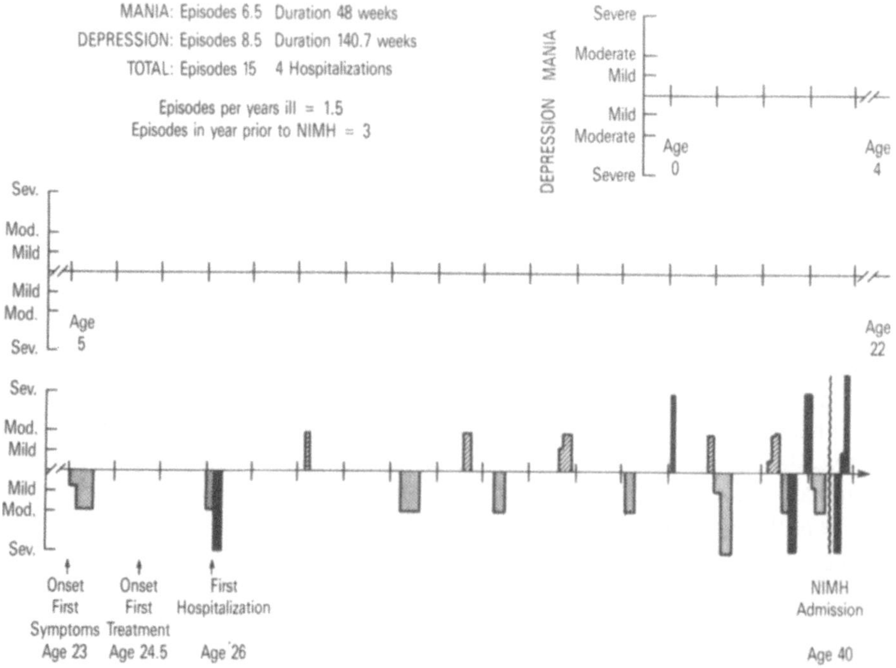

Fig. 1. Median course of affective illness in 82 bipolar manic-depressive patients admitted to NIMH. Note decreasing duration of well intervals between episodes (faster cycling) and increasing severity of illness

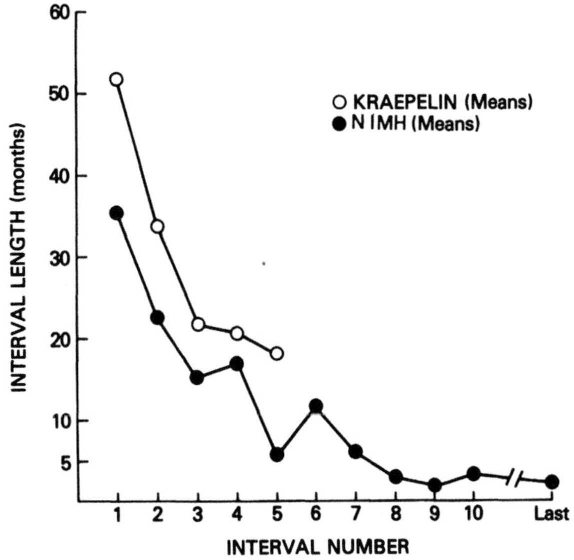

Fig. 2. Decreasing well intervals in recurrent-affective illness. On the average there is a shorter well interval between successive episodes of affective illness as described by Kraeplin (1921) in the pre-psychopharmacological era as well as in our treatment refractory NIMH patients

Table 1. Studies reporting a decreasing well intervall between episodes in manic-depressive illness

Study	Number of subjects
Swift (1907)	105
Kraepelin (1921)	903
Paskind (1930)	633
Lundqvist (1945)	319
Angst and Weis (1967)	388
Bratfos and Haug (1968)	207
Grof et al. (1974)	987
Taschev (1974)	652
Zis et al. (1980)	334
Roy-Byrne et al. (1985)	95
Total	4623

While Fig. 1 illustrates the median course of affective illness in our series of patients admitted to NIMH (often for treatment-refractory illness), it is also clear that there are many exceptions to this profile. Some patients' first episode is that of mania, and in some instances, patients begin rapid cycling almost from the outset of their illness. In the literature, approximately 30% of patients with a first depression do not have a recurrence and the course may never become progressive. Nonetheless, as summarized in Table 2, the median course in patients in the literature does show a characteristic speeding up of episode frequency as a function of recurrence, and this pattern becomes the focus of our current discussion.

Table 2. Parallels in the phenomenology of affective illness to kindling and sensitization

Feature of affective illness	Characteristic of	
	Kindling	Sensitization
1. Genetic component	+	+
2. Early experience predisposes	+	+
3. Mild alterations emerge as full-blown episodes (threshold effects)	+	+
4. Episodes reach plateau and are similar in content and behavior over repeated recurrences	+	+
5. Onset of maximum disturbance occurs earlier in episode with repetitions	+	+
6. Vulnerable to recurrences, cycles may speed up	+	?
7. Early episodes may be precipitated, later ones appear spontaneously	+	?
8. Repeated episodes of mania or depression may lead to emergence of opposite phase (conditioned compensatory reactions)	?	+
9. Lithium carbonate effective prophylaxis	–	(+)
10. Carbamazepine effective prophylaxis	(+)	–

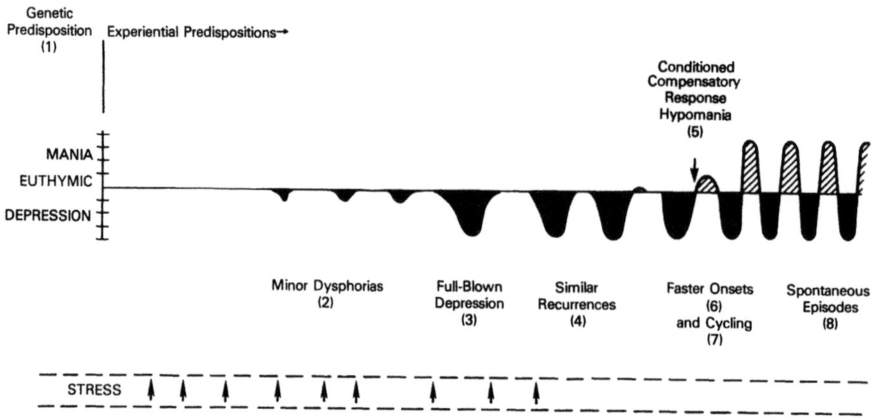

Fig. 3. Schematic illustration of the longitudinal and progressive development of recurrent affective illness (see Post et al. 1986a for details)

2 Behavioral Sensitization Stimulants

We (Post 1975) have postulated, and Antelman and Chiodo (1984) have demonstrated, that stimulant-induced phenomena may, in some instances, mimic behavioral and chemical changes induced by stresses. While considerable work is necessary to delineate which types of stresses show cross-sensitization to stimulant-induced sensitization under what conditions, it is apparent in many cases that stresses in animals will lead to increased behavioral responsivity to psychomotor stimulants and vice versa.

Several aspects of the behavioral sensitization phenomenon are particularly pertinent to our discussion of affective disorders. They emphasize that increased behavioral responsivity can occur following the same pharmacological (or by inference, biochemical) perturbation induced repeatedly. Various elements of behavioral sensitization, including its long-lasting alteration of responsivity, are further summarized in Table 3.

As illustrated in Fig. 4, repeated administration of the same dose of cocaine (10 mg/kg i.p.) evokes increasing motoric responsivity over time. Figure 4 (bottom) emphasizes that this behavioral sensitization can be dependent on the environmental context and, therefore, appears to involve conditioning. That is, animals treated repeatedly with cocaine in the same environmental context are more reactive to the psychomotor stimulant than animals repeatedly given the same enviromental exposures and doses of cocaine but who demonstrate their cocaine-induced hyperactivity in an environment dissimilar to the test apparatus. In fact, in recent studies, we have demonstrated that the degree of similarity of pretreatment environment to test environment is an important determinant of the degree of behavioral sensitization (Weiss et al. 1986a). A single injection of cocaine (40 mg/kg, day 1) sensitizes the animal to a rechallenge with cocaine (10 mg/kg) the next day. As illustrated in Fig. 5, this is entirely context-dependent, as animals injected with cocaine on day 1 in a different environment show no more activity on day 2 than saline-pretreated animals. Moreover, blockade of cocaine-induced behavioral activation on day 1 with neuroleptics or benzodiazepines inhibits the developement of behavioral sensitization to subsequent reexposure to cocaine on day 2 (Fig. 6).

Taken together, these data suggest that the manifestation of cocaine-induced behaviors in the testing environment (or in a similar environment) is important for the phenomena of behavioral sensitization. However, it is also possible, given high enough doses of cocaine and sufficient repetitions, to demonstrate an environmental-context-independent form of sensitization and one that is not blockable by neuroleptics, as illustrated in a study by Gale (1984). These and other data

Table 3. Characteristics of behavioral sensitization to psychomotor stimulants

1. Shorter latency to maximum response following repeated exposure
2. Increased magnitude of behavioral response to same dose
3. Selective increase in some behaviors, not others
4. Alterations in responsivity are long-lasting
5. Intermittent administration produces greater sensitization, as compared to continuous administration
6. Effects are dose-related
7. Genetic component is evident in degree of sensitization
8. Sensitization is not related to alterations in blood levels
9. Sensitization shows environmental context dependency, i. e., is conditionable
10. Cross-sensitization occurs among many stimulants and dopamine agonists
11. Cross-sensitization occurs between stimulants and a variety of stresses (tail-pinch, shock, starvation)
12. Neurotransmitter and receptor changes postulated
13. Hormones and peptides modulate behavioral sensitization

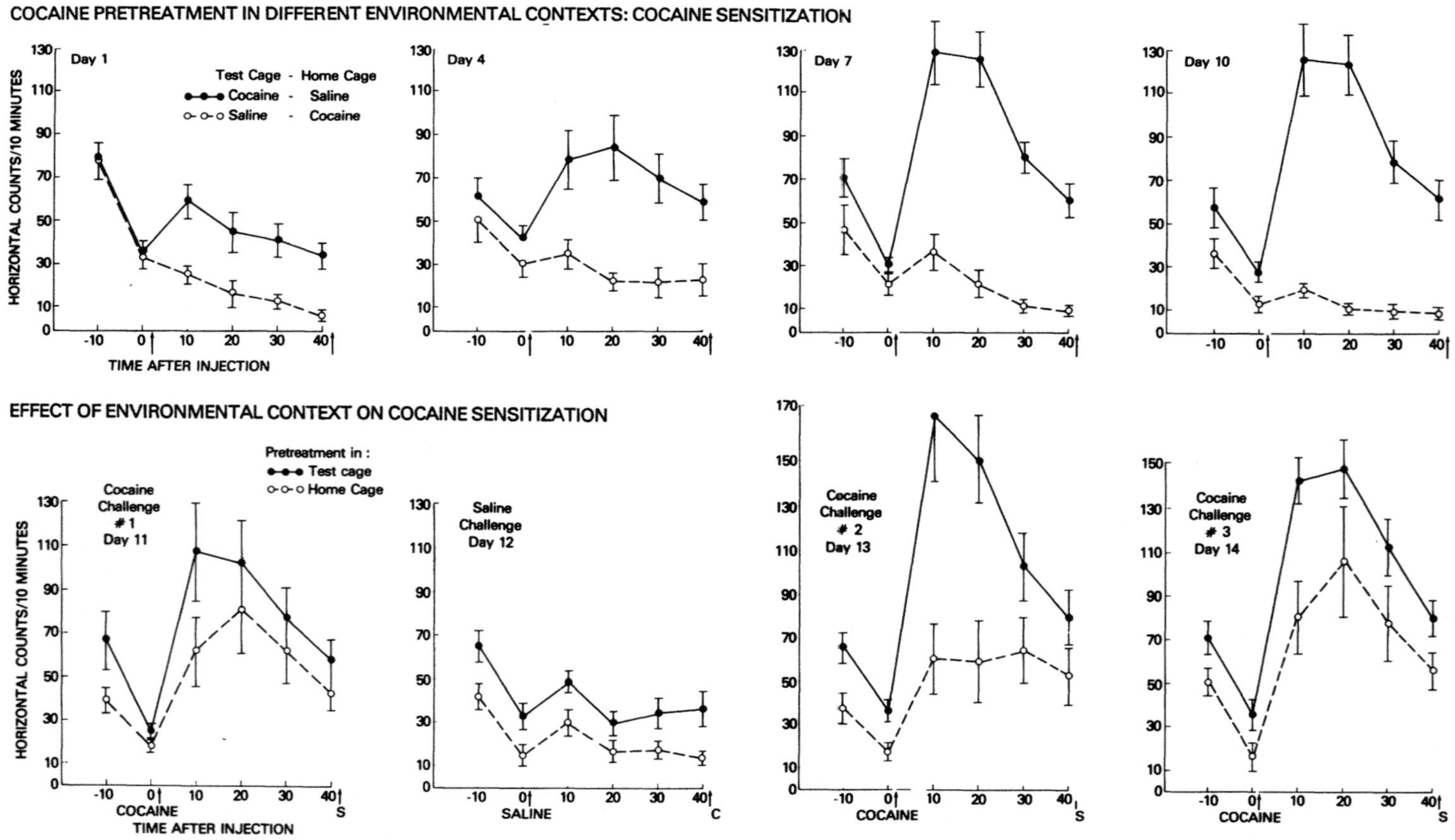

Fig. 4. Cocaine sensitization and environmental context. Repeated cocaine administration results in increased behavioral responsivity only in the group receiving cocaine in the test cage. (From Post et al. 1981)

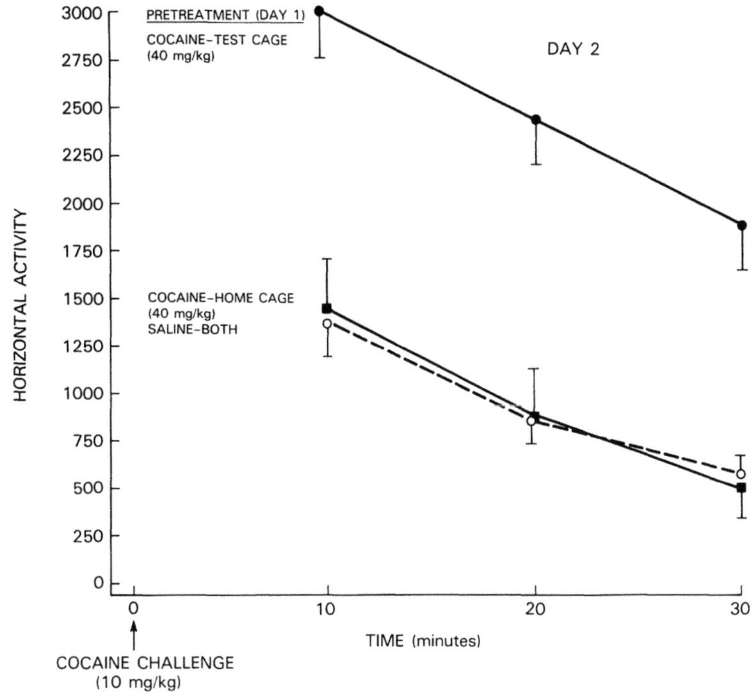

Fig. 5. Influence of environmental context on cocaine-induced behavioral sensitization. Only rats previously treated with cocaine (40 mg/kg) in the test cage (*filled circles*) show increased response to cocaine (10 mg/kg) the next day. Animals receiving cocaine (40 mg/kg) in the home cage (*filled square*) were no more responsive the next day than saline controls

are summarized in Fig. 7, suggesting that there are carefully defined parameters of dose and interval which determine the final output in terms of degree of behavioral sensitization, dependence on context and conditioning principles, and, finally, developement of kindled seizures. Repeated administration of the same dose of a pure local anesthetic stimulant such as lidocaine (Post et al. 1975), or the mixed local anesthetic-psychomotor stimulant cocaine in high doses, may evoke seizures in response to doses which were previously subconvulsant (Post and Kopanda 1976). This "pharmacological" kindling bears a close resemblance to the electrophysiological kindling originally described by Goddard and colleagues in 1969.

Interestingly, while neuroleptics block the development of cocaine- and also amphetamine-induced sensitization, they do not block its *expression* once sensitization has been induced (Beninger and Hahn 1983; Beninger and Herz 1986; Tadokoro and Kuribara 1986; Weiss et al. 1986a). That is, animals previously exposed to stimulants run more in response to rechallenge with stimulants, even if neuroleptics are coadministered (Fig. 8). To the extent that sensitization phenomena are clinically relevant in affective illness or schizophrenia, these data could provide a new conceptual basis for considering neuroleptic nonresponsiveness.

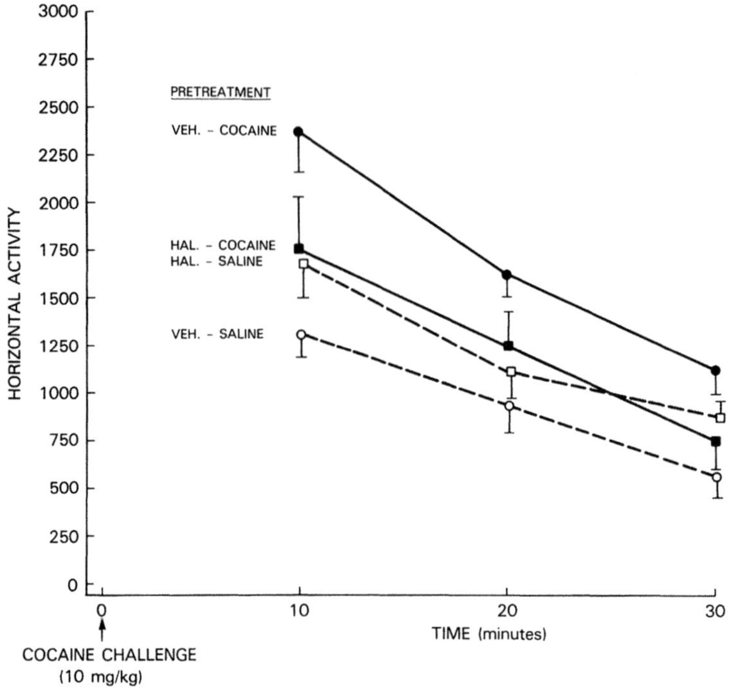

Fig. 6. Blockade of cocaine-induced activity prevents cocaine sensitization. Animals receiving haloperidol prior to cocaine on day one (*filled squares*) were no more active than those receiving haloperidol alone (*open squares*) following the day 2 challenge dose of cocaine

The conditioned components of stimulant-induced sensitization may have become neuroleptic-insensitive! These data and those discussed below suggest the possible importance of initially preventing stimulant-induced and stress-induced abnormal behaviors, in order to prevent sensitization phenomena from occurring.

The fact that behaviors induced by very high doses of cocaine (Fig. 7; Gale 1984) can also produce a sensitization syndrome that cannot be inhibited in its development suggests an additional mechanism for neuroleptic unresponsiveness. In this instance, the increased response does not appear to depend on the previous occurrence of behavior and probably does not involve a dopaminergic substrate.

Fig. 8. Haloperidol blocks the development (*left panel*) but not the expression of cocaine sensitization (*right panel*). In contrast to the effect of haloperidol administered prior to the day one cocaine injection (*left panel*), the same dose of haloperidol (0.5 mg/kg) is unable to block the increase in cocaine-induced activity when it is administered prior to the day two challenge dose of cocaine (*right panel*)

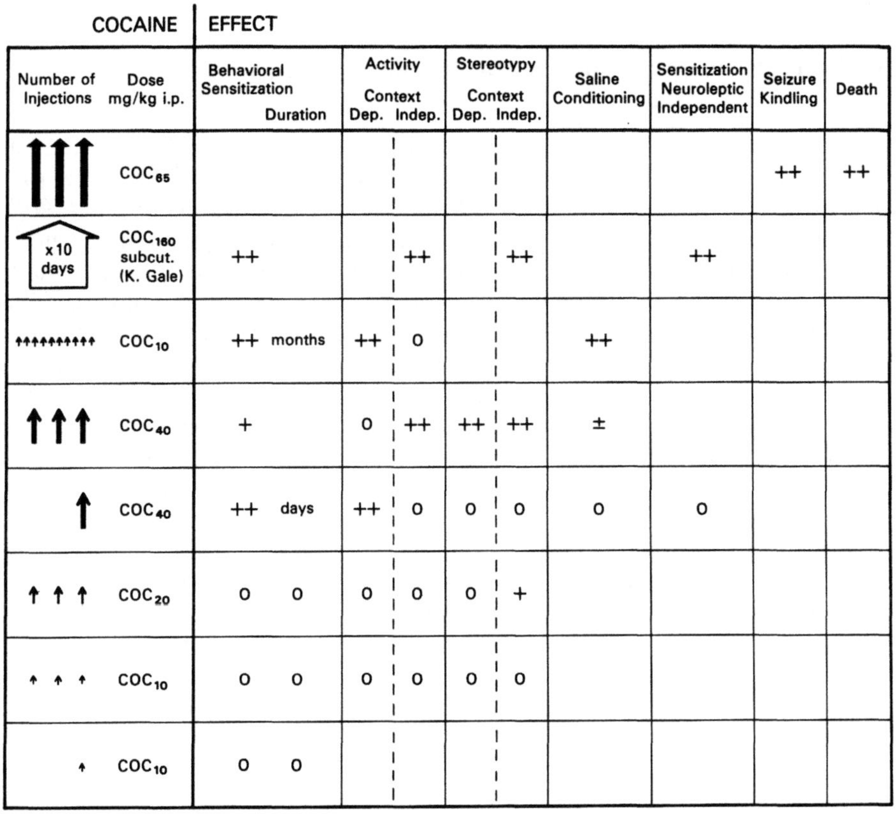

COCAINE: Number of Injections	Dose mg/kg i.p.	EFFECT: Behavioral Sensitization	Duration	Activity Context Dep.	Activity Context Indep.	Stereotypy Context Dep.	Stereotypy Context Indep.	Saline Conditioning	Sensitization Neuroleptic Independent	Seizure Kindling	Death
↑↑↑	COC_{65}									++	++
x10 days	COC_{160} subcut. (K. Gale)	++			++		++		++		
↑↑↑↑↑↑↑↑↑↑	COC_{10}	++	months	++	0			++			
↑↑↑	COC_{40}	+		0	++	++	++	±			
↑	COC_{40}	++	days	++	0	0	0	0	0		
↑ ↑ ↑	COC_{20}	0	0	0	0	0	+				
↑ ↑ ↑	COC_{10}	0	0	0	0	0	0				
↑	COC_{10}	0	0								

Fig. 7. Long term effects of cocaine: function of dose and number of repetitions. Size and number of arrows indicates dose of cocaine and number of administrations. *$COC_{( )}$* Cocaine and dose in mg/kg administered once daily intraperitoneally except in the study of Gale when doses were subcutaneous. *0* no effect; ± equivocal; + moderate effect; ++ marked or definite effect. Effects of cocaine increase and are more persistent with dose and number of repititions. They also become less dependent on environmental context and conditioning. The highest doses may be associated with seizures, kindling, and its associated lethality. Thus, effects shift from behavioral sensitization to kindling as the dose is increased

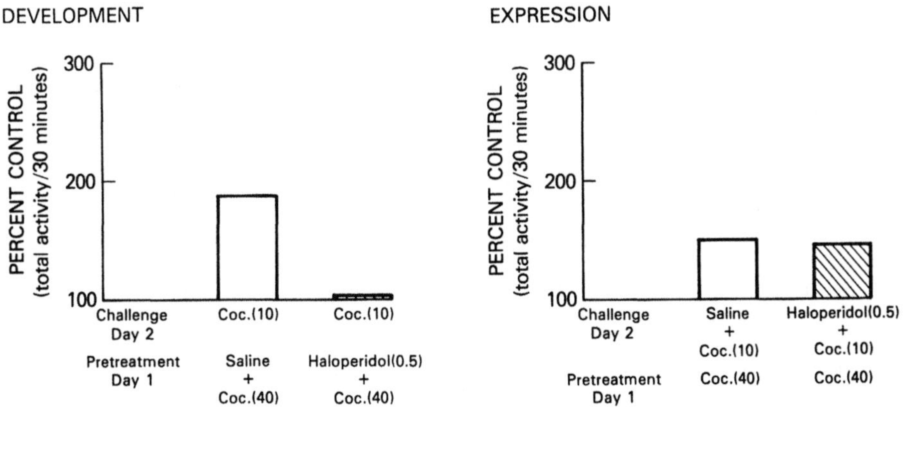

3 Electrophysiological and Pharmacological Kindling

The characteristcs of kindling, summarized in Table 4, involve the repeated but intermittent electrical stimulation of the brain with currents which are initially inadequate to produce behavioral or electrophysiological (spiking) responses. With sufficient repetition, afterdischarges develop, spread throughout many brain structures, and eventually a full-blown "limbic-type" seizure is evoked. If these kindled seizures are repeated sufficiently, the animals develop "spontaneity," a stage in which seizures occur without any exogenous electrophysiological stimulation. Thus, there appear to be three critical stages in the process of kindling: (A) an initial developemental stage in which thresholds for excitability and afterdischarges are altered, and which results in the developement of seizures following sufficient repetitions of the stimulus; (B) an intermediate stage, in which seizures are reliably induced by electrophysiological stimulation; and (C) a stage of spontaneity where no stimulation is required.

It is of considerable interest that the pharmacology of these three stages is quite different (Fig. 9). That is, some drugs such as diazepam, which are effective in inhibiting the development of or completed seizures or the development phase, are ineffective in inhibiting spontaneous seizures of the late phase of kindling (Pinel 1983). Conversely, phenytoin, which is inadequate in the early developmental and completed stages, is effective in inhibiting spontaneous seizures (Pinel 1983). The anticonvulsant carbamazepine does not block the development

Table 4. Electrical kindling: major characteristics

1. Repeated stimulations
   a) Initially subseizure threshold
   b) Intermittent
2. Local afterdischarges and seizure activity
   a) Increase in amplitude, frequency
   b) Increase in duration
   c) Increase in complexity of wave form
   d) Increase in anatomical spread
3. Replicable sequence of seizure stages: behavioral arrest, blinking and masticatory movements, head nodding, opisthotonis, contralateral then bilateral forelimb, clonus, rearing and falling
4. Limbic system kindled more readily than cortex
5. In kindled animals the history of convulsion development is recapitulated as seizure builds
6. Transfer effects to secondary sites; kindling facilitated in other sites even after primary site destroyed
7. Interference: a secondary kindled site interfere with primary site kindling
8. No toxic or neuropathological changes evident: kindling is a transsynaptic process
9. Relatively permanent change in connectivity; a kindled animal will still have seizures after a 1-year seizure-free interval
10. Seizures may develop spontaneously in chronically kindled animals
11. Interictal spikes and spontaneous epileptiform potentials develop

See Goddard et al. (1969) regarding items 1, 4–9; Wada and Sato (1974) regarding 2–3, 6; Wada et al. (1974) regarding 10–11; Racine (1978); Pinel and Rovner (1978a, b)

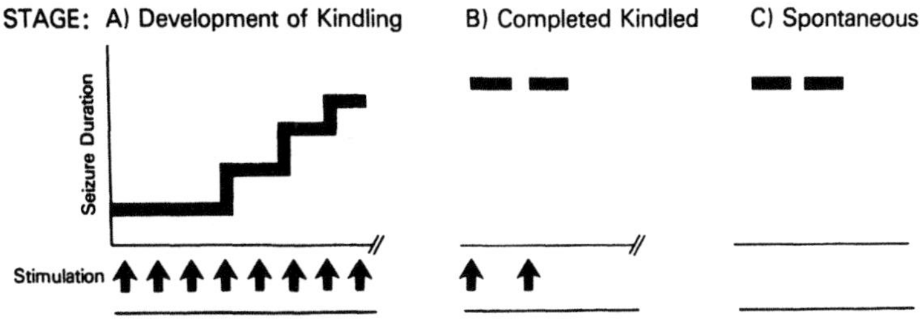

Amygdala Kindling (Electrical)			
Drug	Effectiveness (Rat)		
Diazepam	+++	++	0
Phenytoin	0	+	+++
Carbamazepine	0	+++	?
Lidocaine Kindling (Pharmacological)			
Carbamazepine	+++	0	?
Diazepam	+++	+++	?

Fig. 9. Pharmacology of kindling as a function of stimulus type and stage of development. Anticonvulsants are effective in some stages of amygdala kindling but not others. Moreover, the same anticonvulsant, carbamazepine, is differentially effective in amygdala compared to lidocaine kindling

phase of kindling in the rat (Weiss et al. 1985) as it does in the cat and the baboon (Wada et al. 1976; Wada 1977), but is quite a potent anticonvulsant for completed kindled seizures (Weiss and Post 1987a; Post et al. 1984, 1986b). Conversely, carbamazepine appears to have opposite effects in development-phase and completed seizures kindled by the local anesthetics lidocaine and cocaine (Post et al. 1986b): it is potent in inhibiting the development of these local anesthetic kindled seizures, but is without effect once they are completely manifest.

Thus, it appears to be a general principle that different stages of different types of kindling show differing pharmacological responsivity. These data are schematized in Fig. 9. They further suggest that the biological and biochemical underpinnings of the kindling phenomenon can vary as a function of stage and development. Therefore, we raise the question of whether differences might not also occur in the course of development of clinical conditions, such that pharmacological interventions may vary as a function of course of illness.

This might be particularly important in the consideration of affective disorders, where preliminary evidence suggests that some agents may be effective in one stage of the illness but not in others (Fig. 10). For example, tricyclic and monoamine oxidase antidepressants are effective in the treatment of unipolar patients but may, in some unipolar and bipolar patients, precipitate manic episodes, increase rapidity of cycling, or induce continuous cycles (Bunney et al. 1970; Wehr and Goodwin 1979; Kukopulos et al. 1980; Goodwin and Jamison 1988). An extensive literature also suggests that lithium carbonate, while effective in some 80% of bipolar patients, may be less effective in the later stages of the illness

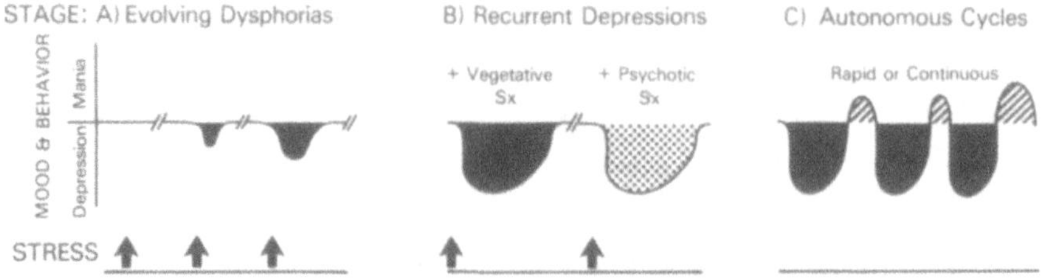

TREATMENT	A) Evolving Dysphorias	B) Recurrent Depressions + Vegetative Sx	B) Recurrent Depressions + Psychotic Sx	C) Autonomous Cycles
Psychotherapy	+++	++		(+)
Benzodiazepines	++	(++)	(+)	
Tricyclics	±	+++	+	−−
MAOI's	±	+++	(+)	−−
ECT		+++	+++	(+)
Tricyclics plus Neuroleptics			+++	
Lithium	±	+	+	(++)
Carbamazepine		(+)	(+)	+++
Lithium plus Carbamazepine				+++

Fig. 10. Psychopharmacotherapy of affective illness as a function of type and stage of development. *MAOIs* monoamine oxidase inhibitors; *ECT* electroconvulsive therapy. Just as different stages of kindling are differentially responsive to pharmacological interventions, we postulate a differential effectiveness of therapeutic agents as a function of the longitudinal course of affective illness (see Post et al. 1986a for details). This schema remains to be directly tested

characterized by rapid cycling or more than four episodes per year (Dunner and Fieve 1974; Hanus and Zapletalek 1984; Prien 1984). It remains to be demonstrated whether this represents an actual shift in pharmacological responsivity as a function of course of illness, or relates to differential subgroups of patients. Nonetheless, these data raise the question of whether pharmacological responsivity may vary as a function of course of illness, and preliminary evidence suggests that some anticonvulsant agents, such as carbamazepine or valproic acid, may be helpful in the very rapidly cycling, lithium-resistant, manic-depressive patients who are often studied later in their course of illness (Post and Uhde 1986; Post et al. 1987).

4 Kindling with Benzodiazepine-Active Ligands: Proconvulsant and Anticonvulsant Effects

It is also of interest that one can kindle seizures with the "peripheral-type" benzodiazepine receptor ligand Ro5-4864 (Fig. 11) (Weiss et al. 1986c). However, these seizures may be occurring through actions at other than the peripheral-type

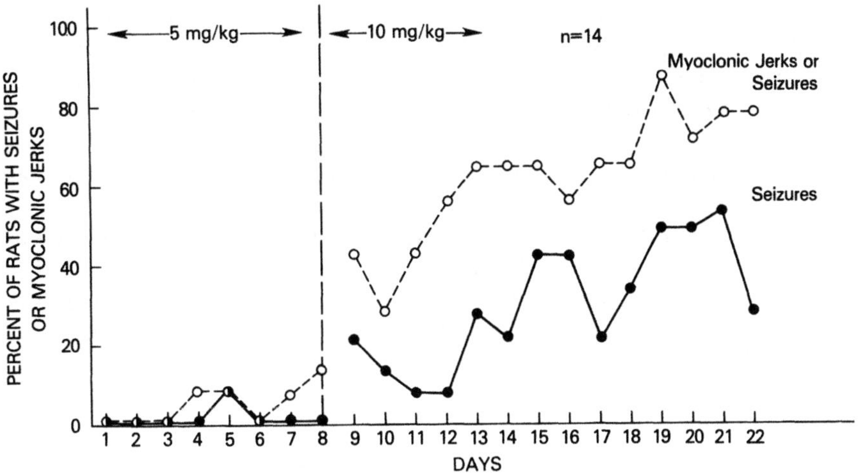

Fig. 11. Repeated injection of Ro5-4864 produces increasing seizures and myoclonic jerks in rats. Repetition of the same once-daily dose produces a pharmacological kindling-like pattern of increased responsivity over time. The high doses of Ro5-4864 may occupy binding sites on the chloride channel of the classical central type benzodiazepam-GABA complex in addition to its effects on the peripheral type binding site

benzodiazepine site, as they are not blocked by PK-11195 or carbamazepine (Weiss et al. 1986c; Weissman et al. 1983). The picrotoxin site may be involved, and seizure development with Ro5-4864 may have some analogies to kindling at the "central-type" benzodiazepine sites with FG-7142 or other beta-carbolines (Little et al. 1984; Nutt and Little 1984). Since putative endogenous ligands for both central-type and peripheral-type benzodiazepine receptors have been described, it is possible that kindling-like responses to repeated activation of these systems could occur in animals.

It does appear that the anticonvulsant effects of carbamazepine are, in part, mediated by the peripheral-type benzodiazepine site (Weiss et al. 1985, 1986c). Carbamazepine displaces binding of [^{3}H]-Ro5-4864 more potently than that of [^{3}H]-BCCE, suggesting a greater biochemical effect at peripheral-type benzodiazepine receptors than at central-type benzodiazepine receptors (Marangos et al. 1983). Moreover, the anticonvulsant effects of carbamazepine, but not diazepam, are reversed by Ro5-4864 in the amygdala kindling model. Conversely, the prototypic central-type benzodiazepine receptor antagonist, Ro15-1788, reverses the anticonvulsant effects of diazepam but not of carbamazepine.

The selective effects of carbamazepine on peripheral-type benzodiazepine receptors are further suggested by the findings that PK-11195 reverses the inhibitory effect of Ro5-4864 (Weiss et al. 1985). In addition, the reversal of carbamazepine's effects by Ro5-4864 is not the result of a nonspecific proconvulsant effect of Ro5-4864, since the similarly proconvulsant beta-carboline, B-CCM, does not reverse the anticonvulsant effects of carbamazepine, but does reverse those of diazepam (Weiss et al. 1986c).

Peripheral-type benzodiazepine receptor mechanisms could also be relevant to other aspects of carbamazepine effects, as they have been implicated in the effects of anxiety (Pellow and File 1984; Mizouli et al. 1985) and stress (Drugan et al. 1986). They also could account for some of the diverse endocrine effects of carbamazepine, including its ability to increase urinary free cortisol secretion (Rubinow et al. 1986b) and to act as a vasopressin-like compound (Gold et al. 1983). Ro5-4864 binding is paticularly dense in endocrine tissues such as the posterior pituitary, adrenal, kidney, and testis. Preliminary data suggest that peripheral-type benzodiazepine receptors are associated with calcium channels (Mestre et al. 1985), in contrast to the central-type benzodiazepine receptors, which are linked to chloride channels and $GABA_A$ receptors (Tallman et al. 1980). The further elucidation of the endocrine effects of peripheral-type benzodiazepine receptors and their role in the mechanisms of action of carbamazepine should provide advances of clinical and basic interest.

5 Peptide Kindling: Opiates and Corticotropin-Releasing Factor

Several peptides can show phenomena akin to behavioral sensitization and kindling. It is of interest that the peptides that have demonstrated this phenomenon most amply are related to the endogenous opiates and to corticotropin-releasing factor (CRF), two peptide systems that have been closely linked to stress responsivity.

At a behavioral level, opiate-related compounds can show phenomena akin to behavioral sensitization. While we usually assume that tolerance develops to repeated opiate administration, Shuster et al. (1975), in an edifying study, demonstrated that, depending on the interval between injections, morphine can induce either sensitization or tolerance to opiate-induced running behavior in mice. With continuous administration or injections repeated at short intervals, tolerance did develop to opiate-induced running activity. However, when the interval was increased, behavioral sensitization and an increased responsivity to morphine were exhibited. Sensitization phenomena have also been demonstrated with a variety of endogenous opiate compounds directly administered into brain regions, such as the ventral tegmental area. Kalivas et al. (1986), for example, have shown that repeated enkephalin administration will produce increased motor activity and that this will show cross-sensitization to the psychomotor stimulant amphetamine. Thus, it is possible that, under conditions of some types of stress which are capable of releasing endogenous opiate compounds, phenomena akin to behavioral sensitization could develop.

Moreover, Cain and collaborators (1984), in a series of studies, have documented that endogenous opiate compounds of the enkephalinergic class are capable of inducing a classical kindling pattern with final manifestation of seizures upon repeated intracerebral (i.c.) administration of previously subconvulsant doses of opiates into the amygdala and hippocampus. These findings have recently been replicated by Tanaka et al. (1986). Cain demonstrated that the opiate-induced kindling was selective to some regions of the amygdala and hippocampus but not others, and was opiate-antagonist-reversible with naloxone.

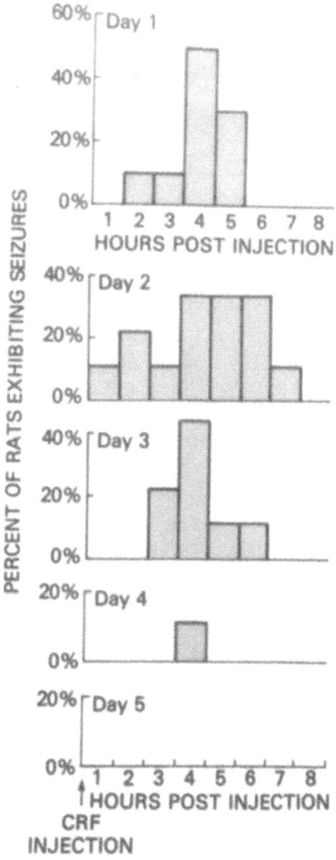

Fig. 12. Temporal distribution of CRF-induced seizures. Seizures generally occur 3–7 h following i.c.v. injection of CRF (100 μg). While all animals have seizures on day 1 or 2 of injection, none do by day 5 suggesting the development of tolerance

We have replicated the original observations of Ehlers et al. (1983) that a single high dose of CRF administered intracerebroventricularly (i.c.v.) will produce the late onset of seizures some 3–7 h after drug administration (Fig. 12; Weiss et al. 1986b). Ehlers et al. (1983) postulated that this involves a pharmacological kindling mechanism that evolves over this long period of time and is associated with repeated episodes of spiking in subcortical structures such as the amygdala. The CRF seizures involve amygdala spiking similar to that seen with electrophysiological kindling; the behavioral seizures themselves are similar to those observed in electrical kindling, and CRF-induced seizures show cross-sensitization to amygdala kindling (Fig. 13; Weiss et al. 1986b). Repeated daily i.c.v. administration of CRF, however, does not show a typical kindling course of increasing response over time. It actually shows the development of tolerance to CRF-induced seizures and to the associated aggressive behavior that is also observed (Fig. 14). That is, all of the animals studied show late onset of seizures and defensive fighting behaviors following i.c.v. CRF on days 1 and 2, but by day 5 of injection, no animals demonstrate seizures. These data are illustrated in Figs. 12 and 14.

In a series of studies with S. R. B. Weiss and A. Pert, we have been unable to localize the origin of CRF-induced seizures with direct i.c. administration of this compound into amygdala, hippocampus, septum, ventral thalamus, anterior hy-

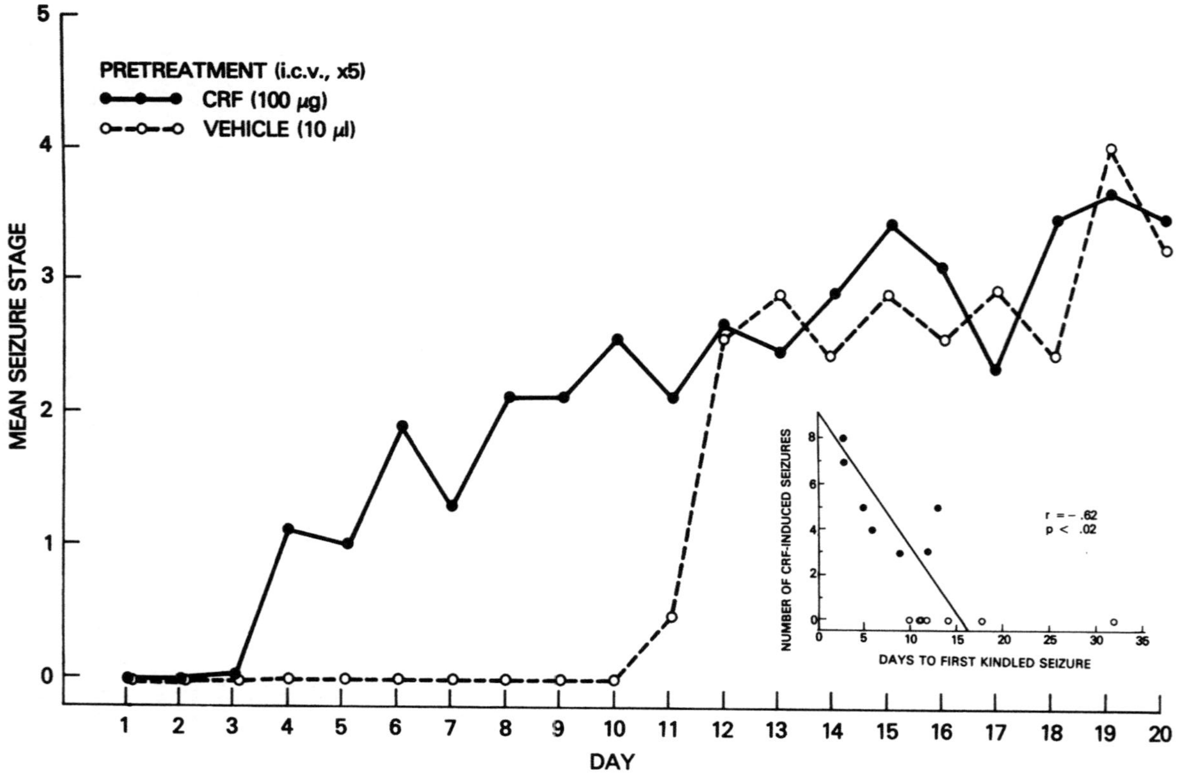

Fig. 13. CRF-induced seizures facilitate the development of amygdala-kindled seizures. Even though tolerance develops to the CRF induced seizures (see Fig. 12) animals pretreated with CRF show significantly faster electrical kindling than vehicle treated controls. Moreover, those rats with greater numbers of CRF seizures, kindle the fastest (insert)

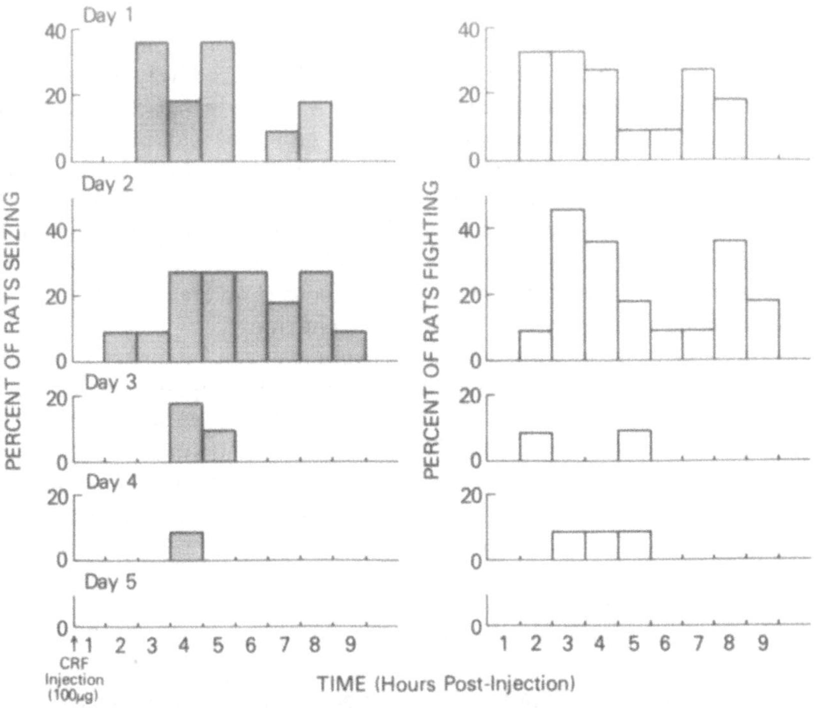

Fig. 14. Temporal distribution of CRF-induced seizures and fighting: development of tolerance to both. The lag in onset in apparent tolerance to CRF-induced seizures is replicated in a second group of animals (*left panel*). The late occurrence of and apparent tolerance to CRF-induced aggression is also illustrated (*open bars, right panel*)

pothalamus, periaqueductal gray, and endopyriform nucleus. Moreover, lesions of amygdala, hippocampus, olfactory tubercle and endopyriform nucleus do not block the development of CRF-induced seizures following i.c.v. administration. Thus, the mechanisms and anatomical regions involved in the late onset of CRF-induced seizures remain to be elucidated.

Nonetheless, the data suggest that, under some conditions, high doses of CRF administered into the ventricular system may be capable not only of inducing a range of behaviors, vegetative and hormonal phenomena that are often observed in affective illness, but also introduce various types of apparent sensitization and tolerance phenomena. Gold and associates, in this volume, have suggested that of CRF administered i.c.v. can mimic many of the phenomena observed in affective illness. These include decreased appetite and sleep, changes in psychomotor activity, anxiety-like behaviors decreased social interaction, cortisol hypersecretion and, at high doses, aggression. Recently, Kalin and colleagues (1985) also demonstrated that i.c.v. CRF in unrestrained primates will induce a huddled posture resembling the posture assumed by primates during despair phases following separation from their mothers or peer groups. Thus, in light of the evidence that cortisol hypersecretion is one of the best documented neuroendocrine concomitants of the affective disorders, and in the light of Gold's data reviewed above, one is tempted to speculate that hypersecretion of CRF could occur in some areas of brain or in the cerebrospinal fluid (CSF) of the affectively ill patients. The initial finding by Nemeroff et al. (1984) that CRF is increased in the CSF of depressed patients has been replicated by that group, but not by Gold et al.; the relevance of these CRF changes to the symptomatology of depression requires further study. These data are also intriguing in light of the report that CRF receptors are decreased in the frontal cortex of suicide victims compared to controls (Nemeroff 1986). The current interest in the effects of repeated peptide administration on behavior and electrophysiology is focused on the possible long-term consequences of altered peptide function on behavior.

Both tolerance and sensitization phenomena can be observed following administration of these peptide substances (endogenous opiates and CRF), depending on initial administration characteristics, such as dose, area of brain involved, and interval between injections. The sensitization and kindling data are raised in this context in order to outline a series of questions regarding possible progressive changes in behavior, and long-lasting consequences following repeated exposure to these and related peptide substances. Obviously, we wish to reemphasize that the kindling model is inadequate for direct application to the affective disorders, as affectively ill patients do not demonstrate seizures.

In this volume, Janowsky reviews the possible alterations in cholinergic mechanisms in changes of mood and behavior. In this regard, it is of considerable interest that cholinergic kindling has been demonstrated in animals. Cain and collaborators reported that administration of a cholinergic agonist into the amygdala eventually elicited seizures to a dose which was previously subthreshold (Cain 1983). This pardigm was atropine-reversible. Similar phenomena have been demonstrated by Wasterlain (1983). While there is controversy as to the extent of the cross-sensitization between cholinergic kindling and electrophysiological kindling, both investigators have documented the existence of cholinergic kindling. Thus, we would raise the possibility not only that a variety of endocrine and peptide substances are altered with stresses and repeated episodes of affective illness, but also that cholinergic alterations might be associated with either sensitization or kindling phenomena.

6 Somatostatin: Relation to Affective Illness, Seizures, and Anticonvulsant Mechanisms

There is considerable evidence linking alterations in somatostatin to aspects of affective disorders, its endocrinology and pharmacology. Gerner and Yamada (1982) and Rubinow et al. (1983) reported decreased somatostatin in CSF of depressed patients. These observations have been replicated by several other investigative groups (Table 5) and extended with the suggestion that, at least in some patient populations, those that demonstrate cortisol hypersecretion or escape from dexamethasone suppression may have significantly lower somatostatin in CSF (Doran et al. 1986). A series of studies demonstrate unequivocally that

Table 5. Low CSF somatostatin in depressed patients

Study	Control	Depressed	Euthymic	Manic	Schizophrenic
Gerner and Yamada (1982)	23 (29)**	15 (28)		26 (10)	19 (13)
Rubinow et al. (1983)	62.1 ± 5.4 (47)**	40.8 ± 4.2 (23) UP 43.7 ± 4.8 (26) BP	63.2 ± 9.1 60.8 ± 6.5 (18)	55.8 ± 6.6 (9)	
Agren and Lindquvist (1984)	–	14.8 ± 4.2 (21)*[a]	18.8 ± 5.8 (25)		
Black et al. (1986)	55.5 ± 2.2 (46)**	19.2 ± 2.4 (18)			
Bissette et al. (1986)	116.1 ± 15.7 (10)*	65.4 ± 7.8 (17)			68.7 ± 7.7 (11)
Doran et al. (1986)		46.9 ± 21.5 (15UP + 4BP)**[b] 44.8 ± 5.1 (15) UP 55.7 ± 14.4 (4) BP*[b]			65.6 ± 26.4 (44)

Results expressed as mean values (pg/ml). Number of subjects in parentheses. UP, unipolar depression (no history of mania); BP, bipolar depression (history of mania)

* $p < 0.05$; ** $p < 0.01$

[a] Values in worst week compared with 2 months later

[b] Significantly lower than in schizophrenic patients

somatostatin is decreased in the brain and spinal fluid of patients with Alzheimer's disease (Wood et al. 1982; Serby et al. 1984; Martin et al. 1984; Price et al. 1984; Soininen et al. 1984). These data, in conjunction with findings of altered somatostatin in several other neuropsychiatric illnesses characterized by cognitive deficits, including parkinsonism (Dupont et al. 1982; Epelbaum et al. 1983) and multiple sclerosis in relapse (Sorensen et al. 1980; Beal et al. 1985), further suggest the possibility that not only some of the behavioral and endocrinological aspects of affective disorders might ultimately be linked to somatostatin alterations, but also the element of depressive pseudodementia. Decreases in somatostatin in various brain regions induced by cysteamine have been associated with behavioral evidence of cognitive defects in animals (Vecsei et al. 1983, 1984; Walsh et al. 1985). Obviously, a considerable amount of work is necessary before somatostatin alterations are linked with changes in cognitive function in depressed patients. However, this appears to be an interesting area for further research.

We have found that the anticonvulsant carbamazepine significantly decreases somatostatin in the CSF of affectively ill patients (Rubinow et al. 1985). This is not observed with a variety of other psychotropic agents, including tricyclic antidepressants and lithium. Carbamazepine has also been reported to decrease brain somatostatin levels in hippocampus and in several other brain regions following its concomitant administration during kindling (Higuchi et al. 1986). We were unable to find decreases in brain somatostatin in rats treated with carbamazepine for 12 days (Weiss et al. 1986 b), but Nagaki et al. (1986) did find decreases in mesolimbic areas and hippocampus after acute and chronic treatment respectively.

There does appear to be a possible link between somatostatin alterations and seizure phenomena. Kato and collaborators (1983) originally reported long-lasting increases in brain somatostatin following a series of three or five kindled seizures. Increases in somatostatin in the amygdala and in perirhinal cortex persisted for 2 months following the last seizure. Nadi et al. (1986) have also reported increases in somatostatin in the human epileptic focus compared with less pathological tissue some 4 cm distant. Moreover, several investigators have demonstrated that inhibition of somatostatin function with depletion by cysteamine or immunoneutralization with somatostatin antibodies, had anticonvulsant effects on electrical or pentylenetetrazol (PTZ)-kindled seizures (Higuchi et al. 1983; Assouline et al. 1984). This group has recently suggested that chronic administration of cysteamine inhibits the development of kindled seizures as well (Higuchi et al. 1986). We have also observed that cysteamine will prevent the development of seizures induced by CRF administration (Weiss et al. 1987, unpublished observations). Somatostatin is also released into CSF after PTZ-induced seizures in the rat (Pitkanen et al. 1986).

Taken together, these data raise the possibilty that somatostatin could play an important role in either the anticonvulsant effects of carbamazepine or some of the clinical or endocrinological aspects of the affective disorders, including mood, cognitive, or motoric alterations and cortisol hypersecretion (Rubinow et al. 1986 a). The possible interactions of somatostatin with seizures and convulsant mechanisms, affective illness, and cognitive impairment each appears to be a complex area deserving further study. Preliminary data from our laboratory suggest that 1 day's exposure to escapable or inescapable shock is insufficient to alter brain somatostatin levels in a variety of regions studied. The ultimate impact on brain somatostatin levels of acute and chronic stresses and other animal models of depression, in addition to learned helplessness, should prove to be an area for fruitful study.

7 Conclusions

Behavioral sensitization and kindling provide interesting paradigms for considering principles and mechanisms underlying the progressive development of increasing behavioral responsivity to the same stimulus over time and its long lasting consequences for behavior. It is apparent that neurotransmitter, endocrine, and peptide substances may, in some instances, be capable of inducing sensitization and kindling-like characteristics. As the pathophysiology of endocrine and peptide involvement in the affective disorders is unraveled, it perhaps behooves us also to consider how they may relate to various aspects of the longitudinal phenomenology of manic-depressive illness. It is apparent that intervention in the sensitization and kindling paradigms with pharmacological agents may differ as a function of stage of development of behavioral pathology (as well as a function of different types of sensitization and kindling). We can ask whether similar phenomena may also be occurring in affective illness. These considerations may not only lead to the development of a new series of psychopharmacological agents based on endocrine and peptide interventions in the affective disorders, but also clarify how these substances may be involved in the phenomenology and physiology as a function of important temporal characteristics of the illness.

References

Ågren H (1984) Low levels of somatostatin in human CSF mark depressive episodes. Psychoneuroendocrinology 9:233

Ambelas A (1979) Psychologically stressful events in the precipitation of manic episodes. Br J Psychiatry 135:15

Angst J, Weiss P (1967) Periodicity of depressive psychoses. In: Brill H, Cole JO, Deniker P et al. (eds) Proceedings of the fifth international medical foundation, international congress series. Excerpta Medica, Amsterdam, pp 702–710

Antelman SM, Chiodo LA (1984) Stress: its effect on interactions among biogenic amines and role in the induction and treatment of disease. In: Iverson LS, Iverson SD, Snyder SH (eds) Handbook of psychopharmacology, vol 18. Plenum, New York, pp 279–341

Assouline G, Barkaie E, Gutnick MJ (1984) Cysteamine suppresses kindled seizures in pentylenetetrazol-kindled rats. Eur J Pharmacol 106:649

Beal MF, Mazurek MF, Black PMcL, Martin JB (1985) Human cerebrospinal somatostatin in neurologic disease. J Neurol Sci 71:91

Beninger RJ, Hahn BL (1983) Pimozide blocks establishment but not expression of amphetamine-produced environment-specific conditioning. Science 220:1304

Beninger RJ, Herz RS (1986) Pimozide blocks establishment but not expression of cocaine-produced environment-specific conditioning. Life Sci 38:1425

Bissette G, Widerlov E, Walleus H, Karlsson I, Eklund K, Forsman A, Nemeroff CB (1986) Alterations in cerebrospinal fluid concentrations of somatostatin-like immunoreactivity in neuropsychiatric disorders. Arch Gen Psychiatry 43:1148 1148
Black PMcL, Ballantine T Jr, Carr DB, Beal MF, Martin JB (1986) Beta-endorphin and somatostatin concentrations in the ventricular cerebrospinal fluid of patients with affective disorder. Biol Psychiatry 21:1077
Bratfos O, Haug FO (1968) The course of manic-depressive psychosis: Follow-up investigation of 215 patients. Acta Psychiatry Scand 44:89
Brown GW, Bhrolchain MN, Harris T (1975) Social class and psychiatric disturbance among women in an urban population. Sociology 9:225
Bunney WE Jr, Murphy DL, Goodwin FK (1970) The switch process from depression to mania: relationship to drugs which alter brain amines. Lancet I:1022
Cain DP (1983) Bi-directional transfer of electrical and carbachol kindling. Brain Res 260:135
Cain DP, Corcoran ME (1984) Intracerebral beta-endorphin, met-enkephalin and morphine: kindling of seizures and handling-induced potentiation of epileptiform effects. Life Sci 34:2535
Doran AR, Rubinow DR, Roy A, Pickar D (1986) CSF somatostatin and abnormal response to dexamethasone administration in schizophrenic and depressed patients. Arch Gen Psychiatry 43:365
Drugan RC, Basile AS, Crawley JN, Paul SM, Skolnick P (1986) Inescapable shock reduces [^{3}H]Ro 5-4864 binding to "peripheral-type" benzodiazepine receptors in the rat. Pharmacol Biochem Behav 24:1673
Dunner DL, Fieve RR, (1974) Clinical factors in lithium prophylactic failure. Arch Gen Psychiatry 30:229
Dupont E, Christensen SE, Hansen AP, Olivarius BdeF, Orskov H (1982) Low cerebrospinal fluid somatostatin in Parkinson's disease: an irreversible abnormality. Neurology 32:312
Ehlers CL, Henriksen SJ, Wang M, Rivier J, Vale W, Bloom FE (1983) Corticotropin releasing factor produces increases in brain excitability and convulsive seizures in rats. Brain Res 278:332
Epelbaum J, Rubert M, Moyse E, Javoy-Agid F, Dubois B, Agid Y (1983) Somatostatin and dementia in Parkinson's disease. Brain Res 278:376
Gale K (1984) Catecholamine-independent behavioral and neurochemical effects of cocaine in rats. In: Sharp CW (ed) National Institue on Drug Abuse research monograph series 54: Mechanisms of tolerance and dependence. U.S. Government Printing Office, Washington DC, pp 323–332
Gerner RH, Yamada T (1982) Altered neuropeptide concentrations in CSF of psychiatric patients. Brain Res 238:298
Goddard GV, McIntire DC, Leech CK (1969) A permanent change in brain function resulting from daily electrical stimulation. Exp Neurol 25:295
Gold PW, Robertson GL, Ballenger JC, Kaye W, Chen J, Rubinow DR, Goodwin FK, Post RM (1983) Carbamazepine diminishes the sensitivity of the plasma arginine vasopressin response to osmotic stimulation. J Clin Endocrinol Metab 5:952
Goodwin FK, Jamison KR (1988) Manic-depressive illness. Oxford, New York (to be published)
Grof P, Angst J, Haines T (1974) The clinical course of depression: practical issues. In: Schattauer FK (ed) Symposia medica hoest, vol 8. Classification and prediction of outcome of depression. Schattauer, New York, pp 141–148
Hanüs H, Zapletalek M (1984) The prophylactic lithium treatment in affective disorders and the possibilities of the outcome prediction. Sb Ved Pr Lek Univ Karlovy 27:5
Higuchi T, Kato N, Noguchi T, Friesen HG, Waga JA (1986) Kindling and somatostatin. In: Wada JA (ed) Kindling III. Raven Press, New York, pp 349–360
Higuchi T, Sikand GS, Kato N, Wada JA, Friesen HG (1983) Profound suppression of kindled seizures by cysteamine: possible role of somatostatin to kindled seizures. Brain Res 288:359

Hoffman DC, Beninger RJ (1985) The effects of pimozide on the establishment of conditioned reinforcement as a function of the amount of conditioning. Psychopharmacology 87:454
Kalin N (1985) Behavioral effects of ovine corticotropin-releasing factor administered to rhesus monkeys. Fed Proc 44:249
Kalivas PW, Richardson-Carlson R, Van Orden G (1986) Cross-sensitization between foot shock stress and enkephalin-induced motor activity. Biol Psychiatry 21:939
Kato N, Higuchi T, Friesen HG, Wada JA (1983) Changes of immunoreactive somatostatin and beta-endorphin content in rat brain after amygdala kindling. Life Sci 32:2415
Kraepelin E (1921) In: Robertson GM (ed) Manic-depressive insanity and paranoia. Livingstone, Edinburgh
Kukopulos A, Reginaldi D, Laddomada P, Floris G, Serra G, Tondo L (1980) Course of the manic-depressive cycle and changes caused by treatments. Pharmakopsychiatria 13:156
Little HJ, Nutt DJ, Taylor SC (1984) Acute and chronic effects of the benzodiazepine receptor ligand FG7162: proconvulsant properties and kindling. Br J Pharmacol 83:951
Lundquist G (1945) Prognosis and course in manic-depressive psychoses. Acta Psychiatry Neurol 35:1
Marangos PJ, Post RM, Patel J, Zander K, Parma A, Weiss S (1983) Specific and potent interactions of carbamazepine with brain adenosine receptors. Eur J Pharmacol 93:175
Martin JB, Beal MF, Uhl G (1984) Neuropeptides and neurology. In: Proceedings of association for research in nervous and mental disease. Inc., symposium: neuropeptides: implications for neurologic and psychiatric diseases, New York, Nov 30 Dec 1
Mestre M, Carriot T, Belin C, Uzan A, Renault C, Dubroeucq MC, Gueremy C, Doble A, Le Fur G (1985) Electrophysiological and pharmacological evidence that peripheral type benzodiazepine receptors are coupled to calcium channels in the heart. Life Sci 36:391
Mizoule J, Gauthier A, Uzan A, Renault C, Dubroeucq MC, Gueremy C, Le Fur G (1985) Opposite effects of two ligands for peripheral type benzodiazepine binding sites, PK11195 and Ro5-4864, in a conflict situation in the rat. Life Sci 36:1059
Nadi NS, Holmes MD, Wyler AR, Porter RJ (1986) Neuropeptides in focal and non-focal epileptic tissue from patients with intractable epilepsy. 16th annual meeting, November 9–14. Society for Neuroscience, Washington DC, p 294 (abstract no 80.4.)
Nagaki S, Kato N, Minatogawa Y, Higuchi T (1986) Effects of anticonvulsants on immunoreactive somatostatin and gamma-aminobutyric acid contents in the rat brain. Epilepsia 27:639
Nemeroff C (1986) TRH and neurotensin during stress and psychiatric illness: Presented at NIH Conference on Mechanisms of Physical and Emotional Stress. Bethesda, Maryland, Nov. 6–8, 1986
Nemeroff CB, Widerlov E, Bissette G, Walleus H, Karlsson I, Eklund K, Kilts CD, Loosen PT, Vale W (1984) Elevated concentrations of CSF corticotropin-releasing factor-like immunoreactivity in depressed patients. Science 226:1342
Nutt DJ, Little HJ (1984) Kindling produced by a benzodiazepine receptor antagonist (inverse agonist). In: Vizi ES, Magyar K (eds) Regulation of transmitter function, basic and clinical aspects. Akademai-Kado, Budapest, pp 443–448
Paskind HA (1930) Manic-depressive psychosis in a private practice. Arch Neurol Psychiatry 23:699
Pellow S, File SE (1984) Behavioural actions of Ro 5-4864: a peripheral-type benzodiazepine? Life Sci 35:229
Pinel JP (1983) Effects of diazepam and diphenylhydantoin on elicited and spontaneous seizures in kindled rats: a double dissociation. Pharmacol Biochem Behav 18:61
Pitkanen A, Jolkkonen J, Honkanen K-L, Riekkinen P (1986) Somatostatin-like immunoreactivity in rat cerebrospinal fluid before and after generalized convulsions induced by pentylenetetrazol. Epilepsia 27:616
Post RM (1975) Cocaine psychoses: a continuum model. Am J Psychiatry 132:225

Post RM (1988) Animal models: clinical relevance. In: Koob G, Ehlers C (eds) Animal models of affective disorders. Birkhauser Inc., Boston
Post RM, Kopanda RT (1976) Cocaine, kindling, and psychosis. Am J Psychiatry 133:627
Post RM, Uhde TW (1986) Anticonvulsants in non-epileptic psychosis. In: Trimble MR, Bolwig TG (eds) Aspects of epilepsy and psychiatry. Wilry, New York, pp 177–212
Post RM, Kopanda RT, Lee A (1975) Progressive behavioral changes during chronic lidocaine administration: relationship to kindling. Life Sci 17:943
Post RM, Weiss SRB, Pert A (1984) Differential effects of carbamazepine and lithium on sensitization and kindling. Prog Neuropsychopharmacol Biol Psychiatry 8:425
Post RM, Rubinow DR, Ballenger SC (1986a) Conditioning and sensitization in the longitudinal course of affective illness. Br J Psychiatry 149:191
Post RM, Lockfeld A, Squillace KM, Contel NR (1981) Drug-environment interaction; context dependency of cocaine-induced behavioral sensitization. Life Sci 28:755
Post RM, Weiss SRB, Szele F, Woodward R (1986b) Differential anticonvulsant effects of carbamazepine as a function of stage and type of kindling. In: Abstracts of 16th annual meeting, society for neuroscience, November 9–14. Washington DC
Post RM, Uhde TW, Roy-Byrne PP, Joffe RT (1987) Correlates of antimanic response to carbamzepine. Psychiatry Res 21:71
Price DL, Struble RG, Whitehouse PJ (1984) Alzheimer's disease: a multisystem disorder. Proceedings of association for research in nervous and mental disease, inc., symposium: neuropeptides: implications for neurologic and psychiatric diseases, New York, Nov. 30–Dec. 1
Prien RF (1984) Five-center study clarifies use of lithium, imipramine for recurrent affective disorders. Hosp Commun Psychiatry 35:1097
Racine R (1978) Kindling: the first decade. Neurosurgery 3:234
Roy-Byrne PP, Post RM, Uhde TW, Porcu T, Davis D (1985) The longitudinal course of recurrent affective illness: life chart data from research patients at the NIMH. Acta Psychiatr Scand 71 [suppl 317]:5
Rubinow DR, Gold PW, Post RM, Ballenger JC, Cowdry R, Bollinger J, Reichlin S (1983) Somatostatin in affective illness. Arch Gen Psychiatry 40:403
Rubinow DR, Post RM, Gold PW, Ballenger JC, Reichlin S (1985) Effects of carbamazepine on cerebrospinal fluid somatostatin. Psychopharmacology 85:210
Rubinow DR, Post RM, Davis CL, Doran AR (1986a) Somatostatin and GHRH: mood and behavioral regulation. In: Goodwin FK, Fragese G (eds) Hypothalamic dysfunction in neuropsychiatric disorders. Raven, New York
Rubinow DR, Post RM, Gold PW, Uhde TW (1986b) Effect of carbamazepine on mean urinary free cortisol excretion in patients with major affective illness. Psychopharmacology 88: 115
Serby M, Richardson SB, Twente S, Siekierski J, Corwin J, Rotrosen J (1984) CSF somatostatin in Alzheimer's disease. Neurobiol Aging 5:187
Shuster L, Webster GW, Yu G (1975) Perinatal narcotic addiction in mice: Sensitization to morphine stimulation. Addict Dis 2:277
Soininen HS, Jolkkonen JT, Reinikainen KJ, Halonen TO, Riekkinen PJ (1984) Reduced cholinesterase activity and somatostatin-like immunoreactivity in the cerebrospinal fluid of patients with dementia of the Alzheimer type. J Neurol Sci 63:167
Sorensen KV, Christensen SE, Dupont E, Hansen AP, Pedersen E, Orskov H (1980) Low somatostatin content in cerebrospinal fluid in multiple sclerosis. Acta Neurol Scand 61:186
Swift HM (1907) The prognosis of recurrent insanity of the manic-depressive type. Am J Insanity 64:311
Tadokoro S, Kuribara H (1986) Reverse tolerance to the ambulation-increasing effect of methamphetamine in mice as an animal model of amphetamine-psychosis. Psychopharmacol Bull 22:757
Tanaka T, Hamasaki Y, Takeshita H, Kawahara R, Hazama H (1986) Chemical kindling with met-enkephalin and changes of brain opioid delta-receptor sensitivity after electrical kindling. Epilepsia 27:596

Tallman JF, Paul SM, Skolnick P, Gallager DW (1980) Receptors for the age of anxiety: pharmacology of benzodiazepines. Science 207:274
Taschev T (1974) The course and prognosis of depression on the basis of 652 patients decreased. In: Schattauer FK (ed) Symposia medica hoest, vol 8: classification and prediction of outcome of depression. Schattauer, New York, pp 156–172
Vecsei L, Bollok I, Telegdy G (1983) Intracerebroventricular somatostatin attenuates electroconvulsive shock-induced amnesia in rats. Peptides 4:293
Vecsei L, Kiraly C, Bollok I, Nagy A, Varga J, Penke B, Telegdy G (1984) Comparative studies with somatostatin and cysteamine in different behavior tests on rats. Pharmacol Biochem Behav 21:833
Wada JA (1977) Pharmacological prophylaxis in the kindling model of epilepsy. Arch Neurol 34:389
Wada JA, Sato M (1974) Generalized convulsive seizures induced by daily electrical stimulation on the amygdala in cats. Neurol 24:565
Wada JA, Sato M, Corcoran ME (1974) Persistent seizure susceptibility and recurrent spontaneous seizures in kindled cats. Epilepsia 15:465
Wada JA, Sato M, Wake A, Green JR, Troupin AS (1976) Prophylactic effects of phenytoin, phenobarbital, and carbamazepine examined in kindled cat preparations. Arch Neurol 33:426
Walsh TJ, Emerich DF, Winokur A, Banki C, Bissette G, Nemeroff CB (1985) Intrahippocampal injection of cysteamine depletes somatostatin and produces cognitive impairments in the rat. Soc Neuroscience Abstr 11:621
Wasterlain CG, Jonec V (1983) Chemical kindling by muscarinic amygdaloid stimulation in the rat. Brain Res 271:311
Wehr TA, Goodwin FK (1979) Rapid cycling in manic-depressives by tricyclic antidepressants. Arch Gen Psychiatry 36:555
Weiss SRB, Post RM (1987a) Carbamazepine and carbamazepine-10,11-epoxide inhibit amygdala-kindled seizures in the rat but do not block their development. Clin Neuropharmacol 10:272
Weiss SRB, Post RM, Patel J, Marangos P (1985) Differential mediation of anticonvulsant effects of carbamazepine and diazepam. Life Sci 36:2413
Weiss SRB, Murman D, Post RM, Pert A (1986a) Conditioning in cocaine-induced behavioral sensitization. In: 16th Annual meeting, November 9–14. Society for Neuroscience. Washington DC, p 914 (abstract no 249.13)
Weiss SRB, Nguyen T, Rubinow DR, Helke CJ, Narang PK, Post RM, Jacobowitz DM (1987b) Lack of effect of chronic carbamazepine on brain somatostatin in the rat. J Neural Transm 68:325
Weiss SRB, Post RM, Gold PW, Chrousos G, Sullivan TL, Walker D, Pert A (1986b) CRF-induced seizures and behavior: interaction with amygdala kindling. Brain Res 372:345
Weiss SRB, Post RM, Marangos PJ, Patel J (1986c) "Peripheral-type" benzodiazepines: behavioral effects and relationship to the anticonvulsant effects of carbamazepine. In: Wada J (ed) Kindling III. Raven, New York, pp 375–392
Weissman BA, Cott J, Hommer D, Quirion R, Paul S, Skolnick P (1983) Pharmacological, electrical, and neurochemical actions of the convulsant benzodiazepine Ro5-4864 (4′-chloridazepam). In: Biggio G, Costa E (eds) Recognition site ligands: biochemistry and pharmacology. Raven, New York, pp 139–152
Wood PL, Etienne P, Lal S, Gauthier S, Cajal S, Nair NPV (1982) Reduced lumbar CSF somatostatin levels in Alzheimer's disease. Life Sci 31:2073
Zis AP, Grof P, Webster M et al. (1980) Prediction of relapse in recurrent affective disorder. Psychopharmacol Bull 16:47

The Mechanisms of Action of Antipsychotics and Antidepressant Drugs *

F. A. Wiesel [1] and L. Träskman-Bendz [2]

Contents

The introduction of neuroleptics and tricyclic antidepressant drugs represents a major breakthrough in the treatment of psychiatric disorders. The changes of central monoamine metabolism induced by these psychoactive compounds are still the basis for current theories on the role of the monoamines in the pathophysiology of psychosis and depression. The depressive disorders have been linked with the functional state of noradrenergic and serotoninergic neurons, schizophrenia mainly with the transmission of dopaminergic neurons. The effects of antipsychotics and antidepressants in psychiatric patients demonstrate the importance of

* The work reported here has been supported by the Swedish Medical Research Council (714/85, 5454 and 7027), Karolinska Institute and the Swedish Medical Society

1 Department of Psychiatry and Psychology, Karolinska Institute, Karolinska Hospital, S-104 01 Stockholm, Sweden

2 Department of Psychiatry, University of Lund, University Hospital, S-221 85 Lund, Sweden

Current Topics in Neuroendocrinology, Vol. 8

the interplay between basic and clinical science in the understanding and the treatment of psychiatric disorders. The mechanism of action of these two major groups of psychopharmacological drugs will be discussed within a clinical framework.

1 Antipsychotic Drugs

1.1 Clinical Aspects

Antipsychotic drugs constitute a diverse group of chemical compounds (Table 1). The reason for grouping them together is their clinical characteristic of reducing psychotic symptoms. The term neuroleptic was introduced by Delay and Deniker (1957) to characterize compounds which (1) had an antipsychotic effect not due to sedation, (2) reduced psychomotor activity, and (3) induced extrapyramidal symptoms like parkinsonism or catalepsy. These compounds are effective in the treatment of both toxic and functional psychosis, which demonstrates that they are suitable for all types of psychosis. The use of the term "antischizophrenic drugs" is therefore inappropriate. The schizophrenics are the major patient group treated with neuroleptics. Neuroleptic treatment does not cure the disease, but the course of the illness has become more favourable since the introduction of neuroleptics (Wing 1966; Bland and Orn 1978; von Sivers 1983). Drug treatment also lends itself to combination with other types of psychiatric treatment (Hogarty et al. 1974; Goldstein et al. 1978; Fallon et al. 1985; Leff et al. 1985).

The clinical efficacy of antipsychotics in the treatment of schizophrenic patients is very well documented by many double-blind placebo studies (Davis and Garver 1978). In acute treatment about 70% of the patients are described as much improved, in line with the treatment of other medical disorders (cf. Davis et al. 1980). A significant reduction in psychotic symptoms occurs within a wide time interval from 1–2 days to 6 months. On average, the major reduction in clinical symptoms is found within 4–6 weeks (Bjerkenstedt et al. 1978; Wode-Helgodt et al. 1978; Davis et al. 1980; Härnryd et al. 1984a). Usually, patients with a long

Table 1. Groups of antipsychotic drugs

Group	Example
Phenothiazines	Chlorpromazine
Thioxanthenes	Flupenthixol
Butyrophenones	Haloperidol
Diarylbutylamines	Pimozide
Dibenzothiazepines	Clothiapine
Dibenzodiazepines	Clozapine
Indoles	Molindone
Benzamides	Sulpiride

prodromal phase or a long duration of psychosis are less sensitive to antipsychotics and their poorer prognosis may be related to a long delay before the start of neuroleptic treatment (May et al 1981; Crow et al. 1986). Schizophrenia is a syndrome rather than a defined disease, and this probably explains the diversity of clinical symptoms; treatment failure with one compound does not exclude the possibility of success with another type of drug. This raises the possibility of classifying drugs according to their clinical profile. This has not been done in a systematic way, and each psychiatrist may have his/her own opinion. In general, antipsychotics are considered to be more efficient in the treatment of positive symptoms (increase in psychic functions: prominent thought disturbances, hallucinations, delusions, delusional mood, feeling of being controlled, ideas of persecution) and less efficient in the treatment of negative symptoms (decrease in psychic functions: poverty and slowness of thinking, affective flattening, anhedonia, avolition, apathy) (cf. Andreasen 1985). Patients with brain abnormalities as shown on computed tomographic images have more negative symptoms and benefit less from neuroleptic treatment (Johnstone et al. 1978; Crow 1980; Weinburger et al. 1980; Andreasen et al. 1982). There is some evidence that selective dopamine D2-blocking compounds are more effective in helping patients with autistic or negative symptoms than conventional neuroleptics (Haase et al. 1974; Elizur et al. 1975; Härnryd et al. 1984a). This difference may be related partly to a lower frequency of lethargic side effects in patients treated with selective compounds. In a subgroup of chronic schizophrenic patients positive psychotic symptoms remain despite treatment with high doses of neuroleptics. Clozapine treatment seems to be of especial value to these severely ill patients (Povlsen et al. 1985; Kuha and Miettinen 1986). It has been proposed that neuroleptics may cause negative or dysmentia-like symptoms (Rifkin et al. 1978; Wilson et al. 1983). Some of these problems are related partly to treatment with too high a dose of neuroleptics (Baldessarini et al. 1984; Manchanda and Hirsch 1986; Sedvall et al. 1986a). However, the amazing thing with the introduction of neuroleptics was how previously retarded, apathetic and socially withdrawn patients were activated (for references Goldberg 1985). The limitations of the therapeutic results should be recognized, but one must differentiate between side effects of neuroleptics and symptoms linked with the course of the disease.

Antipsychotic drugs could also be divided, according to their motor side effects, into typical and atypical neuroleptics, where "atypical" refers to drugs with a low frequency of extrapyramidal side effects (EPS). In rat behavioural studies, atypical neuroleptics block locomotion but not for stereotypes induced by dopamine agonists (cf. Ljungberg and Ungerstedt 1978). Atypical neuroleptics include clozapine, thioridazine, sulpiride and probably also melperone (Borison et al. 1983; Kirkegaard et al. 1981; Honigfeld et al. 1984; Povlsen et al. 1985; Christensson 1988). This group of compounds is also less prone to cause tardive dyskinesia. From a pharmacological point of view this group is very heterogenous. With respect to receptor interactions, sulpiride is a highly selective compound and clozapine is the opposite. The lack of selectivity of clozapine combined with a blockade of serotonin (5-HT) receptors has been suggested to be the background for its clinical profile, i.e. hardly any EPS and partial effectiveness in severely ill schizophrenic patients.

1.2 Biochemical and Receptor Mechanisms

In 1963, Carlsson and Lindqvist suggested that antipsychotic drugs like chlorpromazine and haloperidol block postsynaptic catecholamine receptors and thereby induce compensatory activation of the presynaptic neuron. Later it was demonstrated that most antipsychotic drugs preferentially accelerate the synthesis and turnover of dopamine (Nybäck and Sedvall 1968, 1970). In line with the biochemical findings, electrophysiological studies have demonstrated that different antipsychotic drugs increase the impulse activity of the dopamine neurons, probably by blocking central dopamine receptors (cf. Bunney 1984). Dopamine stimulates an adenylate cyclase in the brain to form cyclic adenosine monophosphate, which may propagate the dopaminergic transmission within the postsynaptic neuron (Kebabian et al. 1972; Siggins et al. 1974). Different classes of antipsychotics were shown to block the stimulatory effect of dopamine on this adenylate cyclase, which was suggested to be related to a dopamine D1 receptor (Clement-Cormier et al. 1974; Karobath and Leitich 1974). However, the in vitro actions of the butyrophenones on the adenylate cyclase were unexpectedly weak in comparison with their in vivo increase of dopamine turnover. In 1975 binding sites for the butyrophenones were identified in the striatum (Creese et al. 1975; Seeman et al. 1975). Kebabian and Calne (1979) called these binding sites D2 receptors, in contrast to the adenylate cyclase-linked D1 receptors. Further subtypes of dopamine receptors have been proposed, but binding sites called D3 and D4 are probably only high-affinity forms of the D1 and D2 receptors (Leff and Creese 1983; Bacopoulos 1984). Most interesting was the observation that the relative affinities of the neuroleptics to the D2 receptors but not to the D1 receptors correlated with their clinical potency in the treatment of schizophrenic patients (Creese et al. 1976; Seeman et al. 1976). The blockade of D2 binding sites seems to be the common effect of neuroleptics in clinical use. Actions on other receptor types, like muscarinic-cholinergic, α-adrenergic, 5-HT and histamine receptors do not correlate with the average clinical antipsychotic potency of neuroleptic drugs (Peroutka and Snyder 1980 b).

Neuroleptics, both typical and atypical, increase brain levels of homovanillic acid (HVA), which is an indirect measure of dopamine turnover (Anden et al. 1964; Sedvall et al. 1973; Wiesel et al. 1973, 1978; Wiesel and Sedvall 1975; Scatton et al. 1977). The inaccessibility of the human brain makes it necessary to determine HVA levels in the cerebrospinal fluid (CSF) in order to obtain an indirect measure of brain dopamine activity. The HVA content in lumbar CSF has been shown to be derived almost exclusively from the brain (Curzon et al. 1971; Post et al. 1973; Sourkes 1973; Gordon et al. 1975), and HVA is the major dopamine metabolite in man (Wiesel 1975; Swahn and Wiesel 1976). These facts make it plausible that HVA levels in CSF at least partly reflect brain dopamine turnover. Therefore the elevation of HVA in CSF following neuroleptic treatment is compatible with an increase of dopamine turnover in the human brain. In contrast to neuroleptics, antidepressants and lithium do not increase HVA levels in CSF (Sedvall et al. 1975). The fact that both typical and atypical neuroleptics increase HVA levels in CSF provides support for the notion of a common action of neuroleptics on dopaminergic systems also in the human brain (Bjerkenstedt et al. 1977;

Table 2. Effects of neuroleptic drug treatment on ^{11}C-raclopride binding to D2 receptors in the putamen of four male schizophrenic patients

Drug	Haloperidol	Flupenthixol	Clozapine	Sulpiride
Dose (mg)	4 b.i.d.	5 b.i.d.	300 b.i.d.	400 b.i.d.
Time after last administration (h)	6	6	6	5
Receptor occupancy (%)	84	74	65	82

Expected specific binding was calculated from the mean Bmax and Kd values for male healthy subjects and from the concentration of ^{11}C-raclopride in the cerebellum (free ligand concentration) in the drug-treated patients (Farde et al. 1988)

Wode-Helgodt et al. 1977; Härnryd et al. 1984b). Until recently it was not possible to demonstrate directly that neuroleptics block dopamine receptors in vivo in man; however, the development of positron emission tomography (PET) has enabled direct visualization of the binding of psychoactive drugs in the living human brain (for review see Sedvall et al. 1986a). Radiolabelled neuroleptics have been shown to bind to human striatum (Wagner et al. 1983; Farde et al. 1985). It has also been possible to demonstrate in the living human brain that the selective D2 receptor-blocking compound, sulpiride, blocks binding sites for D2 receptors but not for D1 receptors (Farde et al. 1987). In patients treated with conventional doses of haloperidol, clozapine, flupenthixol or sulpiride, all drugs were found to block binding sites for D2 receptors (Table 2) (Farde et al. 1986, 1988). These results give evidence that the common mechanism of neuroleptics is a blockade of D2 receptor sites in the brain of schizophrenic patients. It will be possible to correlate the antipsychotic effect with the degree of receptor blockade which should elucidate the importance of D2 receptor blockade for the clinical effect.

1.3 Clinical and Biochemical Mechanisms

Strong clinical support for the importance of dopamine receptor blockade in alleviation of psychotic symptoms is provided by the results of a critical study carried out by Johnstone et al. (1978). They compared two isomeric forms of flupenthixol with placebo in the treatment of acute schizophrenics. The α isomer, which blocks dopamine receptors, reduced psychotic symptoms significantly in comparison with both placebo and β isomer, which does not block dopamine receptors. The β isomer was similar to placebo. The importance of the dopaminergic system is further supported by the finding that α-methyl-tyrosine strengthens the antipsychotic action of subthreshold doses of neuroleptics (Wåhlinder et al. 1976; Larsson et al. 1984). α-Methyl-tyrosine, as an inhibitor of tyrosine hydroxylase, decreases the feedback activation of dopamine neurons following dopamine receptor blockade and therefore strengthens the neuroleptic blockade of dopamine receptors (Carlsson 1974). These findings support the view that the blockade, rather than the increased dopaminergic turnover, is crucial for the clinical effect of neuroleptic treatment.

Another set of clinical data indirectly supporting the importance of dopamine blockade for the reduction of psychotic symptoms is the treatment of amphetamine abusers who have developed psychotic symptoms. Psychotic amphetamine abusers will have a pronounced reduction in psychotic symptoms within a couple of hours following the injection of a neuroleptic compound. Schizophrenic patients whose psychotic symptoms have deteriorated after L-dopa or amphetamine treatment are also improved by neuroleptic administration (Gerlach and Luhdorf 1975; Angrist et al. 1980; Cecarec and Nyman 1985).

Neuroleptic treatment increases brain dopamine turnover in man as measured by HVA levels in CSF. If this effect is relevant for the antipsychotic effect one might expect a correlation between changes in HVA levels and clinical symptoms. Correlations were indeed found between increased HVA levels and reduction in psychotic symptoms (Sedvall et al. 1975; van Praag 1977; Wode-Helgodt 1977; Alfredsson et al. 1984); however, the correlations were fairly weak and not consistent enough to give more than support for the view that neuroleptics may induce antipsychotic effects in schizophrenic patients in a graded fashion proportional to the degree of interaction with central dopaminergic mechanisms.

It has been suggested that HVA levels in plasma may reflect turnover of dopamine in the brain (Bacopoulus et al. 1979; Sternberg et al. 1983). This has motivated the study of plasma HVA in schizophrenic patients. However, neuroleptic treatment has been shown to both increase and decrease human plasma HVA, which indicates that plasma HVA is not a simple measure of central dopaminergic activity (Harris et al. 1984; Pickar et al. 1984; Davis et al. 1985). Furthermore, there is no correlation between HVA in plasma and CSF, as demonstrated in healthy volunteers (Alfredsson et al. 1987). Nevertheless, reduction in plasma HVA following neuroleptic treatment has been found to be correlated to the reduction of psychotic symptoms (Bowers et al. 1984: Pickar et al. 1986). The mechanism for this relationship is not known.

Neuroleptic treatment increases prolactin in serum, an effect which by and large is due to a blockade of D2 receptors in the pituitary. It has been speculated that these D2 receptors may reflect the functional state of central dopamine receptors. This has encouraged investigators to look for relationships between prolactin levels and clinical symptoms after neuroleptic treatment. Generally the results are less convincing than those obtained for HVA levels in CSF (Bjerkenstedt et al. 1979; Wode-Helgodt 1977; Alfredsson et al. 1984; see also Johnstone 1985).

1.4 Site of Action

The data available so far indicate that dopamine receptor blockade is the basic mechanism of neuroleptics for effects on psychotic symptoms. A matter of controversy is the site of action of the dopamine blockade inducing the antipsychotic effect. Using the histochemical fluorescence technique it has been possible to visualize monoaminergic neurons in the brain. The dopamine-containing cell bodies are located in the zona compacta of the substantia nigra (A9) and the adjacent midbrain ventral tegmental area (A10). The A9 neurons project to the striatum, but also to the cingulate cortex, whereas the A10 system innervates various parts of the limbic system, prefrontal, piriform, and entorhinal cortices (for references see Lindwall and Björklund 1978). EPS are due to an interaction between dopamine and muscarinic cholinergic neurons following the blockade of D2 receptors in the striatum. The EPS are not related to the antipsychotic effect, since anticholinergic agents relieve the patients from EPS without any obvious deterioration of the antipsychotic effect. Clozapine is an effective antipsychotic agent which does not induce EPS (Povlsen et al. 1985). This may be due partly to the anticholinergic properties of this compound (Miller and Hiley 1974; Snyder et al. 1974), and may be related to the failure of clozapine treatment to induce depolarization inactivation of A9 dopamine neurons (cf. Bunney 1984). Striatum has been looked upon as a structure important only for motor function, but recent results have demonstrated a close interrelationship of projections between neostriatum and areas of association cortex (Selemon and Goldman-Rakic 1985). This brings up striatum as a possible site of action for the antipsychotic effect.

In behavioural studies of rats, neuroleptics have been found to exhibit potent inhibition of stereotyped behaviour and locomotion. Interestingly enough the selective D2 blocking compound sulpiride fails to inhibit stereotyped behaviour but blocks locomotion induced by dopamine receptor stimulation. Similar findings have been obtained for clozapine, thioridazine and melperone (cf. Ljungberg and Ungerstedt 1978; Christensson 1988). Since locomotion is triggered from limbic areas, the similarity in action of the typical and atypical neuroleptics on locomotion has been considered as evidence of the importance of limbic dopamine blockade for the reduction of psychotic symptoms. Other investigators have suggested that mesocortical dopamine neurons are the major target for the antipsychotic action of neuroleptics. The pharmacology of mesocortical dopamine neurons has been related to the absence of presynaptic dopamine receptors on the nerve terminals and cell bodies (cf. Bannon and Roth 1983). This may explain why neuroleptics do not create tolerance for these dopamine neurons, a finding which has been interpreted in analogy to the lack of tolerance for the antipsychotic effect. If striatal, limbic or mesocortical dopamine neurons are more or less linked to the therapeutic effects of the neuroleptics is for the moment not possible to say. Nevertheless, sulpiride, with a low frequency of EPS, and clozapine, with hardly any EPS, block D2 sites in vivo in human striatum like typical neuroleptics irrespective of their ability to block stereotyped behaviour in experimental animals (Table 2) (Farde et al. 1986, 1988). The argument that the lack of tolerance for the antipsychotic effect should support the view that a blockade of mesocortical dopamine neurons which do not develop tolerance is crucial for the clinical effect may not be relevant. Tolerance is seen for the HVA increase in CSF (the bulk coming from the striatum) following neuroleptic treatment without any indication of deterioration of the psychosis (Goodwin et al. 1975; Bjerkenstedt et al. 1977; Bowers and Heninger 1981; Wode-Helgodt et al. 1977; Härnryd et al. 1984b). In fact, Bowers and Heninger (1981) showed that patients who benefited most from the neuroleptic treatment were those who developed tolerance for the HVA increase. The reduction in plasma HVA following neuroleptic treatment has also been considered to be due to tolerance mechanisms and the reduction in plasma HVA levels were related to clinical improvement (Pickar et al. 1986). Complex compensatory mechanisms must take place in the striatum following dopamine receptor blockade, and developement of tolerance in biochemical measures might in fact indicate that the treatment is effective.

The distribution of dopamine neurons may differ between animals and man. In fact D2 receptors have not been demonstrated in the human frontal cortex (De Keyser et al. 1985), and PET studies with the highly selective D2-blocking compound raclopride demonstrate binding only to the striatum, whereas the ligand for D1 receptors (SCH 23990) also binds outside the striatum (Farde et al. 1985; Sedvall et al. 1986b). However, this does not mean that the antipsychotic effect is directly related to the striatum; instead, one has to consider complex interactions between different neuronal systems in various regions resulting in an antipsychotic effect, even if the primary effect is related to a dopamine blockade in the striatum.

1.5 Dopamine Receptor Supersensitivity

Neuroleptic treatment has been demonstrated to increase the number of dopamine receptors in experimental animals (Burt et al. 1977; Muller and Seeman 1978). This finding has been suggested to be related to the development of tardive dyskinesia in man. In relation to clinical data, such a relationship between tardive dyskinesia and receptor supersensitivity can be seriously questioned (cf. Fibiger and Lloyd 1984; Gerlach 1985). In rats dopamine supersensitivity develops rapidly within weeks, but in man tardive dyskinesia is seen at the earliest after several months of neuroleptic treatment. Tardive dyskinesia in man is affected by dopamine agonists only to a small extent, in contrast to animals, which react with an increase of stereotyped behaviour. It has not been possible to demonstrate an increased number of dopamine receptors in postmortem brains from patients with tardive dyskinesia as compared to patients without tardive dyskinesia (Cross et al. 1985). It has been proposed that tardive dyskinesia is more related to disturbances in GABA neurons following dopamine receptor blockade (for references see Fibiger and Lloyd 1984; Gerlach 1985). There is complex interaction between dopamine and GABA neurons in the striatum, and neuroleptic treatment also results in supersensitive GABA receptors in animals (Gale and Cassu 1981). In monkeys with neuroleptic-induced consistent dyskinesia, a decrease in the activity of glutamic acid decarboxylase in the substantia nigra, the medial part of the globus pallidum and the nucleus subthalamicus was found (Gunne et al. 1984). This gives strong support for the GABA hypothesis and tardive dyskinesia (Gunne et al. 1984).

The significance of an increased number of dopamine receptors following neuroleptic treatment is not clear. Ungerstedt and Ljungberg (1977) suggested that supersensitive dopamine receptors could deteriorate the psychosis, a hypothesis that was suggested to be supported by findings in patients (Chouinard and Jones 1980; Ekblom et al. 1984). This hypothesis was specially coupled to findings with clozapine, but this compound does not develop an increased number of D2 receptors (Jenner et al. 1985). An argument against the existence of supersensitivity psychoses is that the same proportion of patients suffer from relapses at different points in time after discontinuation of neuroleptic treatment (cf. Davis 1985). This does not exclude the possibility that individual subjects may exhibit an increase in psychotic symptoms after sudden discontinuation of clozapine (Ekblom et al. 1984) and other neuroleptics which may be looked upon as a rebound or withdrawal phenomenon. In clinical practice the reduction of the dose of neuroleptics should be made in stepwise fashion over several months and is only meaningful if the patients are symptom-free. Perhaps an increase of dopamine receptors is only one part of a development of a normal tolerance following treatment with conventional neuroleptics. As discussed previously, indications of tolerance of drug-induced changes of dopamine turnover may indicate clinically effective neuroleptic treatment.

1.6 Conclusions

Both clinical and pharmacological data strongly support the view that the basic mechanism of action of antipsychotics is a blockade of D2 receptors. The close neuronal interrelationships between striatum and vast areas of association cortex (Selemon and Goldman-Rakic 1985) and the distribution of D2 receptors (Farde et al. 1986 a) points to the importance of dopamine blockade in the striatum for the antipsychotic effect. In addition to effects on conventional neurotransmitter systems, neuroleptics also interact with neuropeptides, especially methionine-enkephalin, cholecystokinin and neurotensin (cf. Bissette and Nemeroff 1985). However, methodological problems have so far hampered the evaluation of the clinical inportance of neuropeptides regarding the mechanisms of action of antipsychotics. How a psychosis is brought about and how a blockade of dopamine receptors with neuroleptics can relieve the patients from the psychotic symptoms is in principle unknown. The schizophrenic patients are not cured, since they relapse after discontinuation of drug treatment. Moreover, a rather large subgroup of schizophrenic patients is not sufficiently improved by antipsychotics. It is evident that too little is known about the pathophysiological mechanisms of psychotic disorders, which must be the basis for improved treatment in the future.

2 Antidepressant Drugs

2.1 Introduction

The widely quoted hypothesis of monoamine deficiency in depression is based on several observations: (a) Hypertensive patients who received high doses of reserpine sometimes became depressed (Harris 1957), and reserpine was also found to deplete the tissue stores of 5-HT (Shore et al. 1955), noradrenaline (Carlsson and Hillarp 1956) and dopamine (Carlsson et al. 1958). (b) Iproniazid, an antituberculosis drug, in some cases produced elevated mood. Subsequently, iproniazid was serendipitously found to be effective as an antidepressant drug (Loomer et al. 1957). Iproniazid appeared to be an inhibitor of one of the degradating enzymes of monoamines, monoamine oxidase (MAO). Later, the degree of MAO inhibition was related to the antidepressant effect (Ravaris et al. 1976). (c) Imipramine was synthetized as a potential neuroleptic drug but was found to lack antipsychotic properties. Instead, it had favourable effects as an antidepressant (Kuhn 1957). The mode of action of imipramine and similar tricyclic compounds was suggested to be blocking of reuptake of noradrenalin and 5-HT into the presynaptic nerve ending (Axelrod et al. 1961).

2.2 Presynaptic Effects

The most often discussed therapeutic action of tricyclic antidepressants and MAO inhibitors has been the enhancement of the availability of noradrenaline and 5-HT, in the brain either by decreased catabolism due to MAO inhibition or by inhibition of reuptake into the presynaptic nerve ending. It has been suggested that *acute* treatment in some instances induces inhibition of noradrenaline uptake and activates a negative feedback mechanism via presynaptic α_2-autoreceptors (Nybäck et al. 1975). This results in a decrease in impulse flow in noradrenaline neurons, which in turn reduces noradrenaline turnover. Similarly, drugs with 5-HT uptake-inhibiting properties cause decreased 5-HT turnover (Corrodi and Fuxe 1969; Sugrue 1980).

Changes in brain monoamine turnover after *prolonged* treatment with tricyclic antidepressants are not homogeneous. In contrast to the acute administration, a chronic treatment may increase the turnover of noradrenaline even though the firing rate of the neurons is depressed. Tertiary and secondary tricyclic antidepressants seem to act differently on noradrenaline turnover, and there are also heterogeneous actions of nontricyclic antidepressants.

After repeated administration of antidepressants, controversial results have been reported for 5-HT turnover as well. However, the most selective 5-HT inhibitors, such as fluoxethine and zimelidine, were found to decrease 5-HT turnover (Fuxe et el. 1979; Hwang and Van Woert 1980).

2.3 Revision of the Reuptake Hypothesis

With increasing knowledge about antidepressant drugs and their effects on several other neurotransmitters and hormones, as well as effects on postsynaptic receptors, the principal mode of action of antidepressants as reuptake inhibitors or MAO inhibitors has been questioned. The following evidence against the traditional view on antidepressants has been discussed:

1. All antidepressants are not reuptake inhibitors or MAO inhibitors, i.e. mianserin and iprindole.
2. Some reuptake inhibitors are not antidepressants, i.e. cocaine, amphetamine, femoxethine.
3. There is a lag period of 7–14 days between the initiation of reuptake inhibition and/or MAO inhibition and clinical improvement.

These objections were discussed by Carlsson (1984). He stated that all antidepressants need not have the same mechanism of action, as we already know that some are reuptake inhibitors and some are MAO inhibitors. In reuptake inhibitors without antidepressant effects the reuptake inhibition is probably obscured by other actions and phenomena in more prominent biochemical systems. The role of reuptake-inhibiting properties need not necessarily be ruled out by the fact that there is a time period before clinical effect. Instead, this lag period could be due to secondary changes after reuptake inhibition.

The lag period is not adequately documented in clinical studies. Apparent differences in the rapidity of antidepressant action may provide further evidence for the view that there are multiple intervention sites in the CNS for treatment of depression. Biological studies, such as the dexamethasone test, baseline cortisol measures or rapid eye movement (REM) sleep studies might aid to the proper understanding of the lag phenomenon. The action of antidepressants could evidently be enhanced and accelerated by adding L-tryptophan (Wåhlinder et al. 1976a), tyrosine (Van Praag 1984), thyroid hormones (Prange et al. 1970; Wheatly 1972), yohimbine (Johnson et al. 1980), phenoxybenzamine (Paul and Crews 1980) or lithium (De Montigny et al. 1981).

2.4 Postsynaptic Effects

Long-term animal studies now suggest that the antidepressants produce biochemical effects on neurotransmitter receptors which occur with the same time course as does recovery from depression. Therefore the focus of attention has shifted from the biogenic amines themselves toward their receptors. The alleged reduction of neuronal transmission during depression is associated with hypersensitivity (increased responsiveness) of postsynaptic receptors according to the receptor hypothesis of depressive disorder (Vetulani and Sulser 1975).

Modern estimates of receptor activity have improved our understanding of the postsynaptic action of antidepressants. Measures of cyclic AMP production, radioligand receptor binding, behavioural paradigms, and single unit electrophysiological recordings have been used extensively in animal studies during the last decade. Treatment with antidepressants cause decreased firing rates of noradrenaline neurons in locus coeruleus and serotonergic neurons in raphe magnus (Sugrue 1981). Chronic treatment facilitates electrophysiological and behavioural responses to local application of 5-HT a noradrenaline or dopamine in selected brain regions (Modigh 1975; Maj et al. 1979a; Menkes and Aghajanian 1981).

2.5 β-Adrenergic Receptors and 5-HT Receptors

Even though antidepressants seem to exert their main effect on β-adrenergic receptors, they also act on serotonergic sites. Regardless of structure, antidepressants decrease the number of (downregulate) β-adrenergic receptors and 5-HT_2 receptors. The relevance of antidepressant-induced 5-HT_2 receptor downregulation may be questioned because of the observations that amphetamine produces this effect and that ECT increases 5-HT_2 receptors (Kellar et al. 1981). In spite of these observations, there is a reasonably coherent picture of *enhanced* 5-HT receptor function and *enhanced* transmission during antidepressant treatment. This is in line with the indole amine hypothesis but not with the receptor hypothesis of depression, as the latter states that there is receptor supersensitivity during depression due to decreased transmission. The MAO inhibitors are exceptions, as they apparently *reduce* 5-HT receptor function. It is also possible that synaptic transmission is reduced during MAO inhibitor treatment (Willner 1985).

The importance of 5-HT neurotransmission was recently put forward by Brunello et al. (1982) and Janowsky and coworkers (1982), who showed that an intact 5-HT neuronal system is a prerequisite for the decrease in β-adrenergic binding sites during long-term administration of tricyclic antidepressant.

2.6 α_1- and α_2-Adrenergic Receptors

It has been suggested that antidepressant action could depend on concomitant downregulation of β-receptors and upregulation of presynaptic α_1-adrenergic receptors (Vetulani and Pilc 1982). The role of α_1- and α_2-receptors during antidepressant treatment is not clearly understood. Animal data favour increased responses mediated by α_1-receptors. Similarly, both biochemical and electrophysiological studies indicate that desipramine produces a reduction in the sensitivity of the α_2-adrenergic autoreceptor (Crews and Smith 1978). These reports on the action of the α_2-receptor could not, however, be verified by Sugrue (1982).

2.7 Other Receptors

The action of antidepressants on α_1-adrenergic, histaminergic (H_1) and muscarinic receptors produce a variety of side effects such as somnolence, orthostatic hypotension, dry mouth, constipation, blurred vision and urinary retention.

Increased density of muscarinic receptors has been reported during antidepressant treatment. Jones (1980) suggested an inhibitory interaction of catecholamines on muscarinic receptors. Other authors found no effects of chronic antidepressant treatment on muscarinic receptors (Peroutka and Snyder 1980a; Menkes and Aghajanian 1981).

The role of dopamine during antidepressant treatment has been ignored. Serra et al. (1979) found an attenuated response to apomorphine, a dopamine agonist, in dopamine-containing cell bodies following chronic administration of antidepressants. Subsensitivity of dopamine autoreceptors has also been reported (Leonard 1984). This subsensitivity could be due to a regulatory action of 5-HT turnover upon dopamine turnover, as suggested by Ågren et al. (1986).

2.8 New Receptor Theories

The premise of the receptor hypothesis of antidepressant action is that the therapeutic action is a consequence of changes in the sensitivity of monoamine systems. One synapse may have more than one transmitter. It is possible that β-adrenergic receptors and 5-HT_2 receptors reside on the same cells and thus influence each other, because they are both activated by environmental factors such as stress (Stone 1983).

Possibly, peptides are cotransmitters and act as neuromodulators of synaptic neurochemistry. The function of such a cotransmitter would be to regulate the fine tuning of the sensitivity of receptors for the primary transmitter. One working hypothesis is therefore that antidepressant drugs either mimic or antagonize cotransmitters (Costa 1982).

2.9 Human Receptor Studies

Receptors in human brain can be directly studied using positron emission tomography (PET) or nuclear magnetic radiography (NMR). Such studies have not yet been performed with antidepressants as far as we know. Instead, peripheral markers may serve as models for neurotransmitter receptors in brain. Lymphocytes contain β-receptors as well as noradrenaline-stimulated adenylate cyclase. Lerer et al. (1981) showed that a possible antidepressant agent, salbutamol, decreased β-adrenergic receptor sensitivity in lymophocytes. Garcia-Sevilla et al. (1981) reported induction of platelet α_2 subsensitivity during treatment with tricyclic antidepressant drugs.

2.10 Biochemical Subgroups of Depressed Patients and Antidepressant Efficacy of Some Antidepressant Drugs

The original monoamine deficiency hypothesis has been investigated in studies of monoamine metabolites in urine and lumbar cerebrospinal fluid (CSF). As antidepressant drugs could be subdivided into those which mainly prevent noradrenaline uptake, 5-HT uptake or both, Sjöqvist (1971) proposed that depressive states could be treated differentially depending on their biochemistry: noradrenaline or 5-HT deficiency.

The reuptake inhibition of tricyclic compounds has been documented by measuring noradrenaline or 5-HT reuptake in vitro during antidepressant treatment using patient plasma. Changes in the monoaminergic systems has also been estimated from levels of monoamine metabolites in urine, blood or CSF. All such investigations have revealed reduction in the noradrenaline metabolite 3-methoxy-4-hydroxyphenyl glycol (MHPG). Certain antidepressants, including monoamine oxidase inhibitors the also show effects on the 5-HT metabolite 5-hydroxyindole acetic acid (5-HIAA) in the CSF (Potter et al. 1985).

2.10.1 Urinary 3-Methoxy-4-hydroxyphenyl Glycol

Several studies have shown that depressed patients with *low* levels of MHPG in their pretreatment 24-h urine samples respond better to antidepressants which are regarded as noradrenaline uptake inhibitors than do patients with high MHPG levels (Maas et al. 1972; Beckmann and Goodwin 1975; Schatzberg et al. 1981). There is also evidence that patients with low urinary MHPG respond more rapidly and need lower doses of antidepressants than do patients with high MHPG levels. A recent report from the NIMH collaborative study shows that favourable antidepressant responses of imipramine and amitriptyline were seen in bipolar (but not unipolar) patients with low to normal urinary and CSF MHPG (Maas et al. 1984).

The data on antidepressants in patients with *high* pretreatment urinary MHPG levels are more contradictory: early studies on amitriptyline indicated that this drug was favourable for patients with high MHPG levels (Beckmann and Goodwin 1975), but these findings were not confirmed in later studies (Spiker et al. 1980: Modai et al. 1979). Investigations with amitriptyline, desipramine and phenelzine have shown no consistent relationship between change in urinary MHPG and clinical response (Beckmann and Goodwin 1975; Cobbin et al. 1979; Beckmann and Murphy 1977).

2.10.2 CSF 5-Hydroxyindole Acetic Acid

Studies of 5-HIAA in the CSF of depressed patients have revealed that two biochemical subgroups may exist. Endogenously depressed patients (melachonics) may be divided into those with low and those with high levels of CSF 5-HIAA (Van Praag and Korf 1971; Åsberg et al. 1973; Cowdry and Goodwin 1978). Studies have indicated that noradrenaline uptake-inhibiting parent drugs such as nortriptyline (Åsberg et al. 1973) or maprotiline and drug metabolites such as demethyl clomipramine (Träskman et al. 1979) act more or less selectively in patients with *high* CSF 5-HIAA. According to our studies the clinical effect is related to the change of CSF MHPG and not CSF 5-HIAA (Träskman et al. 1979). During clomipramine treatment there were no relationships between clinical effect and biochemical change in *low* CSF 5-HIAA patients.

The NIMH collaborative study showed that unipolar, but not bipolar, patients with low CSF 5-HIAA responded better to imipramine and amitriptyline (Maas et al. 1984) than did high CSF 5-HIAA patients. The investigators in this study suggested in their report that the interaction between noradrenaline and serotonin rather than these two transmitters per se should be used when subgrouping patients.

In another study (Wägner et al. 1987) a serotonergic estimate, imipramine binding to platelets, was measured before and during antidepressant treatment. Patients with low pretreatment imipramine binding usually had a poorer clinical outcome than those with high imipramine binding.

2.11 Relationships Between Plasma Drug Level and Biochemical Effect

Even though significant relationships between plasma concentrations of drugs and biochemical effects are rarely described, higher drug concentrations seem to cause less selective effects on noradrenaline or 5-HT. This might be the reason for the curvilinear relationship between nortriptyline plasma levels and clinical effect reported by Åsberg et al. (1973). Another reason for this drug-response pattern could be that high plasma levels of certain antidepressants cause postsynaptic receptor blockade, which in turn affects the clinical outcome. When clomipramine was administered chronically to depressed patients or obsessive-compulsive patients, there was a U-shaped association between change of 5-HIAA in the CSF and clomipramine levels in plasma. High levels of clomipramine did not cause a substantial decrease of CSF 5-HIAA. This could also be a sign of postsynaptic receptor blockade (Träskman-Bendz et al. 1981). Trazodone, a novel psychotropic drug, has similar dose-dependent pharmacological actions. Small doses of the drug act as a central 5-HT receptor antagonist while large doses seem to produce a profile of a 5-HT receptor agonist (Maj et al. 1979 b).

2.12 Toward a New Generation of Antidepressant Drugs

Second-generation antidepressants have been developed in order to reduce cardiotoxic side effects and to make drugs more specific for the noradrenaline and serotonergic systems, respectively. Some of these new drugs successfully present with less side effects.

All current antidepressants, regardless of structure, function as facilitators of central noradrenergic transmission. Further studies of 5-HT-specific drugs need to prove that these have antidepressant properties. The most extensively studied 5-HT uptake blocker, zimelidine, seems to have a therapeutic potency similar to that of standard tricyclic antidepressants (Claghorn and Mathew 1981; Åberg-Wistedt 1982). Zimelidine, however, does not act selectively on the 5-HT system, since it also has an effect on the noradrenergic system (Bertilsson et al. 1980). Thus, the new generation of 5-HT uptake blockers are not as selective as had been hoped.

Little is known about the proper mechanism of action of antidepressant drugs. Elucidating the lag problem might further our understanding. In future, more direct studies of the receptor action of antidepressants are needed, i.e. by using PET or NMR. Perhaps better knowledge of antidepressant action in a genetically determined site *beyond* the β-adrenergic and 5-HT_2 receptor will solve the riddle of the mechanism of action.

Acknowledgements. The authors are grateful to Ms. Birgit Lönn and Ms. Eva Skagert for preparing the manuscript.

References

Åberg-Wistedt A (1982) A double-blind study of Zimelidine, a serotonin uptake inhibitor, and desipramine, a noradrenaline uptake inhibitor in endogenous depression. I Clinical Findings. Acta Psychiatr Scand 66:50–65

Ågren H, Mefford IN, Rudorfer MV, Linnoila M, Potter WZ (1986) Interacting neurotransmitter systems. A non-experimental approach to the 5HIAA-HVA correlation in human CSF. J Psychiatr Res 20(3):175–193

Alfredsson G, Bjerkenstedt L, Edman G, Härnryd C, Oxenstierna G, Sedvall G, Wiesel F-A (1984) Relationships between drug concentrations in serum and CSF, clinical effects and monoaminergic variables in schizophrenic patients treated with sulpride or chlorpromazine. Acta Psychiatr Scand 69 [suppl 311]:49–74

Alfredsson G, Wiesel F-A, Tylec A (1988) Relationships between glutamate and monoamine metabolites in cerebrospinal fluid and serum. Biol Psychiatry (to be published)

Anden N-E, Roos B-E, Werdinius B (1964) Effects of chlorpromazine, haloperidol and reserpine on the levels of phenolic acid in rabbit corpus striatum. Life Sci 3:149–158

Andreasen NC (1985) Positive vs. negative schizophrenia: a critical evaluation. Schizophr Bull 11:380–389

Andreasen NC, Olsen SA, Dennert JW, Smith MR (1982) Ventricular enlargement in schizophrenia: relationship to positive and negative symptoms. Am J Psychiatry 139:297–302

Angrist B, Rotrosen J, Gershon S (1980) Responses to apomorphine, amphetamine, and neuroleptics in schizophrenic subjects. Psychopharmacology 67:31–38

Åsberg M, Bertilsson L, Tuck D, Cronholm B, Sjöqvist F (1973) Indole amine metabolites in the cerebrospinal fluid of depressed patients before and during treatment with nortriptyline. Clin Pharmacol Ther 14:227–286

Axelrod J, Whitby LG, Hertting G (1961) Effect of psychtropic drugs on the uptake of ^{3}H-norepinephrine by tissues. Science 133:383–384

Bacopoulos NG (1984) Dopaminergic ^{3}H-agonists receptors in rat brain: new evidence on localization and pharmacology. Life Sci 34:307–315

Bacopoulos NG, Hattox SE, Roth RH (1979) 3,4-Dihydroxyphenylacetic acid and homovanillic acid in rat plasma: possible indicators of central dopaminergic activity. Eur J Pharmacol 56:225–236

Baldessarini RJ, Katz B, Cotton P (1984) Dissimilar dosing with high-potency neuroleptics. Am J Psychiatry 141:748–752

Bannon MJ, Roth RH (1983) Pharmacology of mesocortical dopamine neurons. Pharmacol Rev 35:53–68

Beckmann H, Goodwin FK (1975) Antidepressant response to tricyclics and urinary MHPG in unipolar patients. Arch Gen Psychiatry 32:17–21

Beckmann H, Murphy DL (1977) Phenelzine in depressed patients: effects on urinary MHPG excretion in relation to clinical response. Neuropsychobiology 3:49–55

Bertilsson L, Tuck JR, Siwers B (1980) Biochemical effects of zimelidine in man. Eur J Clin Pharmacol 18:483–487

Bissette G, Nemeroff CB (1985) Do neuropeptide systems mediate some of the effects of antipsychotic drugs? Prog Clin Biol Res 192:349–353

Bjerkenstedt L, Gullberg B, Härnryd C, Sedvall G (1977) Monoamine metabolite levels in cerebrospinal fluid of psychotic women treated with melperone or thiothixene. Arch Psychiatr Nervenkr 224:107–118

Bjerkenstedt L, Härnryd C, Grimm V, Gullberg B, Sedvall G (1978) A double-blind comparison of melperone and thiothixene in psychotic women using a new rating scale, the CPRS. Arch Psychiatr Nervenkr 226:157–172

Bjerkenstedt L, Gullberg B, Härnryd C, Sedvall G (1979) Relationships between clinical and biochemical effects of melperone and thiothixene in psychotic women. Arch Psychiatr Nervenkr 227:181–192

Bland RD, Orn H (1978) 14 year outcome in early schizophrenia. Acta Psychiatr Scand 58:327–338

Borison RL, Hitri A, Blowers AJ, Diamond BI (1983) Antipsychotic drug action: clinical, biochemical and pharmacological evidence for site specificity of action. Clin Neuropharmacol 6:137–150
Bowers MB, Heninger GR (1981) Cerebrospinal fluid homovanillic acid patterns during neuroleptic treatment. Psychiatry Res 4:285–290
Bowers MB, Swigar ME, Jatlow PI, Goicoecha N (1984) Plasma catecholamine metabolites and early response to haloperidol. J Clin Psychiatry 6:248–251
Brunello N, Chuang DM, Costa E (1982) Use of specific brain lesions to study the site of action of antidepressants. Adv Biosci 40:141–145
Bunney BS (1984) Antipsychotic drug effects on the electrical activity of dopaminergic neurons. TINS 7:212–215
Burt DR, Creese I, Snyder SH (1977) Antischizophrenic drugs: chronic treatment elevates dopamine receptor binding. Science 196:326–327
Carlsson A (1974) Antipsychotic drugs and catecholamine synapses. J Psychiatr Res 11:57–64
Carlsson A (1984) Current theories on the mode of action of antidepressant drugs. Adv Biochem Psychopharmacol 39:213–221
Carlsson A, Hillarp NÅ (1956) Release of adrenaline from the adrenal medulla of rabbits produced by reserpine. Kgl Fysiogr Sällsk Lund Förh 26(8)
Carlsson A, Lindqvist M (1963) Effect of chlorpromazine or haloperidol on formation of 3-methoxytyramine and normetanephrine in mouse brain. Acta Pharmacol Toxicol 20:140–144
Carlsson A, Lindqvist M, Magnusson T, Waldeck B (1958) On the presence of 3-hydroxytyramine in brain. Science 127:471
Cesarec Z, Nyman AK (1985) Differential response to amphetamine in schizophrenia. Acta Psychiatr Scand 71:523–538
Chouinard G, Jones BD (1980) Neuroleptic induced supersensitivity psychosis: clinical and pharmacologic characteristics. Am J Psychiatry 137:16–21
Christensen E (1988) Pharmacologic data of the atypical neuroleptic compound melperone. Acta Psychiatr Scand [suppl] (to be published)
Claghorn JL, Mathew RJ (1981) Clinical trials with new bicyclic and tetracyclic compounds: zimelidine and maprotiline. Psychopharmacol Bull 17:23–26
Clement-Cormier YC, Kebabian JW, Petzold GL, Greengard P (1974) Dopamine-sensitive adenylate cyclase in mammalian brain: a possible site of action of antipsychotic drugs. Proc Natl Acad Sci USA 71:1113–1117
Cobbin DM, Requin-Blow B, Williams LR, Williams WO (1979) Urinary MHPG levels and tricyclic antidepressant drug selection: a preliminary communication on improved drug selection in clinical practice. Arch Gen Psychiatry 36:1111–1115
Corrodi H, Fuxe K (1969) Decreased turnover in central 5-HT nerve terminals induced by antidepressant drugs of the imipramine type. Eur J Pharmacol 7:56–59
Costa E (1982) Perspectives in the molecular mechanisms of antidepressant action. In: Costa E, Racagni G (eds) Typical and atypical antidepressants: molecular mechanisms. Raven, New York, pp 21–26
Cowdry RW, Goodwin FK (1978) Amine neurotransmitter studies and psychiatric illness: toward more meaningful diagnostic concepts. In: Spitzer RL, Klein DF (eds) Critical issues in Psychiatric diagnosis. Raven, New York, pp 281–304
Creese I, Burt DR, Snyder SH (1975) Dopamine receptor binding: differentiation of agonist and antagonist states with ^{3}H-dopamine and ^{3}H-haloperidol. Life Sci 17:993–1002
Creese I, Burt DR, Snyder SH (1976) Dopamine receptor binding predicts clinical and pharmacological potencies of antischizophrenic drugs. Science 192:481–483
Crews FT, Smith CB (1978) Presynaptic alpha-receptor subsensitivity after long-term antidepressant treatment. Science 202:322–324
Cross AJ, Crow TJ, Ferrier IN, Johnson JA, Johnstone EC, Owen F, Owens DGC, Poulter M (1985) Chemical and structural changes in the brain in patients with movement disorder. In: Casey DE, Chase TN, Christensen AV, Gerlach J (eds) Dyskinesia – research and treatment. Springer, Berlin Heidelberg New York (Psychopharmacology Supplementum 2, pp 104–110)

Crow TJ (1980) Molecular pathology of schizophrenia: more than one disease process? Br Med J 280:66–68
Crow TJ, MacMillan JF, Johnson AL, Johnstone EC (1986) A randomized controlled trial of prophylactc neuroleptic treatment. Br J Psychiatry 148:120–127
Curzon B, Gumpert EJW, Sharpe DM (1971) Amine metabolites in the lumbar cerebrospinal fluid of humans with restricted flow of cerebrospinal fluid. Nature 231:189–191
Davis JM (1985) Maintenance therapy and the natural course of schizophrenia. J Clin Psychiatry 11:18–21
Davis JM, Garver DL (1978) Neuroleptics: clinical use in psychiatry. In: Iverson LL, Iverson SD, Snyder SH (eds) Handbook of psychopharmacology. Neuroleptics and schizophrenia, vol 10. Plenum, New York
Davis JM, Schaffer CB, Killian GA, Kinard C, Chan C (1980) Important issues in the drug treatment of schizophrenia. Schizophr Bull 6:70–87
Davis KL, Davidson M, Mohs RC, Kendler KS, Davis BM, Johns CA, DeNegris Y, Horvath TG (1985) Plasma homovanillic acid concentrations and the severity of schizophrenic illness. Science 227:1601–1602
De Keyser J, De Backer J-P, Ebinger G, Vauquelin G (1985) Regional distribution of the dopamine D_2 receptors in the mesotelencephalic dopamine neuron system of human brain. J Neurol Sci 71:119–127
Delay J, Deniker O (1957) Caracteristiques psycho-physiologiques des medicaments neuroleptiques. In: Garattini S, Ghetti V (eds) The psychotropic drugs. Elsevier, Amsterdam, pp 485–501
De Montigny C, Grunberg F, Mayer A, Deschenes JP (1981) Lithium induces rapid relief of depression in tricyclic antidepressant non-responders. Br J Psychiatry 138:252–256
Elizur A, Davidson S, Psych FRC (1975) The evaluation of the anti-autistic activity of sulpiride. Curr Ther Res 18:578–584
Ekblom E, Eriksson K, Lindström LH (1984) Supersensitivity psychosis in schizophrenic patients after sudden clozapine withdrawal. Psychopharmacology 83:293–294
Falloon JRH, Boyd JL, McGill CW, Williamson M, Razani J, Moss HB, Gilderman AM, Simpson GM (1985) Family management in the prevention of morbidity of schizophrenia. Arch Gen Psychiatry 42:887–896
Farde L, Ehrin E, Eriksson L, Greitz T, Hall H, Hedström C-G, Litton J-E, Sedvall G (1985) Substituted benzamides as ligands for visualization of dopamine receptor binding in the human brain by positron emission tomography. Proc Natl Acad Sci USA 82:3863–3867
Farde L, Hall H, Ehrin E, Sedvall G (1986) Quantitative analysis of dopamine D_2 receptor binding in the living human brain by positron emission tomography. Science 231:258–261
Farde L, Halldin C, Stone-Elander S, Sedvall G (1987) PET analysis of human dopamine receptor subtypes using ^{11}C-SCH 23390 and ^{11}C-raclopride. Psychopharmacology 92:278–284
Farde L, Wiesel F-A, Halldin C, Sedvall G (1988) Central D_2-dopamine receptor occupancy in schizophrenic patients treated with antipsychotic drugs. Arch Gen Psychiatry 45:71–76
Fibinger HC, Lloyd KG (1984) Neurobiological substrates of tardive dyskinesia: the GABA hypothesis. TINS 462–464
Fuxe K, Ögren SO, Agnati LF (1979) The effects of chronic treatment with the 5-hydroxytryptamine uptake blocker zimelidine on central 5-hydroxytryptamine mechanism. Evidence for the induction of a low affinity binding site for 5-hydroxytryptamine. Neurosci Lett 13:307–312
Gale K, Casu M (1981) Dynamic utilization of GABA in substantia nigra: regulation by dopamine and GABA in the striatum, and its clinical and behavioural implications. Mol Cell Biochem 39:369–405
Garcia-Sevilla JA, Zis AP, Zelnik TC, Smith CB (1981) Tricyclic antidepressant drug treatment decreases alpha-2 adrenoreceptors on human platelet membranes. Eur J Pharmacol 69:121–123

Gerlach J (1985) Pathophysiological mechanisms underlying tardive dyskinesia. In: Casey E, Chase T, Christensen AV, Gerlach J (eds) Dsykinesia – research and treatment. Springer, Berlin Heidelberg New York (Psychopharmacology suppl 2, pp 98–103)

Gerlach J, Lühdorf K (1975) The effect of L-Dopa on young patients with simple schizophrenia, treated with neuroleptic drugs. A double-blind cross-over trial with Madopar and placebo. Psychopharmacologia 44:105–110

Goldberg SC (1985) Negative and deficit symptoms in schizophrenia do respond to neuroleptics. Schizophr Bull 11:453–456

Goldstein MJ, Rodnick EH, Evans JR, May PRA, Steinberg MR (1978) Drug and family therapy in the aftercare of acute schizophrenics. Arch Gen Psychiatry 35:1169–1177

Goodwin FK, Post RM, Sack RL (1975) Clinical evidence for neurochemical adaption to psychotropic drugs. In: Mandall EJ (ed) Neurobiological mechanisms of adaption and behavior. Raven New York, pp 33–45

Gordon E, Perlow M, Oliver J, Ebert M, Kopin I (1975) Origins of catecholamine metabolites in monkey cerebrospinal fluid. J Neurochem 25:347–349

Gunne L-M, Häggström J-E, Sjöquist B (1984) Association with persistent neuroleptic induced dyskinesia of regional changes in brain GABA synthesis. Nature 309:347–349

Haase HJ, Floru L, Ulrich F (1974) Klinisch, neuroleptische Untersuchung des *N*-(1-Äthyl-Pyrrolidin-2-yl-Methyl)-2 Methoxy-5-Sulfamoyl-Benzamide-Neurolepticums Sulpirid (Dogmatil) in akut erkrankten Schizophrenen. Int Pharmacopsychiatry 9:77–94

Härnryd C, Bjerkenstedt L, Björk K, Gullberg B, Oxenstierna G, Sedvall G, Wiesel F-A, Wik G, Åberg-Wistedt A (1984a) Clinical evaluation of sulpiride in schizophrenic patients – a double-blind comparison with chlorpromazine. Acta Psychiatr Scand [suppl] 69:17–30

Härnryd C, Bjerkenstedt L, Gullberg B, Oxenstierna G, Sedvall G, Wiesel F-A (1984b) Time course for effects of sulpiride and chlorpromazine on monoamine metabolite and prolactin levels in cerebrospinal fluid from schizophrenic patients. Acta Psychiatr Scand [suppl] 69:75–92

Harris PQ, Brown SJ, Friedman J, Bacopoulos NG (1984) Plasma drug and homovanillic acid levels in psychotic patients receiving neuroleptics. Biol Psychiatry 19:849–860

Harris TH (1957) Depression induced by *Rauwolfia* compounds. Am J Psychiatry 113:950

Hogarty GE, Goldberg SC, Scholler NR (1974) Drug and sociotherapy in the aftercare of schizophrenic patients. Arch Gen Psychiatry 31:609–618

Honigfeld G, Patin J, Singer J (1984) Clozapine: antipsychotic activity in treatment-resistant schizophrenics. Adv Ther 1:77–97

Hwang EC, Van Woert MH (1980) Acute versus chronic effects of serotonin uptake blockers on potentiation of the "serotonin syndrome". Commun Psychopharmacol 4:161–167

Janowsky A, Okada F, Mahler DH, Applegate CD, Sulser F, Steranka LR (1982) Role of serotonergic input in the regulation of the beta-adrenergic receptor-coupled adenylate update system. Science 218:900–901

Jenner P, Rupniak NMJ, Marsden CD (1985) Differential alteration of striatal D-1 and D-2 receptors induced by the long-term administration of haloperidol, sulpiride or clozapine to rats. In: Casey DE, Chase TN, Christensen AV, Gerlach J (eds) Dyskinesia – research and treatment, suppl 2. Springer, Berlin Heidelberg New York, pp 174–181

Johnson RW, Reisine T, Spotnitz S, Wiech N, Ursillo R, Yamamura HI (1980) Effects of desipramine and yohimbine on α_2 and β-receptor sensitivity. Eur J Pharmacol 67:123–127

Johnstone EC (1985) Acute schizophrenia. Br J Hosp Med 34:198–201

Johnstone EC, Crow TJ, Frith CD, Carney MWP, Price JS (1978) Mechanism of the antipsychotic effect in the treatment of acute schizophrenia. Lancet I:848–851

Jones RSG (1980) Long-term administration of atropine, imipramine and viloxazine alters responsiveness of rat cortical neurones to acetylcholine. Can J Physiol Pharmacol 58:531–535

Karobath M, Leitich H (1974) Antipsychotic drugs and dopamine-stimulated adenylate cyclase prepared from corpus striatum of rat brain. In: Proc Natl Acad Sci USA 71:2915–2918

Kebabian JW, Calne DB (1979) Multiple receptors for dopamine. Nature 277:93–96

Kebabian JW, Petzold GL, Greengard P (1972) Dopamine-sensitive adenylate cyclase in caudate nucleus of rat brain, and its similarity to the "dopamine receptor". Proc Natl Acad Sci USA 69:2145–2149

Kellar KJ, Cascio CS, Butler JA, Kurtzke RN (1981) Differential effects of electroconvulsive shock and antidepressant drugs on serotonin-2 receptors in rat brain. Eur J Pharmacol 69:515–518

Kirkegaard A, Kirkegaard G, Geismar L, Christensen I (1981) Additional studies on side effects of melperone in long-term therapy for 1 to 15 years in psychiatric patients. Arzneimittelforschung 31:737–740

Kuha S, Miettinen E (1986) Long-term effect of clozapine in schizophrenia. Nord Psykiatr Tidsskr 40:225–230

Kuhn R (1957) Über die Behandlung depressiver Zustände mit einem Iminodibenzylderivat (G 22355). Schweiz Med Wochenschr 87:1135–1140

Larsson M, Öhman R, Wallin L, Wålinder J, Carlsson A (1984) Antipsychotic treatment with alfa-methyltyrosine in combination with thioridazine: prolactin response and interaction with dopaminergic precursor pools. J Neural Transm 60:115–132

Leff ST, Creese I (1983) Dopaminergic D-3 binding sites are not presynaptic autoreceptors. Nature 306:586–589

Leff J, Kuipers L, Berkowitz, Sturgeon D (1985) A controlled trial of social intervention in the families of schizophrenic patients: to year follow-up. Br J Psychiatry 146:594–600

Leonard BE (1984) Pharmacology of new antidepressants. Prog Neuropsychopharmacol Biol Psychiatry 8:97–108

Lerer B, Ebstein RP, Belmaker RH (1981) Subsensitivity of human beta-andrenergic adenylate cyclase after salbutamol treatment of depression. Psychopharmacology 75:169–172

Lindwall O, Björklund A (1978) Organization of catecholamine neurons in the rat central nervous system. In: Iversen LL, Iversen SD, Snyder SH (eds) Handbook of psychopharmacology, vol 9. Plenum, New York, pp 139–231

Ljungberg T, Ungerstedt U (1978) Classification of neuroleptic drugs according to their ability to inhibit apomorphine-induced locomotion and gnawing: evidence for two different mechanisms of action. Psychopharmacology 56:239–247

Loomer HP, Saunders JC, Kline NS (1957) A clinical and pharmacodynamic evaluation of iproniazid as a psychic energizer. Psychiatr Res Rep 8:129–141

Maas R, Fawcett JA, DeKirmenjian H (1972) Catecholamine metabolism, depressive illness and drug response. Arch Gen Psychiatry 26:252–262

Maas JW, Koslow SH, Katz MM, Bowden CL, Gibbons RL, Stokes PE, Robins E, Davis JM (1984) Pretreatment neurotransmitter metabolite levels and response to tricyclic antidepressant drugs. Am J Psychiatry 141(10):1159–1171

Maj J, Mogilnicka E, Kordecka A (1979a) Chronic treatment with antidepressant drugs: potentiation of apomorphine-induced aggressive behaviour in rats. Neurosci Lett 13:337–341

Maj J, Palider W, Rawlow A (1979b) Trazodone, a central serotonin antagonist and agonist. J Neural Transm 44:237–248

Manchanda R, Hirsch SR (1986) Low dose maintenance medication for schizophrenia. Br Med J 293:515–516

May PRA, Tuma AH, Dixon WJ, Yale C, Thiele DA, Kraude WH (1981) Schizophrenia: a follow-up study of the results of five forms of treatment. Arch Gen Psychiatry 38:776–784

Menkes DB, Aghajanian GK (1981) α_1-Adrenoceptor-mediated responses in the lateral geniculate nucleus are enhanced by chronic antidepressant treatment. Eur J Pharmacol 74:27–36

Miller RJ, Hiley CR (1974) Anti-muscarinic properties of neuroleptics and drug-induced parkinsonism. Nature 248:596–597

Modai I, Apter A, Golomb M, Wijsenbeek H (1979) Response to amitriptyline and urinary MHPG in bipolar depressive patients. Neuropsychobiology 5:181–184

Modigh K (1975) Electroconvulsive shock and postsynaptic catecholamine effects: increased psychomotor stimulant action of apomorphine and clonidine in reserpine pretreated mice by repeated ECS. J Neural Transm 36:19–32

Muller P, Seeman P (1978) Dopaminergic supersensitivity after neuroleptics: time course and specificity. Psychopharmacology 60:1–11

Nybäck H, Sedvall G (1968) Effect of chlorpromazine on accumulation and disappearance of catecholamine formed from tyrosine-^{14}C in brain. J Pharmacol Exp Ther 162:294–301

Nybäck H, Sedvall G (1970) Further studies on the accumulation and disappearance of catecholamines formed tyrosine-^{14}C in mouse brain. Effect of some phenothiazine analogues. Eur J Pharmacol 10:193–205

Nybäck HV, Walters JV, Aghajanian GK, Roth RH (1975) Tricyclic antidepressants. Effects on the firing rate of brain noradrenergic neurons. Eur J Pharmacol 32:302–312

Paul SM, Crews FM (1980) Rapid desensitization of cerebral cortical β-adrenergic receptors induced by desmethylimipramine and phenoxybenzamine. Eur J Pharmacol 62:349–350

Peroutka SJ, Snyder SH (1980a) Long-term antidepressant treatment decreases spiroperidol-labeled serotonin receptor binding. Science 210:88–90

Peroutka SJ, Snyder SH (1980b) Relationship of neuroleptic drug effects at brain dopamine, serotonin, alfa-adrenergic, and histaminc rcceptors to clinical potency. Am J Psychiatry 137:1518–1522

Pickar D, Labarca R, Linnoila M, Roy A, Hommer D, Everett D, Paul SM (1984) Neuroleptic-induced decrease in plasma homovanillic acid and antipsychotic activity in schizophrenic patients. Science 225:954–957

Pickar D, Labarca R, Doran AR, Wolkowitz OM, Roy A, Breier A, Linnoila M, Paul SM (1986) Longitudinal measurement of plasma homovanillic acid levels in schizophrenic patients. Arch Gen Psychiatry 43:669–676

Post RM, Goodwin FK, Gordon E, Watkin DM (1973) Amine metabolites in human cerebrospinal fluid. Effect of cord transection and spinal fluid block. Science 179:897–899

Povlsen UJ, Noring U, Fog R, Gerlach J (1985) Tolerability and therapeutic effect of clozapine. Acta Psychiatr Scand 71:176–185

Potter WZ, Scheinin M, Golden RN, Rudorfer MV, Cowdry RW, Calil HM, Ross RJ, Linnoila M (1985) Selective antidepressants and cerebrospinal fluid. Lack of specificity on norepinephrine and serotonin metabolites. Arch Gen Psychiatry 42:1171–1177

Prange AJ, Wilson IC, Knox A, McClane TK, Lipton MA (1970) Enhancement of imipramine by thyroid stimulating hormone: clinical and theoretical implications. Am Psychiatry 127:191–199

Ravaris CL, Nies A, Robinson DS, Ives JO, Lamborn KR, Korson A (1976) A multiple dose controlled study of phenelzine in depression anxiety states. Arch Gen Psychiatry 33:347–350

Rifkin A, Quitkin F, Klein DF (1978) Are prophylactic antiparkinson drugs necessary? Arch Gen Psychiatry 32:672–674

Scatton B, Bischoff S, Dedek J, Korf J (1977) Regional effects of neuroleptics on dopamine metabolism and dopamine-sensitive adenylate cyclase activity. Eur J Pharmacol 44:287–292

Schatzberg AF, Rosenbaum AH, Orsulak PJ, Rohde WA, Maruta T, Kruger ER, Cole JO, Schildkraut JJ (1981) Toward a biochemical classification of depressive disorders. III. Pretreatment urinary MHPG levels as predictors of response to treatment with maprotiline. Psychopharmacology 75:34–38

Sedvall G, Mayevsky A, Fri C-G, Sjöquist B, Samuel D (1973) The use of stable oxygen isotopes for labelling of homovanillic acid in rat brain in vivo. Adv Biochem Psychopharmacol 7:57–68

Sedvall G, Alfredsson G, Bjerkenstedt L, Eneroth P, Fryö B, Härnryd C, Swahn C-G, Wiesel F-A, Wode-Helgodt B (1975) Selective effects of psychoactive drugs on levels

of monoamine metabolites and prolactin in cerebrospinal fluid of psychiatric patients. In: Airaksinen H (ed) Proceedings of the sixth international congress of pharmacology, vol 3. Forssan Kirjapaino, Oy, pp 255–267
Sedvall G, Farde L, Persson A, Wiesel F-A (1986a) Imaging of neurotransmitter receptors in the living human brain. Arch Gen Psychiatry 43:995–1005
Sedvall G, Farde L, Stone-Elander S, Halldin C (1986b) Dopamine D1 receptor binding in the living human brain. In: Breese GR, Creese I (eds) Neurobiology of central D1-dopamine receptors. Adv Exp Med Biol 204:119–124
Seeman P, Chau-Wong M, Tedesco J, Wong K (1975) Brain receptors for antipsychotic drugs and dopamine binding assay. In: Proc Natl Acad Sci USA 72:4376–4380
Seeman P, Lee T, Chau-Wong M (1976) Antipsychotic drug doses and neuroleptic/dopamine receptors. Nature 261:717–719
Selemon LD, Goldman-Rakic PS (1985) Longitudinal topography and interdigitation of corticostriatal protections in the rhesus monkey. J Neurosci 5:776–794
Serra G, Argiolas A, Klimek V, Fadda F, Gessa GL (1979) Chronic treatment with antidepressants prevents the inhibitory effect of small doses of apomorphine on dopamine synthesis and motor activity. Life Sci 25:415–424
Shore PA, Silver SL, Brodie BB (1955) Interaction of reserpine, serotonin and lysergic acid diethylamide in brain. Science 122:284–285
Siggins GR, Hoffer BJ, Ungerstedt U (1974) Electrophysiological evidence for involvement of cyclic adenosine monophosphate in dopamine responses of caudate neurons. Life Sci 15:779–784
Sjöqvist F (1971) A pharmacokinetic approach to the treatment of depression. Int Pharmacopsychiatry 6:147–169
Snyder SH, Greenberg D, Yamamura HI (1974) Antischizophrenic drugs and brain cholinergic receptors: affinity for muscarinic sites predicts extrapyramidal effects. Arch Gen Psychiatry 31:58–61
Sourkes TL (1973) On the origin of homovanillic acid (HVA) in the cerebrospinal fluid. J Neurol Transm 34:153–157
Spiker DG, Edwards D, Hanin I, Neil J, Kupfer DJ (1980) Urinary MHPG and clinical response to amitriptyline in depressed patients. Am J Psychiatry 137:1183–1187
Sternberg DE, Heninger GR, Roth RH (1983) Plasma homovanillic acid as an index of brain dopamine metabolism: enhancement with debrisoquin. Life Sci 32:2447–2452
Stone EA (1983) Problems with current catecholamine hypotheses of antidepressant agents: speculations leading to a new hypothesis. Behav Brain Sci 6:535–577
Sugrue MF (1980) Changes in rat brain monoamine turnover following chronic antidepressant administration. Life Sci 26:423–429
Sugrue MF (1981) Current concepts on the mechanism of action of antidepressant drugs. Pharmacol Ther 13:219–247
Sugrue MF (1982) A study of sensitivity of rat brain alpha$_2$-adrenoreceptors during chronic antidepressant treatments. Naunyn-Schmiedebergs Arch Pharmacol 320:90–96
Swahn C-G, Wiesel F-A (1976) Determination of conjugated monoamine metabolites in brain tissue. J Neural Transm 39:281–290
Träskman L, Åsberg M, Bertilsson L, Cronholm B, Mellström B, Neckers L, Sjöqvist F, Thorén P, Tybring G (1979) Plasma levels of chlorimipramine and its demethyl metabolite during treatment of depression. Clin Pharmacol Ther 26:600–610
Träskman-Bendz L, Åsberg M, Bertilsson L (1981) Serotonin and noradrenaline uptake inhibitors in the treatment of depression – relationship to 5-HIAA in spinal fluid. Acta Psychiatr Scand [suppl 290] 63:209–218
Ungerstedt U, Ljungberg T (1977) Behavioral patterns related to dopamine neurotransmission: effect of acute and chronic antipsychotic drugs. In: Costa E, Gessa GL (eds) Advances in biochemical psychopharmacology, vol 16. Raven, New York, pp 193–199
Van Praag HM (1977) The significance of dopamine for the mode of action of neuroleptics and the pathogenesis of schizophrenia. Br J Psychiatry 130:463–474
Van Praag HM (1984) Studies in the mechanism of action of serotonin precursors in depression. Psychopharmacol Bull 20:599–602

Van Praag HM, Korf J (1971) Endogenous depressions with and without disturbances in the 5-hydroxytryptamine metabolism: a biochemical classification? Psychopharmacology 19:148–152
Vetulani J, Pilc A (1982) Postdecapitation convulsions in the rat measured with an animex mobility meter: relation to α-adrenoreceptors. Eur J Pharmacol 85:269–275
Vetulani J, Sulser F (1975) Action of various antidepressant treatments reduces reactivity of noradrenergic cyclic AMP-generating system in limbic forebrain. Nature 257:495–496
Wåhlinder J, Skott A, Nagy A, Roos BE (1976a) Potentiation of the antidepressant action of clomipramine by tryptophan. Arch Gen Psychiatry 33:1384–1389
Wåhlinder J, Skott A, Carlsson A, Roos B-E (1976b) Potentiation by metyrosine of thioridazine effects in chronic schizophrenics. Arch Gen Psychiatry 33:501–505
Wägner A, Åberg-Wistedt A, Åsberg M, Bertilsson L, Mårtensson B, Montero D (1987) Effects of antidepressant treatments on platelet imipramine binding in major depressive disorders. Arch Gen Psychiatry 44:870–877
Wagner HN, Burns HD, Dannals RF, Wong DS, Långström B, Duelfer T, Frost JJ, Ravert HT, Links JM, Rosenbloom SB, Lukas SE, Kramer AV, Kuhar MJ (1983) Imaging dopamine receptors in the human brain by positron tomography. Science 221:1264–1266
Weinberger DR, Bigelow LB, Kleinman JE, Klein ST, Rosenblatt JE, Wyatt RJ (1980) Cerebral ventricular enlargement in chronic schizophrenia: an association with poor response to treatment. Arch Gen Psychiatry 37:11–13
Wheatley D (1972) Potentiation of amitriptyline by thyroid hormone. Arch Gen Psychiatry 26:229–233
Wiesel F-A (1975) Massfragmentographic determination of acid dopamine metabolites in human cerebrospinal fluid. Neurosci Lett 1:219–224
Wiesel F-A, Sedvall G (1975) Effect of antipsychotic drugs on homovanillic acid levels in striatum and olfactory tubercle of the rat. Eur J Pharmacol 30:364–367
Wiesel F-A, Fri C-G, Sedvall G (1973) Determination of homovanillic acid turnover in rat striatum using a monoamine oxidase inhibitor. Eur J Pharmacol 23:104–106
Wiesel F-A, Bjerkenstedt L, Skett P (1978) Effect of melperone, two of its metabolites and thiothixene on central monoamine metabolism and prolactin levels in rodents. Acta Pharmacol Toxicol 43:126–136
Willner P (1985) Antidepressant and serotonergic transmission: an integrative review. Psychopharmacology 85:387–404
Wilson JC, Garbutt JC, Lanier CF, Moylan J, Nelson W, Prange AJ (1983) Is there a tardive dysmentia? Schizophr Bull 9:187–191
Wing JK (1966) Five year outcome in early schizophrenia. Roc Rodh Soc Med 59:17–18
Wode-Helgodt B (1977) Clinical and biochemical effects of chlorpromazine in psychotic patients. Doctoral thesis, Karolinska Institute, Stockholm
Wode-Helgodt B, Fyrö B, Gullberg B, Sedvall G (1977) Effect of chlorpromazine treatment on monoamine metabolite levels in cerebrospinal fluid of psychotic patients. Acta Psychiatr Scand 56:129–142
Wode-Helgodt B, Borg S, Fyrö B, Sedvall G (1978) Clinical effects and drug concentrations in plasma and cerebrospinal fluid in psychotic patients treated with fixed doses of chlorpromazine. Acta Psychiatr Scand 58:149–173

Catecholamines and Mood: Neuroendocrine Aspects

N. Matussek

Contents

The results of pharmacological-clinical studies in humans, e.g. those using psychomotor stimulants and other substances, leave no doubt that catecholaminergic neurons in the brain are involved in mood regulation. However, according to present knowledge, we have to remember that, given the close connections between the different neurotransmitter systems and the influence of diverse hormones on these systems, it is not possible to determine which brain process is finally responsible for the feelings of sadness, depression and happiness, for example, and their conscious perception. Bearing in mind these restrictions, I will attempt in this article to reveal the relations of the three catecholaminergic systems – the dopamine (DA), noradrenaline (NA) and adrenaline system – with regard to some aspects of neuroendocrinology and mood. We know today that dys-

Psychiatric Hospital, Universitiy of Munich, Nussbaumstrasse 7, D-8000 Munich 2, FRG

Current Topics in Neuroendocrinology, Vol. 8

regulation in many hormone systems, such as thyroid, adrenal cortex and gonads, may lead to mood swings (Bleuler 1979) and that in the regulation mechanisms of these hormones the catecholamines are markedly involved (Fuxe et al. 1979a; Tuomisto and Männisto 1985). But on the other hand, thyroid (Kunos 1981), ketosteroid and sex hormones (Motulsky and Insel 1982) have strong influence on different catecholaminergic neurotransmitter systems. Since hypercortisolism and thyrotropin-releasing hormone (TRH) are reviewed in separate chapters in this volume, I will concentrate mainly on catecholamine challenge tests, especially in relation to growth hormone (GH), but also with some results concerning cortisol and prolactin. I will restrict myself to neuroendocrine tests in connection with the different types of depression since they are especially suitable for investigating these aspects of mood swings. For excellent reviews of this topic see also Checkley and Arendt (1984), Terry (1984), and Siever and Davis (1985).

The first to speculate about relations between catecholamines and depression were Everett and Toman in 1959, on the basis of animal experimental studies with the reversal of the reserpine sedation with Dopa. Some years later the NA deficit

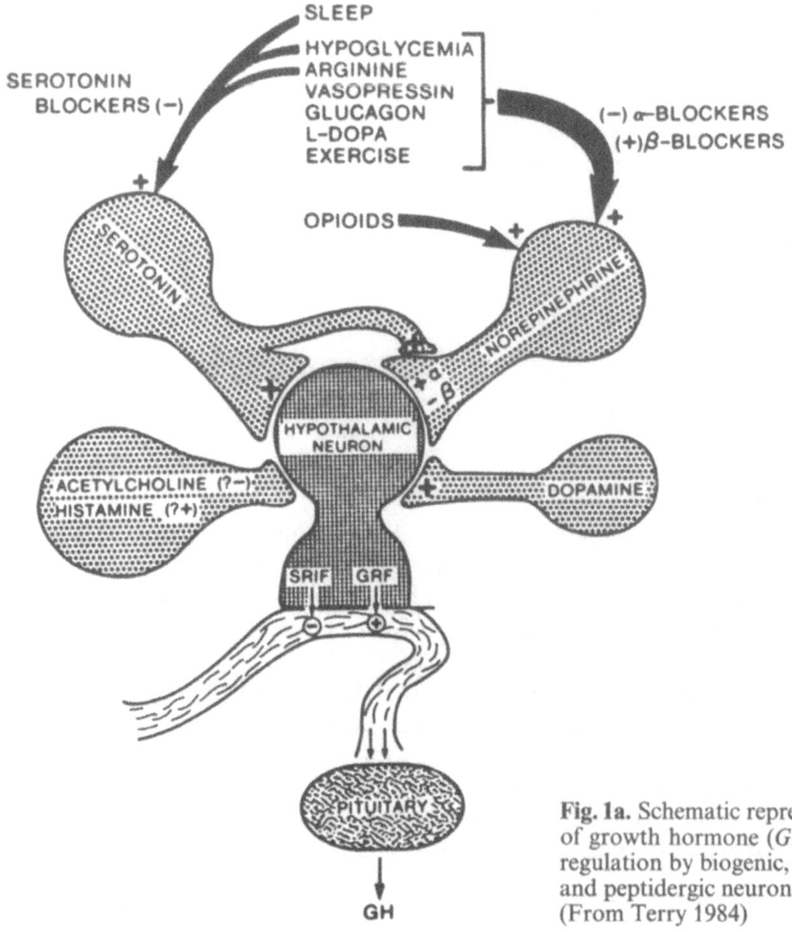

Fig. 1a. Schematic representation of growth hormone (*GH*) regulation by biogenic, aminergic and peptidergic neuronal systems. (From Terry 1984)

hypothesis of depression was formulated in more detail by several groups working independently (Bunney and Davis 1965; Schildkraut 1965; Matussek 1966). This hypothesis was based on the model of depression after resperine administration and the mechanism of action of the tricyclic thymoleptics and monoamine oxidase (MAO) inhibitors. Using "reserpine depression" as model for all forms of depressions was too simple from today's point of view. An NA deficit in cerebrospinal fluid (CSF), blood, urinary output or postmortem studies found by measuring NA, 3-methoxy-4-hydroxyphenylglycol (MHPG) or vanillic mandelic acid (VMA) shows no clear-cut results with regard to common NA deficit related to depressive states (review Matussek 1980).

The introduction of neuroendocrinological methods into biological psychiatry research at the end of the 1960s and beginning of the 1970s also opened up the possibility of using insulin or other variously specific drugs with influence on catecholaminergic neurons as tools for exploring neuroendocrine functions. We have known for some time that GH release is stimulated by catecholaminergic mechanisms among others (Fig. 1 a). The regulation of GH secretion is a fascinating, complex mechanism, even today not completely understood, in which many transmitter and cotransmitter systems, steroids and thyroid hormones, are involved (Fig. 1 a). Recent studies have introduced the so-called medianosome concept. "The medianosome is defined as an integrative unit, which consists of well

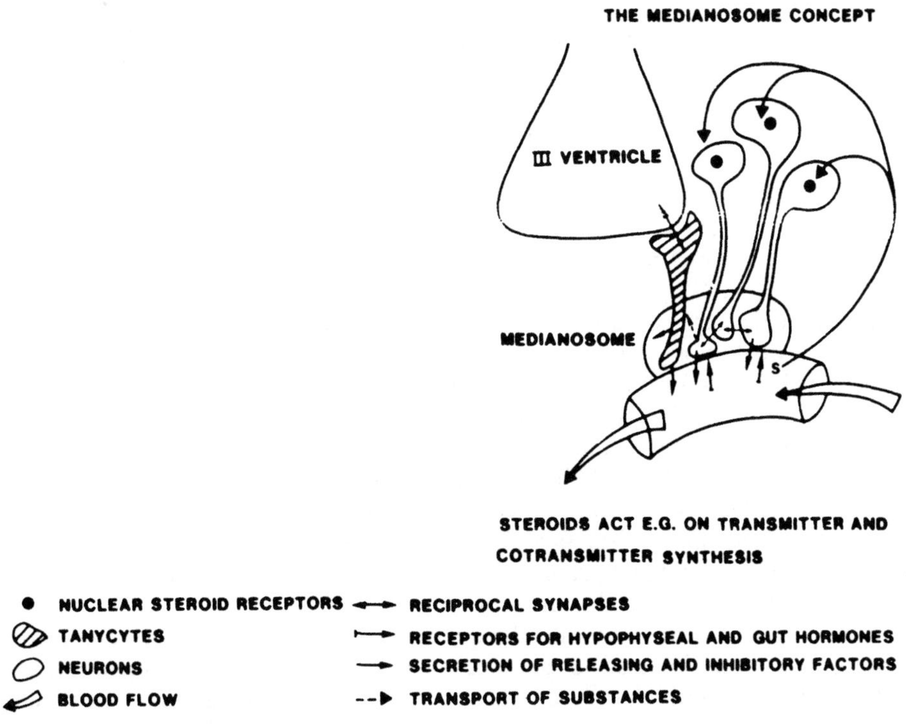

Fig. 1b. Schematic illustration of the medianosome concept and of the sites of action of peptide, steroid and thyroid hormones. (From Fuxe et al. 1986 a)

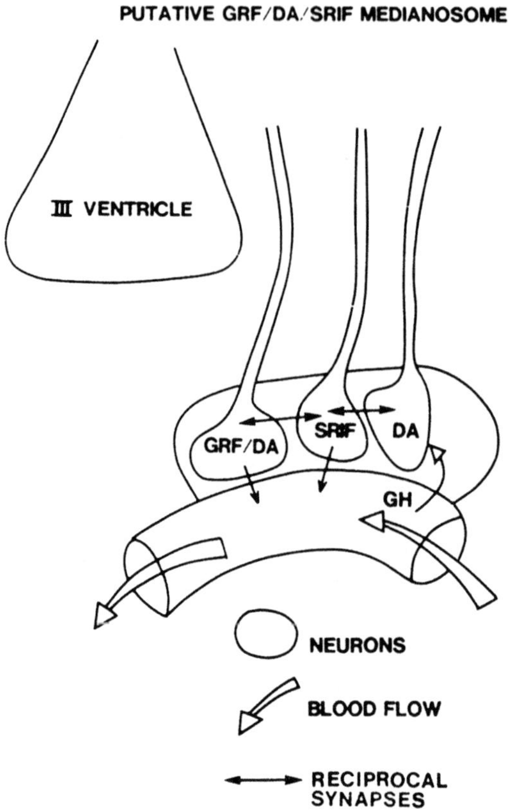

Fig. 1c. Schematic illustration of the medianosome controlling growth hormone (*GH*) secretion. (From Fuxe et al. 1986a). *SRIF*, somatostatin; *GRF*, growth hormone-releasing factor; *DA*, dopamine

defined aggregates of transmitter identified nerve terminals interacting with one other in the external layer of the median eminence" (Fuxe et al. 1986a, Fig. 1b). Probably, some GH receptors are located within the somatotropin release-inhibiting factor (SRIF), growth hormone-releasing factor (GRF)/DA and DA medianosomes (Fig. 1c). α_2-Adrenoceptors are also strongly involved in GH stimulation, but we do not yet know where these receptors are located in the brain (see Fuxe et al. 1986a for further references). But recent studies with inhibitors of adrenaline synthesis show that, as well as from the NA system, the central adrenaline system has a major function in the generation of the GH secretory rhythm (review Terry 1986, 1988). Other mechanisms with influence on GH secretion will be discussed later in relation to special clinical studies.

I should first like to discuss different catecholamine challenge tests with influence on GH, cortisol and prolactin which have been used in the past in affective disorders. Next, biological neuroendocrine differences between endogenous and nonendogenous depression and the question of state and trait markers for depression will be discussed. Some mechanisms influencing GH stimulation will then be examined which should be considered in the interpretation of certain GH challenge results. Finally, I will raise two topical problems in the relation between α-adrenoceptors and GH stimulation which are perhaps of some importance also for affective disorders.

1 Catecholamine Challenge Tests

1.1 Insulin Tolerance Test

The first GH stimulation test used in affective disorders was the insulin tolerance test (ITT). It is included in this chapter because the glucoreceptor involved in GH stimulation is an α-adrenoceptor. When this α-adrenoceptor is blocked with phentolamine, there is no GH stimulation after insulin. The GH response is dependent on the hypoglycemic response to insulin. In the late 1960s and in the 1970s many studies were carried out with the help of this test, all finding that particularly unipolar depressive patients, in comparison to controls and, in some studies, also to nonendogenous depressives, showed a blunted GH response (Mueller et al. 1969a; Endo 1970; Sachar et al. 1971; Endo et al. 1974; Gruen et al. 1975; Casper et al. 1977; Gregoire et al. 1977; Czernik et al. 1980; Czernik 1982). It seemed a reasonable assumption that a result confirmed by so many biological depression research groups from different countries would be a solid one. However, in 1982 two studies were published which showed no significant differences between endogenous depressive patients and controls (Koslow et al. 1982) or neurotic depressive patients (Berger et al. 1982). In Koslow's multicenter study, unipolar depressives were found to differ from bipolar depressive patients and controls in respect to the glucose decrease after insulin which had also been found before by Mueller et al. (1969b), Wright et al. (1978), and Czernik (1982). But this difference was not found in the study by Berger et al. (1982). Possibly the insulin insensitivity in depression is caused by the influence of cortisol on insulin receptor sites (Kahn et al. 1976; Kendler and Davis 1977; Flier et al. 1979). Nathan et al. (1981) found a significant association between cortisol secretion and insulin resistance in endogenous depressives. Thus, the results of the ITT in endogenous depression are perhaps not directly connected with the α-adrenoglucoreceptor but more with the insulin receptor insensitivity, which could explain the rate of glucose utilization (Mueller et al. 1969b), or insulin resistance, at least in a subgroup of endogenous depressive patients.

1.2 L-Dopa

L-Dopa is the precursor of the three catecholamines DA, NA and adrenaline. But serotonergic neurones, too, are influenced by Dopa administration (Chalmers et al. 1971, with further references).

Boyd et al. (1970) have shown that L-Dopa causes a significant GH stimulation, and a decrease of plasma prolactin level was found by Kleinberg et al. (1971) in man. The prolactin effect is probably due mainly to the DA mechanism, with influence exerted also by other neurotransmitter systems in the brain and at the pituitary level (Tuomisto and Männisto 1985). The GH response to L-Dopa is inhibited by phentolamine (Martin 1973), which indicates that mostly α-adrenoceptors are involved in GH stimulation. With the L-Dopa test depressive patients were for the first time found to have reduced GH response compared to healthy controls (Sachar et al. 1972), but later this difference was demonstrated to disappear, if "once the effect of the menopause is taken into account" (Sachar et al. 1975). Subsequently, Gold and Goodwin (1977) found with L-Dopa, more pronounced prolactin suppression and GH stimulation in bipolar depressive patients than in unipolar patients and healthy subjects. These results could not be confirmed by Linkowski et al. (1983), but in their study, too, GH response to L-Dopa was more blunted in postmenopausal than in premenopausal women. Basal prolactin was significantly lower in unipolar and bipolar premenopausal patients than in controls. Prolactin response to L-Dopa was less inhibited in postmenopausal than in premenopausal healthy subjects. Linkowski et al. (1983) conclude that there seems to be a dysfunction in subgroups of affective disorders and emphasize "the importance of considering the menopausal status in neuroendocrine studies of psychiatric disorders".

1.3 Amphetamine and Growth Hormone

First, we in our group began investigating the influence of amphetamine on GH secretion. As is known, amphetamine releases DA and NA from the respective nerve endings and inhibits the reuptake of these amines. It is therefore termed an indirectly acting psychomimetic drug. Since both DA and NA mechanisms are involved in the stimulation of GH, it was to be expected that amphetamine would lead to a GH stimulation in man. We found that endogenous depressives showed a significantly lower GH response to amphetamine than nonendogenous reactive depressives and healthy subjects (Langer et al. 1975, 1976). This result was interpreted as a dysfunction in DA and/or NA transmission in endogenous depressive patients. On the basis of today's knowledge we have to be very careful with our interpretations, since in our study, more of the endogenous depressive patients than of the reactive depressive group were postmenopausal. Later on, Arató et al. (1983) confirmed our results. They likewise found a significant difference between endogenous depressives and healthy subjects. But in their study, too, 6 of 20 female patients but only 2 of 13 controls were postmenopausal women. They showed that the GH response to amphetamine of the same patients was not correlated to the dexamethasone suppression test (DST) or to the thyroid-releasing hormone (TRH) thyroid-stimulating hormone (TSH) test. Halbreich et al. (1982), on the other hand, did not find with the amphetamine GH test a significant difference between healthy postmenopausal women and depressive postmenopausal women and other depressive patients, who mostly showed rather low GH maximum values (generally below 6 ng/ml). Neither did they find a significant difference between a group of five healthy young men and two male endogenous depressives. These male groups were too small for any conclusions to be drawn therefrom.

A very interesting study with dextroamphetamine was done by Nurnberger et al. (1982). These authors searched for behavioral excitation and assessed some neuroendocrinological and biochemical parameters after amphetamine in mono- and dizygotic twins and in "well-state" bipolar patients. In monozygotic twins the behavioral excitation, measured with a special rating scale (Nurnberger et al. 1981), was significantly correlated, as were baseline plasma MHPG, GH, and prolactin response (Nurnberger et al. 1981). Not concordant were cortisol, pulse rate, blood pressure and amphetamine plasma level; this was interpreted by the authors as resulting from environmental influences. Equally interesting was the finding that behavioral excitation was blocked by haloperidol without any influence on cortisol and GH secretion, showing the strong influence of DA on excitation response. In none of the parameters measured, however, was a significant difference found between bipolar depressives in remission and the healthy twins, indicating that a blunted GH response to amphetamine is not a trait marker for bipolar patients.

1.4 Amphetamine and Cortisol

A similar stimulatory drug, methylamphetamine, which also releases DA and NA, was used by Checkley and Crammer (1977) for the purpose of checking the cortisol and GH response in endogenous depressive patients during the depressive phase and after recovery. Cortisol secretion was shown to be significantly lower in the depressive phase than in the recovered state. There were no significant differences in GH stimulation, but the maximum median GH concentrations were 6 ng/ml during depression and 3.5 ng/ml after recovery, within the same range as in our above-mentioned study (Langer et al. 1976). We must, however, keep in mind that a rather high percentage of postmenopausal women were included in both studies. Later, using the same methodology, Checkley (1979) compared endogenous depressives with reactive depressives and with patients suffering from other functional psychoses and psychiatric illnesses. Only the endogenous group showed significantly reduced cortisol stimulation compared to the other groups. The authors interpreted their findings to indicate that the reduced cortisol stimulation in endogenous depressives was due to a functional deficiency at α-adrenergic receptors in these patients during the time they were depressed, "because only the cortisol response to methylamphetamine is blocked by thymoxamine, an α-adrenoceptor antagonist" (Rees et al. 1970). Checkley's (1979) findings concerning the cortisol response in endogenous depressives in comparison with healthy controls were confirmed later by Sachar et al. (1980) with dextroamphetamine (0.1 mg/kg i.v.) in patients fulfilling the Research Diagnostic Criteria (RDC) for major depressive disorders. These authors, in contrast to Checkley (1979), found not a blunted, but a decreased cortisol response to amphetamine in depressives, compared to cortisol stimulation in controls. They interpreted their results as follows: "Since NA normally inhibits hypothalamic corticotropin releasing hormone factor (and adrenocorticotropic hormone) secretion, a noradrenergic deficit may account for cortisol hypersecretion in depression; amphetamine may transiently "correct" this deficit in depressed patients, thereby reducing their cortisol secretion" (Sachar et al. 1980). Later, Sachar et al. (1985) performed the same investigation with 0.15 mg/kg dextroamphetamine in a larger group of 79 endogenous depressives and 64 normal adult volunteers and compared the DST results and cortisol plasma concentrations from 1 p.m. to 4 p.m. in the same patient. Cortisol response to amphetamine was the best test for depression, with sensitivity of 72% and specificity of 88%. Combining all three tests, Sachar et al. obtained a specificity of 100%. However, as often in biological psychiatric research, these results were not confirmed by others, in this case Feinberg et al. (1981). These authors, at 8 a.m., found neither a blunted nor a decreased cortisol response in endogenous depressives using the same dosage of amphetamine as by Sachar et al. (1980). Yet we have to consider that Feinberg et al. (1981) did *not* find an *increase* in cortisol with amphetamine in the nonendogenous group, as did Checkley (1979), or in normal subjects, as did Brown et al. (1978). Feinberg and coworkers tried to explain this discrepancy in terms of a lower dosage in their study. We have to keep in mind that the studies with depressive patients were done at different times of day. Checkley (1979) and Sachar et al. (1985) performed the test at 4 p.m., whereas Sachar et al. (1980) carried it out at 9.30 a.m. and Feinberg et al. (1981) at 8 a.m. We know that cortisol secretion, in particular, exhibits a strong circadian rhythm.

1.5 Apomorphine Test

The importance of disturbances in the dopaminergic system for affective disorders is still under discussion (Randrup et al. 1975; Randrup and Braestrup 1977). In order to check whether there is reduced postsynaptic DA receptor sensitivity in depressive patients, the apomorphine GH stimulation test has been used. Apomorphine is a rather specific DA agonist and thus could be expected to lead to GH stimulation, as DA neurons, too, are involved in GH release (Fig. 1 a). Moreover, prolactin release is also regulated by dopaminergic mechanisms (Brown et al. 1979).

Three studies with 0.75 mg apomorphine were published in the 1970s. None of them found a significant difference between depressive patients and healthy controls in GH response to apomorphine (Frazer 1975; Casper et al. 1977; Maany et al. 1979). Later, too, Meltzer et al. (1984), working with 0.75 mg apomorphine, found no significant difference in GH response between major depressive disorder patients and healthy controls. However, in a large group of patients with different psychotic disorders, compared to controls, these authors found that the degree of prolactin suppression was significantly correlated to depression and anxiety, the connection being even more pronounced in patients with affective disorders. There was a significant inverse correlation between prolactin baseline levels and GH response to apomorphine.

Furthermore, Jimerson et al. (1984) likewise found no difference in GH response to apomorphine (0.75 mg) between patients with major depressive disorders and healthy volunteers. These authors, in contrast to the above-mentioned study, found lower baseline and, in agreement therewith, lower postapomorphine prolactin levels in depressives than in healthy control subjects. They could, however, find no difference in percent prolactin decrease between patients and controls. Furthermore, chronic treatment for about 26 days with the dopaminergic agonist piribedil did not influence neuroendocrine responses in either group. Jimerson et al. concluded that "decreased baseline prolactin levels in depressed patients do not reflect altered postsynaptic receptor sensitivity in the tuberoinfundibular dopamine system". The baseline prolactin levels in depressed patients were, very controversially, either decreased, increased, or normal (review Jimerson et al. 1984). The authors emphasize the importance of the menstrual status also for prolactin studies – this factor had not been taken into account in many previous studies.

With 0.5 mg apomorphine, however, Ansseau et al. (1984) found a blunted GH response in five of eight primary depressive patients and somewhat higher GH response in another four secondary depressives. And in a new study with the same dosage, seven major depressives showed a significantly lower GH response than seven minor depressives (Ansseau et al. 1987). These authors are of the opinion that lower apomorphine dosages are necessary to uncover dopaminergic disturbances. With 5 ng/kg apomorphine, a rather low dose, three of eight endogenous depressives also showed a blunted GH response (Corn et al. 1984 b). Further studies are needed to prove whether or not a subgroup of depressive patients have reduced postsynaptic DA receptor sensitivity that can be distinguished by means of the GH response to low apomorphine doses.

1.6 Desmethylimipramine Test

It has long been well known that desmethylimipramine (DMI) blocks NA uptake in the nerve endings. Laakmann et al. (1977, 1986 a) of our clinic demonstrated that DMI and other tricyclic antidepressants show dose-dependent GH stimulation which is blocked by yohimbine, an α_2-, and phentolamine, an α_1- and α_2-adrenoceptor antagonist, and potentiated by β-blocking agents. On the other hand, prazosin, a rather specific α_1-antagonist, has no effect. The authors therefore conclude, correctly, that the GH stimulation after DMI is mediated by α_2-postsynaptic adrenoceptors. In the DMI stimulation test only male, not female, endogenous depressive patients showed significantly reduced GH maxima compared to neurotic depressives and controls (Laakmann et al. 1986 b). In this study menopausal status was not considered. Similar results were found by Sawa et al. (1982), but the menopausal status was not mentioned in their study either. On the other hand, Meesters et al. (1985) found only in premenopausal females a significant difference between endogenous and nonendogenous depressive patients in comparison to controls, but not in males. The lack of agreement between the results obtained by Laakmann et al. (1986 b) in males and those of Meesters et al. (1985) may be due to the fact that different diagnostic systems (RDC vs ICD) were applied. Furthermore, in the study by Laakmann et al. (1986 b) only endogenous patients with three previous phases were included. Moreover, in a small group of patients, five depressive compared to three euthymic patients showed a blunted GH response to DMI (Calil et al. 1984). Interestingly, the cortisol response to DMI of endogenous depressives was significantly decreased compared to normal controls (Asnis et al. 1986), similar to the above-mentioned amphetamine cortisol test (Sect. 1.4).

1.7 Clonidine Test

On the basis of our amphetamine study mentioned above, showing a blunted GH stimulation in endogenous depressives, we thought that the blunted response might be due to a catecholamine (CA) deficit in the nerve endings or to reduced CA receptor sensitivity. For the purpose of assessing the CA receptor sensitivity we chose clonidine, a preferentially α_2-adrenoceptor agonist acting on pre- and postsynaptic adrenoceptor sites (review Kobinger 1979). Today we must assume that the GH stimulation after clonidine is regulated via postsynaptic α_2-adrenoceptors, because α_2-antagonists, such as piperoxane or yohimbine, completely block GH stimulation to clonidine in baboons (McWilliam and Meldrum 1983) and in rats (Krulich et al. 1982), whereas prazosin, an α_1-adrenoceptor blocker, had no such effect in baboons. In dogs, presynaptic α_2-adrenoceptors might regulate GH stimulation with clonidine (Cella et al. 1984).

1.7.1 Endogenous Depressives Versus Healthy Controls

In clonidine stimulation tests, endogenous depressive patients showed a blunted growth hormone response compared to normal controls and nonendogenous depressive patients (Matussek et al. 1980). The clonidine GH results of endogenous depressives and healthy subjects have been confirmed by seven other studies with different dosages and application forms (Table 1). An analysis of variance in 45 depressed patients has shown that "impairment of GH response to clonidine is associated with measures of endogeneicity of depression, and not with a range of other patient variables," such as sex, age, baseline plasma GH level, length of drug-free period, weight, weight loss, anxiety, and severity of depression as measured by the total Hamilton depression score (Checkley and Corn 1986; Checkley et al. 1986).

All groups finding a blunted GH response to clonidine challenge interpreted this effect with a reduced postsynaptic α_2-adrenoceptor sensitivity. However, the lower sensitivity might just as well have been caused by various other mechanisms influencing α_2-adrenoceptors. This point will be discussed later (Sects. 4.1–4.7).

Table 1. Clonidine GH stimulation studies in depressive patients and controls

Endogenous depressives (*n*)	Controls (*n*)	Dose and route	Authors
10	32	150 μg i.v.	Matussek et al. (1980)
10	10	2 μg/kg i.v.	Checkley et al. (1981)
19	20	2 μg/kg i.v.	Siever et al. (1982a)
15	12	5 μg/kg i.v.	Charney et al. (1982)
12	12	150 μg i.v.	Boyer et al. (1982)
50	46	2.5 μg/kg i.v.	Lechin et al. (1985a, b)
16	24	2 μg/kg i.v.	Höhe et al. (1986)
11	11	2 μg/kg i.v.	Uhde et al. (1986)

All studies are listed in which patients with endogenous depression or major depressive disorders were compared with control subjects. In these studies only those patients were included who had been free of medication for at least 14 days prior to the investigation. In every study GH stimulation was significantly lower in endogenous depressivs than in controls

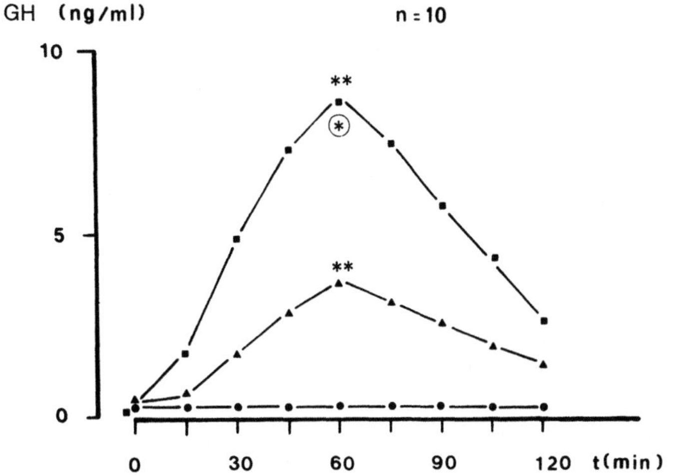

Fig. 2. Dose-response curve of growth hormone (*GH*) stimulation to 1.5 μg/kg body weight (b.w.) i.v. (▲) and 2 μg/kg b.w. i.v. (■) clonidine in comparison to placebo (○) in healthy male subjects. The test was carried out 3 times in the same subject. ⊛$p<0.05$, 2.0 μg/kg b.w. clonidine vs placebo; ** $p<0.01$, 1.5 μg/kg b.w. vs 2.0 μg/kg b.w. clonidine. There was no significant difference between placebo and 1.5 μg/kg b.w. clonidine

In this context it is curious, and we should keep the fact in mind when making interpretations, that in a preliminary study, guanfacine, an even more specific α_2-adrenoceptor agonist than clonidine, showed no difference in GH response between 11 depressive patients (DSM-III: major depressive disorder) and a sex- and age-matched group of controls (Eriksson 1985). I will come back to this point in my summary (Sect. 6).

There are some methodological comments I would like to make in connection with the clonidine GH stimulation. Two routes of clonidine administration are possible, oral or i.v. (Hunt et al. 1986 with a review of 14 different clonidine studies). As was to be expected, GH peak values appeared 90 min after oral administration, but 45–60 min after i.v. administration.

Following a dose-response curve drawn on the basis of these results, we preferred i.v. application of 2 μg/kg (to be published; Fig. 2). We must, however, bear in mind that GH response to clonidine under the same experimental conditions is not an absolutely stable marker, even in young healthy male subjects. Of 12 such subjects tested by us four times at 3-week intervals, only five always showed a GH response of >5 ng/ml (Fig. 3), whereas five had a rather instable GH response; for example, twice >5 ng/ml and twice <5 ng/ml (to be published). In our view, we have to be very careful in interpreting GH response to clonidine on the basis of only one test. We do not yet know the reason for this instability in healthy male subjects.

Likewise, the blunted GH response to clonidine in endogenous depressives does not seem correlated to changes in the circadian rhythm, because, at least between 9 a.m. and 6 p.m., there was no difference in GH stimulation to clonidine;

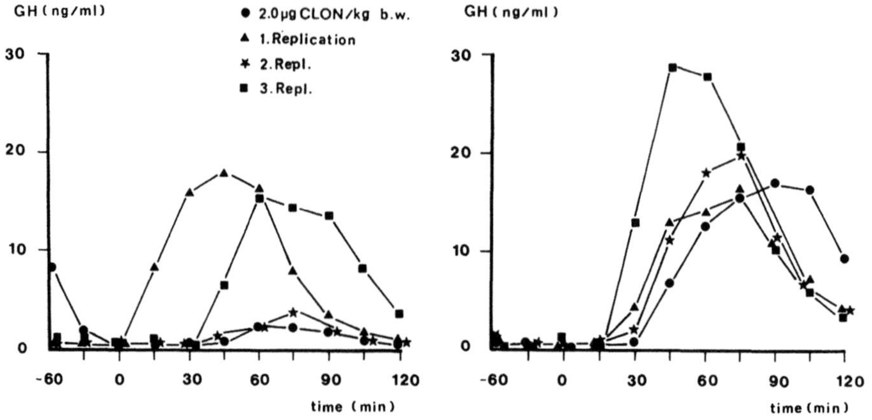

Fig. 3. Four clonidine growth hormone (*GH*) stimulation tests (2.0 µg/kg b.w. i.v.) in two young healthy male controls at 3-week intervals

at 9 p.m. much higher baseline levels were found and thus no stimulation could be calculated. The clonidine effect on blood pressure and alertness was also similar at the three times studied (Honer et al. 1984).

1.7.2 Clonidine Test in Other Psychiatric Illnesses

There are a number of findings equally indicating that the clonidine GH stimulation test is not specific for endogenous depression, since schizoaffective patients (Matussek et al. 1980) and patients with obsessive-compulsive (Siever et al. 1983) and panic (Uhde et al. 1986) disorders, i.e., with illnesses that are related to depression, have been shown also to have a blunted GH response to clonidine. In another study, panic disorder patients were shown to have a higher GH response to clonidine than depressive patients (Schittecatte et al. 1988). In my view, it is most important that unmedicated manics, like major depressives, show a significantly reduced GH response to clonidine compared to minor depressives (Ansseau et al. 1987). In an excellent study and review, various biological markers, such as DST, sleep EEG, GH response to clonidine, and platelet ^{3}H-imipramine binding, were compared in obsessive-compulsive and affective disorders (Insel et al. 1984). It was shown that "each of the studies suggested similarities between patients with obsessional disorder and those with affective illness," and the authors conclude, correctly, that "the two syndromes have 'something' in common. Just what that 'something' is, remains to be discovered". Female migraine patients (Weizmann et al. 1984) as well as acute schizophrenics (Matussek et al. 1980; Ackenheil et al. 1985) display a particularly high GH response to clonidine. Acute neuroleptic treatment does not alter the GH stimulation in schizophrenic patients, but long-term neuroleptic treatment significantly reduces it (Ackenheil et al. 1985). A normal GH response to clonidine was found in anorexia nervosa (Brambilla et al. 1987), whereas malnourished patients (Kaye et al. 1985, with further references) and starved healthy subjects (Fichter and Pirke 1986) showed elevated GH baseline values. Further studies will be needed to evaluate the disease specificity of the clonidine GH stimulation test.

1.7.3 Comparison with Other Neuroendocrine Tests

Today other biological markers are also being discussed with regard to depression, e.g., the DST, the TRH-TSH test and REM latency. However, in most of the studies done so far, only one test was applied, since one cannot leave a patient who has a marked depressive syndrome without treatment for too long. In addition, some tests, like the TRH-TSH test and also the DST, may influence GH stimulation. Ansseau et al. (1984), in a study mentioned above in connection with apomorphine, applied four biological markers in the same patient in a group of 12 endogenous depressives including eight primary and four secondary (Table 2) depressions. The sensitivity with regard to apomorphine was 42%, to dexamethasone suppression 50%, to REM latency 67% and to clonidine-induced GH stimulation 75%. The authors found no correlation between clonidine or apomorphine challenge and nonsuppression after dexamethasone or shortened REM latency.

In another study, GH responses to clonidine and apomorphine were compared in the same patient. The authors (Corn et al. 1984 b) used doses of clonidine (1.3 μg/kg) and apomorphine (5 ng/kg) which induce similar GH responses in normal subjects. In eight drug-free patients with endogenous depression (six females, some of them probably postmenopausal) the GH response to clonidine was significantly smaller than to apomorphine. The authors conclude a "defect in the adrenergic, but not in the dopaminergic regulation of GH in patients with endogenous depression".

Table 2. Summary of results of four biological markers for 12 endogenous depressives: eight primary (nos. 1–8) and four secondary (nos. 9–12). (From Ansseau et al. 1984)

Patient number	Nights with RL 50 (%)	Shortest RL (min)	DST	CLO	APO	No. of abnormal markers
1	100	0	S	–	–	3
2	25	27	NS	–	–	4
3	75	0	NS	–	–	4
4	0	80	S	–	–	2
5	67	2.5	NS	–	–	4
6	33	9.5	S	–	+	2
7	50	15.5	NS	–	+	3
8	33	17	NS	–	+	3
9	0	52	S	+	+	0
10	0	58.5	S	–	+	1
11	0	51	NS	+	+	1
12	100	29	S	+	+	1

RL, REM latency; DST, dexamethasone suppression test; S/NS, suppression/nonsuppression in DST; CLO, clonidine test; APO, apomorphine test; +/–, normal/blunted response

Furthermore, Siever et al. (1985) compared clonidine with amphetamine in the same patient in a group which included eight premenopausal females with major depressive disorders, some of them with secondary depression. There was a significant correlation between GH response to clonidine and amphetamine in the group. Six of the eight subjects had a blunted GH response to clonidine and three to amphetamine. The authors conclude that "clonidine may provide a more specific index of the responsiveness of the α_2-adrenergic receptor". α_2-Receptors seem also to be responsible for GH stimulation after amphetamine, because DA blockade with neuroleptic drugs did not influence GH response to amphetamine in men (Nurnberger et al. 1982).

2 Neuroendocrine Differences Between Endogenous and Nonendogenous Depressive Patients

Even a psychiatrist with long experience is often not able at the first manifestation of a depressive syndrome, based only on the symptomatology, to differentiate between endogenous and nonendogenous depression. In my point of view, however, this is a very important question for every patient as regards both treatment and prognosis.

In this context, I would like to emphasize the differences found between endogenous and nonendogenous depressive patients in various neuroendocrinological studies, i.e., in the methylamphetamine cortisol test and the amphetamine, clonidine and DMI GH stimulation tests and also in the ITT (Table 3). The results of the different tests are not uniform. With the clonidine GH stimulation test, significant differences between the two groups of depressive patients were found by three groups (Matussek et al. 1980; Boyer et al. 1982; Checkley et al. 1984), but not by Dolan and Calloway (1986) and Katona et al. (1986). But in the study by Dolan and Calloway (1986) a TRH-TSH test had been carried out 3 h prior to the clonidine GH stimulation test, and this may have influenced the GH response to clonidine. It is known that TRH in lower concentrations has a stimulating effect and in high concentrations an inhibitory effect, on GH release depending on the concentration of growth hormone-releasing factor (GRF) (Denef et al. 1986). It is most interesting that, in the study of Katona et al. (1986), dexamethasone non-suppressors showed a significantly lower GH response than dexamethasone suppressors.

Impressive are the findings of Laakmann et al. (1986 b) with the DMI test, because in this study only, male endogenous and nonendogenous depressive patients were compared to an age- and sex-matched group of healthy subjects. In this sample the estrogen status cannot be responsible for the difference between both groups. Likewise, the mean GH maxima of nonendogenous depressives were found to be higher, though not significantly so, than in healthy subjects. This was often seen also in other GH stimulation tests, such as those using insulin, clonidine, or amphetamine. Similar results, obtained in a smaller group, were reported by Sawa et al. (1982). I cannot explain with certainty why Meesters et al. (1985) did not find a difference between endogenous and nonendogenous depres-

Table 3. Neuroendocrine studies comparing major depressive disorder or endogenous with nonendogenous depressive patients

Endogenous		Non-endogenous		Signif-icance[a]	Dose and route	Author(s)
n	(F)	*n*	(F)			
1. Insulin tolerance test						
6	(6)	5	(1)	*	0.1 IU/kg s.c.	Mueller et al. (1969a)
24	(24)	24	(24)	*	0.1 IU/kg s.c.	Czernik (1982)
10	(10)	19	(10)	n.s.	0.1 IU/kg s.c.	Berger et al. (1982)
2. Amphetamine GH stimulation test						
9	(5)	7	(4)	*	0.1 mg/kg i.v.	Langer et al. (1976)
3. Methylamphetamine cortisol stimulation test						
20	(14)	6	(3)	*	15 mg/kg i.v.	Checkley (1979)
4. DMI GH stimulation test						
13	(?)	4	(?)	*	50 mg i.m.	Sawa et al. (1982)
30	(0)	10	(0)	n.s.	75 mg i.m.	Meesters et al. (1985)
10	(10)	8	(8)	n.s.	75 mg i.m.	Meesters et al. (1985)
15	(15)	12	(12)	n.s.	75 mg i.m.	Meesters et al. (1985)
16	(0)	11	(0)	*	75 mg i.m.	Laakmann et al. (1986b)
5. Clonidine GH stimulation test						
10	(8)	12	(7)	*	150 μg i.v.	Matussek et al. (1980)
12	(9)	12	(9)	*	150 μg i.v.	Boyer et al. (1982)
10	(3)	10	(3)	*	1.3 μg/kg i.v.	Checkley et al. (1984)
7	(?)	7	(?)	n.s.	5.0 μg/kg p.o.	Dolan and Calloway (1986)
7	(4)	7	(4)	*	150 μg/kg i.v.	Ansseau et al. (1987)
12	(?)	11	(?)	n.s.	1.3 μg/kg i.v.	Katona et al. (1986)

[a] * significant; n.s. not significant
[b] Premenopausal women
[c] Postmenopausal women

sive males, in contrast to Laakmann et al. (1986b), but in Laakmann's study the endogenous depressives had at least three previous depressive phases or were bipolar. Meesters et al. (1985) made their diagnoses according to the RDC and did not mention the number of depressive phases. They found lower GH responses to DMI in the endogenous groups of pre- and postmenopausal females than in the nonendogenous depressives, but this difference did not reach statistical significance.

As regards the GH response in the ITT, endogenous patients were also found to have blunted stimulation and nonendogenous depressives in contrast, often to have a much higher GH stimulation (Mueller et al. 1969a; Czernik et al. 1980). We should remember (Sect. 1.1) that in both studies endogenous depressives showed insulin resistance compared to nonendogenous depressives. Why Berger et al. (1982) found no significant difference between the two groups of depressive patients is not clear. Besides the ICD-8 and RDC systems, they also used for diagnosis the Newcastle scale, which should differentiate between endogenous and nonendogenous depressives. In addition, they mentioned the episode length, but not the number of depressive phases. In a small group of six bipolar patients and four secondary depressives, the prolactin response to insulin-induced hypoglycemia was significantly reduced in the primary, but not in the secondary depressives (Grof et al. 1983).

Although in the different tests up to now there is no complete agreement about the results, differentiation between endogenous and nonendogenous groups of depressives appears – at least to me – a very important topic, perhaps giving hints as to essential biological differences between the two groups of depressive patients. For nonendogenous depressive patients often show much higher GH stimulation after clonidine, DMI or in the ITT than controls. This can perhaps be explained by higher sensitivity of the postsynaptic α-adrenoceptor caused by a lower presynaptic NA release. Ashcroft et al. (1972) were the first to differentiate, on theoretical considerations, between a lower output depression and a lower sensitivity depression, i.e., between presynaptic and postsynaptic depression, since both a pre- and a postsynaptic defect could induce the same behavioral disturbances. Thus, nonendogenous depression could be a lower output or presynaptic depression and endogenous depression a lower sensitivity or postsynaptic depression (Matussek 1980). However, presynaptic depression could theoretically be caused by an NA deficit in the storage pool, such as after reserpine and similarly acting drugs, or by a more sensitive presynaptic α_2-adrenoceptor, which could also be responsible for a lower NA release. On the basis of these results and speculations, I would like to emphasize that in all biological studies with depressive patients it is most important to differentiate between endogenous and nonendogenous depressives. Thus, in order to arrive at the right diagnosis, especially for monopolar patients, it seems to me vital also to consider the course of the illness and the number of depressive phases. This has been pointed out before by many famous psychiatrists, e.g. Kraepelin, but these factors are given no consideration in the DSM-III and some other diagnostic systems.

3 Neuroendocrine Response: State or Trait Marker?

The blunted cortisol response to methylamphetamine found by Checkley and Crammer (1977) appears to be a state-dependent marker. Of the other amphetamine-cortisol studies mentioned above, none is a systematic investigation of this question. But since the DST and hypercortisolemia are state-dependent in depressive patients, the disturbances found in such studies should disappear after remission of the depressive phase.

Another question is whether the reduced GH responses in endogenous depressives to insulin, amphetamine, DMI and clonidine are either state- or trait-dependent. The results of the ITT show that after improvement of the endogenous depressive phase there is generally an increase in GH response (Mueller et al. 1969a; Endo et al. 1974; Czernik 1982), probably because the insulin receptor insensitivity disappears. In contrast, neurotic-reactive depressive patients with a higher GH response in the ITT were, with improvement of the depressive symptomatology, found to have a decreased response to insulin (Mueller et al. 1969a; Czernik 1982) as well as to DMI (Laakmann 1980). The same result was found in two of three nonendogenous depressive patients in the clonidine GH stimulation test who had rather high GH maxima in the depressive phase (Matussek et al. 1986). This reduction of GH maxima during treatment may be explained by drug effects (see below).

The GH response to amphetamine, at least in bipolar patients in the free interval, shows normal GH values (Nurnberger et al. 1982), whereas in our earlier study (Langer et al. 1976) two male unmedicated unipolar depressive patients showed a blunted GH response also in the free interval.

On the other hand, in most of the studies done so far with the clonidine GH stimulation test, the blunted GH response in endogenous depressives was still found after different antidepressant treatment, independent of improvement (Slade and Checkley 1980; Siever et al. 1982b; Charney et al. 1982b, 1984; Corn et al. 1984a; Möller et al. 1984; Price et al. 1986). Only in two studies with depressive patients (Glass et al. 1982; Matussek et al. 1986) and in one study with DMI treatment in healthy subjects was an increased GH response to clonidine found after treatment (Corn et al. 1984c). As yet, it is not possible to explain these divergent results. From animal experiments we would expect a reduced GH response to clonidine after longer treatment with most antidepressants, since chronic treatment with most of the antidepressant drugs leads to a decrease of α_2-adrenoceptor sensitivity. This is not the case with DMI (Maj et al. 1984); this was also shown in a study by Brambilla et al. (1987), in which DMI treatment for 30 days did not reduce GH response to clonidine. On the other hand, we know that many antidepressant drugs, as well as electroconvulsive treatment lead to a β-adrenoceptor downregulation (Sulser 1982). When β-adrenoceptors are blocked, however, the GH stimulation increases (Brown et al. 1979). This mechanism will be discussed in more detail later (Sect. 4.5).

In their studies with antidepressant treatment inducing no change in the GH response to clonidine, Charney et al. (1984) and Corn et al. (1984a) mentioned that the blunted GH response to clonidine might be a trait marker for endogenous depression. But in order to give a clear answer to the question of whether it is either a state or trait marker, the clonidine test must be applied to unmedicated endogenous depressive patients in the free interval. Yet, after the introduction of the effective prophylactic lithium treatment for uni- and bipolar depressive patients, it is difficult to find unmedicated endogenous depressive patients in the free interval, particularly so since lithium treatment has to be excluded because it reduces the GH response to clonidine (Catalano et al. 1984). In our group we have so far tested 15 endogenous depressives without any drug for at least 14 days, and only one patient showed a GH response of >5 ng/ml (see also preliminary report, Höhe et al. 1986) (Fig. 4). Siever et al. (1986) reported that in six of eight endogenous depressive patients there was also a blunted GH response to clonidine (<5 ng/ml). These authors, too, consider this to be a trait marker for endogenous or major depressive disorder. In this context it is of great interest that in a very careful longitudinal study done in Pittsburg, USA, during sleep, patients with major depression (DSM III) as compared to controls were found to have significantly lower GH values in the first part of the night, also in remission (Jarrett and Kupfer, personal communication). In my opinion, however, further studies with more drug-free patients unmedicated for longer periods are necessary to answer this question. In addition, studies in families with affective disorders and in con-

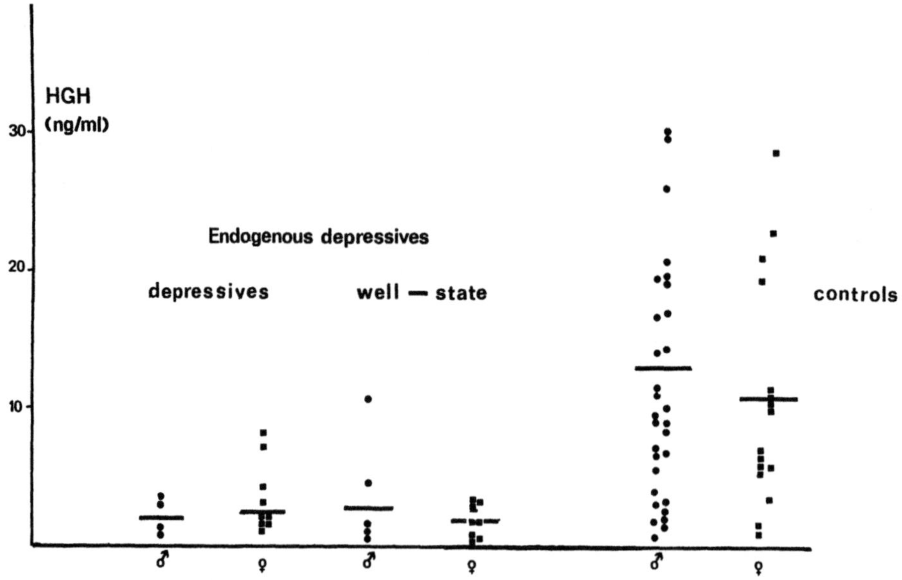

Fig. 4. Growth hormone (*GH*) maxima to clonidine challenge of 13 endogenous depressives in the depressive phase (*left*), 15 endogenous depressives in the free interval (*well-state*; *middle*) and 47 healthy controls (*right*). The patients had been free of medication, including lithium, for at least 14 days, free interval patients even longer. They were bipolar or had at least two previous depressive phases. All females were premenopausal. The test was done between the 10th and the 20th day of the menstrual cycle. The difference between controls and endogenous depressives was significant ($p<0.001$) (Matussek and Höhe, in preparation)

cordant and discordant monozygotic twins with endogenous depression will perhaps help to clear up this problem.

Instead of defining a blunted GH response to clonidine as a trait marker for endogenous depression, I would rather speak of a vulnerability or risk marker with regard to depression. A blunted GH response to clonidine is often found in postmenopausal women who have frequent depressions and in people with regular alcohol intake who are not alcoholics (Matussek et al. 1984), and about 25% of young psychically healthy volunteers display the same result. From the work of Winokur (1986) we know that a close relationship exists between alcoholism and depression.

4 Influence of Other Mechanisms on GH Stimulation and α_2-Adrenoceptors

4.1 Estrogens and Age

From many studies we have known for some time that estrogens exert a strong influence on GH stimulation. In the postovulatory stage of the menstruation cycle GH response to the different stimuli is more pronounced than during menstruation (Frantz and Rabkin 1965; Merimee and Fineberg 1971; Genazzani et al. 1975). In animal experiments with clonidine and GRF the influence of sex steroids in GH stimulation was clearly demonstrated (Eriksson and Modigh 1984; Eriksson 1985). We, too, could show in the clonidine GH stimulation test that during the menstruation period there was a blunted and during the ovulatory stage a significantly higher GH response to clonidine (Matussek et al. 1984). Furthermore, we found, with a mood self-rating scale (v. Zerssen 1981), that the score during the menses was significantly lower than at ovulation, which indicates that during menstruation our females felt slightly better. Actually, we do not know whether the GH response in the clonidine or other GH stimulation tests may be correlated to the premenstrual tension syndrome. Considering all the studies mentioned, it seems absolutely necessary not to do a GH stimulation test during the time of menstruation. Halbreich et al. (1982) likewise emphasized the importance of the menstrual cycle in connection with the GH response. Yet in many GH stimulation studies done with females, this stricture has not been properly observed, and this seems to me one important reason for the discrepancy in the results of some neuroendocrine GH stimulation studies.

Another important aspect regarding GH stimulation tests in females, apart from the menstruation cycle, is the postmenopausal stage. It is well known that the GH response in elderly subjects to physiological and pharmacological stimuli is reduced as regards clonidine, in females at an earlier age than in males (Gil-Ad et al. 1984). Likewise, Siever et al. (1982a) and we (Matussek et al. 1980) were able to show a significantly reduced GH response to clonidine in postmenopausal compared to premenopausal females. On the other hand, Siever et al. (1982a) found in postmenopausal endogenous depressive females a GH response to clonidine significantly lower than in postmenopausal controls. Laakmann et al. (1986b) found that the response in the DMI GH stimulation test also decreased with age, the age dependency in females not so clearly demonstrable as in healthy male subjects. In view of the relation of GH response to α_2-adrenoceptor sites it is interesting that brain adrenergic and dopaminergic receptor binding in rats (Misra et al. 1980), and also platelet α_2-adrenoceptors in humans, show an age-dependent decrease (Brodde et al. 1982). In many of the above-mentioned GH stimulation studies postmenopausal women were included or postmenopausal status was not mentioned. In one study with the ITT, for instance, the female patients must have been postmenopausal (Koslow et al. 1982). I do not think this is the correct methodology for GH stimulation tests, because age and sex hormone status reduce GH stimulation in postmenopausal women. Therefore, only premenopausal women were included in the female group of our clonidine interval study and the challenge test was carried out between the 10th and 20th day of the menstruation cycle.

Whether the above-mentioned lower GH response to clonidine in females than in males (Gil-Ad et al. 1984) is correlated to a higher vulnerability for depression in females requires further investigation. Recent epidemiological studies also reveal a much higher rate of major depression and dysthymia in women than in men, as well as with increasing age (Weissman et al. 1986; Klerman 1986).

4.2 Cortisol

We have seen above that cortisol hypersecretion is probably responsible for the insulin insensitivity in endogenous depressives, since cortisol influences the insulin receptor (Sect. 1.1). Thus, the question arises of whether in other GH stimulation tests, too, hyperactivity of the HPA axis might be responsible for the reduced GH response in endogenous depression. Adrenergic nerves innervate corticotropin-releasing factor (CRF)-immunoreactive cell bodies in the medial parvocellular part of the paraventricular nucleus of the hypothalamus (Fuxe et al. 1976; Agnati et al. 1985), which "probably have an inhibitory role in the regulation of CRF release" (Härfstrand et al. 1986). When this adrenergic mechanism is blocked with phenylethanolamine-*N*-methyl transferase (PNMT) inhibitors, the result is a marked increase of corticosterone secretion (see Fuxe et al. 1979a; Roth et al. 1981), probably through stronger CRF release. Furthermore, CRF centrally inhibits the release of GRF in the rat (Rivier and Vale 1984). These results demonstrate a strong inhibitory influence of the HPA axis on GH secretion. But on the other hand, corticosterone shows very interesting opposite effects on α_2-adrenoceptors within the hypothalamic paraventricular and the supraoptic nucleus with a diurnal rhythm of up- and down-regulation (Jhanwar-Uniyal and Leibowitz, 1986, with further references).

Some clinical studies also point in the same direction. Recently it was demonstrated that eight patients with major depression showed significantly higher ($p<0.001$) CRF plasma levels than ten normal controls (Widerlöv et al. 1986). Fairly often, endogenous depressive patients in particular have high plasma cortisol levels (Halbreich et al. 1985a, b) and higher urinary free cortisol excretion (v. Zerssen et al. 1984). But in the clonidine and DMI studies in which both GH and cortisol plasma levels were measured (Matussek et al. 1980; Lechin et al. 1985a, b; Laakmann, personal communication), including the first publication of the clonidine GH stimulation test in healthy subjects (Lal et al. 1975), no negative correlation was found between GH and cortisol plasma levels or the DST (Ansseau et al. 1984), either in depressive patients or in controls.

But we must keep in mind that Dolan and Calloway (1986) found a significant negative correlation between the GH stimulation after clonidine and free cortisol in a 24 h urine collection period (Sect. 2). With the 24 h collection period the hyperactivity of the HPA axis can be better recorded than with some plasma cortisol levels. Hence, studies determining 24 h cortisol excretion or, better, CRF plasma concentration should be done in connection with measuring GH levels.

A very interesting study (Siever et al. 1984) hints at the close relation between GH stimulation with clonidine and the HPA axis. The authors found a sixfold higher cortisol decrease after clonidine in patients with major depressive disorders than in healthy subjects. In this study, there is a significant, negative correlation between baseline plasma cortisol levels and the reduction of the plasma concentration of MHPG. The authors speculate about "the possibilty that abnormalities in the responsiveness of the α_2-noradrenergic system may be associated with the HPA dysfunction found in depressed patients. The noradrenergic inhibitory influence may be apparent only in the subjects with a hyperactive HPA axis observed in most of our depressed patients". But in our clonidine study, too (Matussek et al. 1980), the endogenous depressives showed a more pronounced cortisol decrease than did healthy subjects. This difference, however, attained no statistical significance.

Furthermore, it has been shown that with increasing age there is a highly significant increase of 24 h plasma cortisol levels in women which is not found in men (Halbreich et al. 1984). Is this higher cortisol level then correlated to the reduced GH response to clonidine and DMI with age, as demonstrated above, and to a higher vulnerability in females with age (Klerman 1986; Weissman et al. 1986)?

On the basis of all the results mentioned we have to reflect upon the possibiltiy that hyperactivity of the HPA axis might be responsible for or associated with the blunted GH response in the different challenge tests. But according to the clinical studies with depressive patients done so far, HPA hyperacivity seems to be a state rather than a trait marker for depression. However, as discussed above (Sect. 3), at least the blunted GH response to clonidine is probably a trait marker, because in the free interval GH response is still blunted and HPA hyperactivity cannot be assumed.

But there is no question that the very interesting relationships between GRF and CRF mechanisms have to be cleared up in both animal and clinical studies. In the regulation of these two neuroendocrine systems neuropeptide Y is also strongly involved (Härfstrand et al. 1986); I will come back to this later (Sects. 4.7 and 5).

4.3 Opioids

Above, when interpreting the blunted GH response to clonidine and DMI in endogenous depressives, I mentioned that a reduced sensitivity of α_2-adrenoceptors or *of structures that are related to them* might be responsible for this. Recently we have learned, for instance, that α_2-adrenoceptors are probably very closely connected to μ-opiate receptors (Aghajanian 1985). Most likely this is the explanation for the efficacy of clonidine treatment after opiate withdrawal. Furthermore, Eriksson et al. (1981) showed in rats that morphine stimulates GH release and activates GH response to clonidine. But it is also very interesting that the morphine-enhanced GH secretion is completely blocked by adrenaline synthesis inhibition (Terry 1986, 1988), indicating again the importance of the central adrenaline system for GH secretion.

In view of the close connections between α_2-adrenoceptors and μ-opiate receptors we investigated the effect of opiates on GH, prolactin, and cortisol secretion as well as on NA and adrenaline plasma levels in healthy volunteers and depressive patients (Höhe et al. 1984; Höhe 1986). Especially with fentanyl (0.1, 0.2, 0.25 μg/70 kg), a rather specific μ-opiate agonist, we obtained in healthy volunteers a dose-dependent increase of prolactin and GH secretion, as well as, for the first time, a marked, dose-dependent increase of NA plasma levels. In a preliminary study, 16 endogenous depressives showed a significantly lower GH response to fentanyl than 15 healthy controls (Fig. 5; to be published). If this result could be further confirmed by us and other groups we would have to assume that μ-

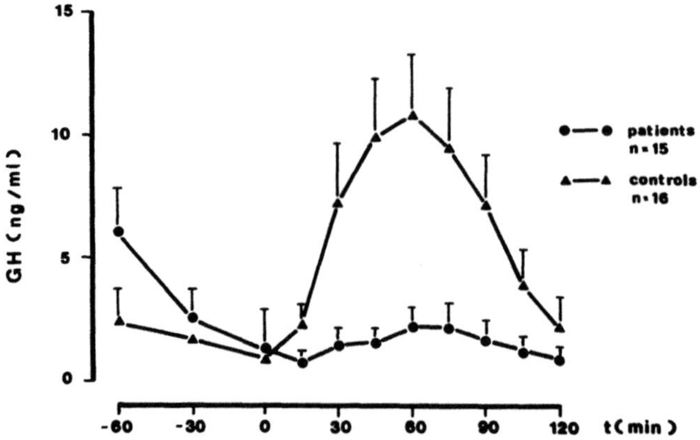

Fig. 5. Course of growth hormone (*GH*) plasma concentration after fentanyl (0.2 mg/70 kg) in endogenous depressive patients and controls

receptor subsensitivity could also be responsible for the blunted GH response to clonidine. Furthermore, it would be important to clarify whether the reduced GH response to fentanyl is either state- or trait-dependent.

4.4 Cholinergic Mechanisms

α_2-Adrenoceptors and cholinergic neurons also have an interesting relationship (review Vizi 1983). Some time ago, Janowsky et al. (1972) postulated a cholinergic hyper- and noradrenergic hyposensitivity in depression and, vice versa, in manic states (see Janowsky in this volume). In this connection it is remarkable that Siever et al. (1981) found, in a preliminary study, that depressive patients or healthy subjects with a blunted GH response to clonidine showed a depressive mood reaction to arecoline or physostigmine, in contrast to probands with a normal GH stimulation to clonidine. These authors interpret their results as supporting the cholinergic-adrenergic balance theory, according to which a blunted GH response to clonidine should be an indication of cholinergic hyperactivity. This interpretation is supported by Delitala et al. (1983), since pirenzepine, a cholinergic muscarinic agonist, completely blocks the clonidine- and apomorphine-induced GH stimulation in healthy subjects. This means that a cholinergic hyperactivity blocks the GH stimulation by noradrenergic challenge.

4.5 β-Adrenoceptors

With many different GH stimuli it has been shown that β-blockade increases and α-blockade decreases GH stimulation (review Brown et al. 1979). In the DMI GH stimulation test (Laakmann et al. 1986a, b), too, the mutual influence of β-adrenoceptor downregulation and α-adrenoceptor upregulation could be seen. This balance between β- and α-adrenoceptors is also seen in other tissues or organs

(Kunos 1980; Kunos et al. 1985). Likewise, thyroid hormones, which play an important role in affective disorders (see chapter by Prange, this volume), are markedly involved in the balance between α- and β-adrenoceptors (Kunos 1981). On the other hand, in hepatic tissue for instance, the inverse regulation is found, controlled by phospholipase activity, between α_1- and β-receptors (Kunos et al. 1985), and β-agonists induced a selective increase in α_2-(H)-affinity states in cortex membranes (U'Prichard 1984). However, as explained above, postsynaptic α_2-adrenoceptors, and not α_1-receptors, are involved in GH stimulation. Today we do not know the molecular biological mechanisms responsible for the regulation of α_1-, α_2-, and β-adrenoceptors. I think that more knowledge of these processes is needed to understand the essential biological mechanisms in affective disorders, including the mechanisms responsible for the therapeutic action of antidepressant treatment. At present, β-downregulation is believed to be a very important adaptation step toward the therapeutic effect of antidepressant treatment and closely related to the serotonergic input (Sulser 1986). We should keep in mind, however, that in most of the studies using the clonidine-GH stimulation test under antidepressant treatment no increase of GH response was found after recovery. An increase might have been expected if – via β-downregulation – antidepressant treatment had the same mechanism of action as β-adrenoceptor blockers. But most antidepressant drugs, besides reducing β-, also decrease α_2-adrenoceptor sensitivity (review Maj et al. 1984).

4.6 Somatostatin, Growth Hormone-Releasing Factor

Somatostatin, the counterpart of the GH secretion, does not seem to be responsible for the reduced GH response in endogenous depressive patients to various drugs. In contrast, Rubinow et al. (1983) and Ågren and Lindqvist (1984) found significantly lower somatostatin levels in the CSF of depressive patients. Whether the blunted GH response in the various tests in endogenous depression is due to GRF deficit remains to be cleared up. CSF levels of GRF in bipolar patients were found to be the same as in controls (Berrettini et al. 1987), but in preliminary studies the GH response to GRF was not uniform. On the one hand, in 11 (Eriksson 1985), and in 14 (Krishnan et al. 1986) patients with major depressive disorders the peak GH response was found to be significantly higher than in controls. Krishnan et al. discussed this finding in relation to the above-mentioned lower CSF somatostatin concentration in depressed patients and concluded that no GRF deficit could be responsible for the reduced GH response in major depressive disorders. However, Risch et al. (1986), in the first 15 min post injection of hGRF, found a significantly attenuated log GH concentration in five patients with major depressive disorders compared to 11 age- and sex-matched normal controls. Furthermore, Laakmann et al. (1986c, personal communication), in nine endogenous depressive patients, did not find the GH response to GRF (100 μg i.v.) to be higher than to DMI (75 mg i.m.) in the same patients, although they found a blunted GH response to DMI in such patients (Sect. 1.6). The GH responses in to guanfacine and GRF are also highly significantly correlated in depressed patients and healthy controls (Eriksson 1985). More careful GRF studies appear necessary to clear up the very important question of whether an α_2-adrenoceptor subsensitivity or a GRF deficit contributes to the blunted GH response in endogenous depressive patients.

4.7 Neuropeptide Y

One further peptide, neuropeptide Y (NPY), appears to be very important for the regulation of GH secretion and also, according to first studies, in affective disorders. Like clonidine, NPY, via its action on postsynaptic α_2-adrenoceptors at adrenaline synapses in particular brain areas, leads to a reduction of arterial blood pressure. This effect might be due to the ability of NPY to increase α_2-adrenoceptors (see Fuxe et al. 1986b with further references). But in contrast to clonidine, "the cardiovascular effects of NPY are not reversed by α_2-adrenergic receptor blockade, indicating the existence of NPY receptors which can produce their biological effects independently of the α_2-adrenergic receptor agonist site" (Fuxe et al. 1986c; Härfstrand et al. 1984). These authors proposed a very interesting model according to which there exists a special integrative mechanism between NPY and α_2-adrenergic receptors (Fuxe and Agnati 1985; Fuxe et al. 1986c) (Fig. 6). They assume "that one biochemical error in spontaneous hypertensive rats may be an abnormality in the receptor-receptor interaction at some adrenaline synapses" (Fuxe et al. 1986b). The interaction between the two receptors may be disturbed "both at the level of the coupling device and the level of the proteins carrying recognition sites" (Fuxe et al. 1986c). The close relationships between α_2-adrenoceptors and NPY are also demonstrated by the elevation of NPY in cardiac nerves after clonidine treatment (Nagata et al. 1986).

Due to the clonidine and NPY effects on α_2-adrenoceptors, one might expect that NPY would increase secretion, if mechanisms similar to those in the blood pressure-regulating areas of the brain also exist in the GH-releasing brain areas. But in contrast to clonidine, NPY, at a dose of 250 pg i.v.t., decreases GH secretion and increases adrenocorticotropic hormone (ACTH) level. At lower doses NPY leads to no significant changes in GH or ACTH secretion (Härfstrand et al. 1987). This demonstrates that the action of NPY on GH secretion is contrary to that of clonidine. Is the mechanism similar to that of the above-mentioned effect on the cardiovascular system, i.e., that NPY acts directly at NPY receptors, independently of the α_2-adrenoceptor agonist site?

In this context, the first results reported by Widerlöv et al. (1986) appear of special interest. These authors demonstrated in patients with major depression ($n=37$), in comparison to normal controls ($n=20$), a highly significantly ($p<0.001$) reduced NPY-like immunoreactivity in CSF, which in 10 of the patients did not differ during the recovery phase. From these results Widerlöv et al. assumed that CSF-NPY could be "a trait marker for depressive vulnerability". But in contrast to somatostatin (Rubinow et al. 1983) and other transmitters or metabolites, CSF-NPY shows no rostrocaudal gradient (Berrettini et al. 1986), so the reduced CSF NPY values do not have to correlate with a CNS NPY deficit. NPY was found in the spinal cord of humans (Allen et al. 1984) and animals (Gibson et al. 1984) and in the autonomic nerve terminals and fibers of human skin

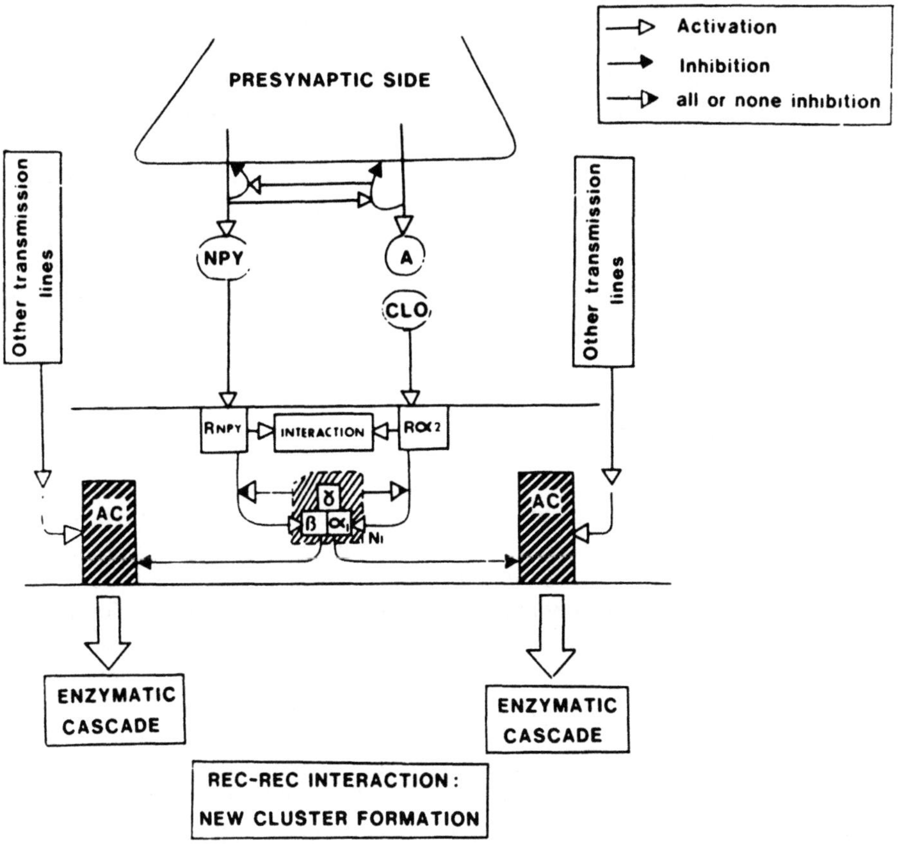

Fig. 6. Schematic representation of interactions between neuropeptide Y (*NPY*) and α_2-adrenergic receptors at the level of the pre- and postsynaptic membrane of NPY/adrenaline (*A*) synapses of the medulla oblongata of the rat. *CLO*, clonidine; *R*, receptor; *Ni*, inhibitory guanine nucleotide binding coupling proteins; *AC*, adenylatcyclase catalytic moiety; α, β, γ, subunits of N proteins. (From Fuxe et al. 1986c)

(Johansson 1986). If the lower CSF levels of NPY found by Widerlöv et al. (1986) are related to a brain deficit of NPY, which cannot be clearly asserted, then the blunted GH response of depressive patients in the different challenge tests cannot be related to the low NPY level, because NPY reduces GH secretion (Härfstrand et al. 1987).

Low NPY levels probably cannot be responsible either for the higher plasma CRF levels in major depressive disorders (Widerlöv et al. 1986) mentioned above (Sect. 4.2), because NPY increases ACTH. I believe, however, that the interaction between NPY receptors and α_2-adrenoceptors could have a strong influence on GH secretion mechanisms, not only in hypertension, but also in affective disorders. The connection to depression is further emphasized by the finding that imipramine and zimelidine treatment in rats for 21 days elevate NPY concentration in the cerebral cortex, but not in hypothalamus, pons and striatum. Bulbectomy, a well-known animal model for depression, reduces the cortical NPY concentration (Widerlöv et al. 1986).

The mechanisms mentioned in Sects. 4.1–4.7 are all involved in the regulation of adrenoceptor sensitivity and GH secretion. Many of them have a strong influence on mood in human beings. They may provoke a depressive syndrome, like β-blockers or cholinomimetic drugs, they may trigger depressive or manic states, like cortisol, or cause feelings of well-being, like opioids (review Matussek and Holsboer 1987). Which mechanism including the peptides might be responsible for the postulated reduced α_2-adrenoceptor sensitivity and blunted GH response in endogenous depressives is unknown. This problem will be discussed from the genetic standpoint in the next section.

5 Adrenoceptors and Hypertension

Changes in adrenoceptor sensitivity, in the opposite direction to those postulated in endogenous depression, are a very important topic too in connection with essential hypertension. Both of these disorders have a strong genetic determination. At present, on the basis of animal studies, higher adrenoceptor binding sites in brain stem and hypothalamus of spontaneously hypertensive rats and in platelets of humans with essential hypertension are being discussed. Excellent reviews with further references to these problems have recently been published by Eriksson (1985) and Middeke (1985; see also Cantor et al. 1981; Wirz-Justice et al. 1983). There are also differences in platelet binding and cyclic AMP production between depression and hypertension (Kafka et al. 1985). Furthermore, GH response to clonidine is significantly higher in spontaneous hypertensive rats than in Wistar-Kyoto rats (Eriksson et al. 1980). In this context it is of interest that in a study comparing healthy mono- and dizygotic twins, Bmax of α_2-adrenoceptors in human platelets was shown to be genetically controlled (Propping and Friedl 1983; Bondy, to be published). Furthermore, we have seen from another twin study that the GH response to dextroamphetamine is also genetically determined (Nurnberger et al. 1981) (Sect. 1.3). But, as mentioned above (Sect. 4.7), recent studies have shown that NPY is also closely related to hypertension, and we have seen that CSF levels of NPY are significantly lower in major depression (Sect. 4.6) than in healthy controls.

Many animal studies comparing spontanecus hypertensive rats and Wistar-Kyoto rats have shown that in such behavioral and experimental conditions as locomotor activity (Leaton et al. 1983), anticipation of electroshock (McCarty et al. 1978), neophobia (Delini-Stula and Hunn 1985), corticosterone response to "mild" stress only in the light phase (Kräuchi et al. 1983), and behavioral suppressant effects of clonidine (Tilson et al. 1977), great differences exist between the two strains in the sense of enhanced adrenergic activity in spontaneous hypertensive rats. Furthermore, as mentioned above, Eriksson et al. (1980) demonstrated a markedly higher GH response to clonidine in spontaneous hypertensive than in Wistar-Kyoto rats. But in a study with hypertensive subjects (diastolic blood pressur >90 mmHg) investigating the age effect of GH response to clonidine, no differences were found in comparison with normotensive subjects (Gil-Ad et al. 1984). The authors did not, however, specifically search for essential hypertension. But in another study, patients with primary hypertension showed a blunted GH response to clonidine in contrast to normotensive subjects matched for sex, age, and body weight (Barbieri et al. 1980). Essential hypertension and endogenous depression, both of them strongly genetically determined, are perhaps due either to high or low adrenoceptor density, or to related mechanisms like NPY, in specific brain regions. But if this assumption were correct and genetic determination were involved both in cardiovascular and in "depressiogenic" brain areas in the reverse direction, patients with essential hypertension could not become endogenously depressed and vice versa.

As regards schizophrenia, one group (Seidel 1977) has reported an association between hypertension and paranoid states and another (Lake et al. 1980) has described a slight increase in diastolic blood pressure in a combined group of paranoid, nonparanoid, and schizoaffective schizophrenic patients compared with controls. If these findings were confirmed this could perhaps give further hints at the postulated catecholamine hyperactivity hypothesis of schizophrenia, because both DA *and* NA hyperactivity seem to be important, at least for the paranoid-hallucinatory type (Matussek 1982). As mentioned above (Sect. 1.7.2), schizophrenics have a particularly high GH response to clonidine (Matussek et al. 1980; Ackenheil et al. 1985) as well as to low doses (6 µg/kg) of apomorphine (Müller-Spahn et al. 1985) in comparison to healthy subjects. Hyperactivity of the DA (Sowers et al. 1982) and NA (Lake et al. 1981) systems are assumed to be involved also in hypertension. Van Valkenburg and Winokur (1984), however, could not confirm the results of Seidel (1977) and Lake et al. (1980), but "this difference could be accounted for by use of matched inpatient control groups" in the van Valkenburg and Winokur studies in contrast to those of Lake et al., which used healthy volunteer subjects.

To my knowledge there are as yet only four studies investigating the relation between essential hypertension and endogenous depression. Slater (1938) reported in a large family study of manic depressives that "the cause of death in these families shows a clearly high frequency of heart diseases". Moreover, Weeke (1979) found a highly significant mortality due to cardio-vascular diseases in manic depressives. But cardio-vascular diseases are often found in hypertensive patients. Seidel (1977) found lower blood pressure in affective disorder (nonparanoid) than in paranoid state patients.

Thus we wanted to evaluate in a preliminary study whether the group of 304 depressive patients who came to our clinic in 1985 showed a difference in respect to diastolic respectively systolic blood pressure between endogenous ($n=182$) and nonendogenous ($n=122$) depressives, such as neurotic or reactive depressives. We also compared the endogenous ($n=23$) and nonendogenous ($n=187$) groups of depressive patients and the healthy controls of a long-term population study with 2334 subjects in Bavaria (Fichter et al. 1987) in terms of differences in blood pressure. We applied the WHO criteria for high blood pressure, i.e., systolic $\geqq 160$ mmHg and diastolic $\geqq 95$ mmHg. We did not find lower blood pressure values in age-corrected endogenous depressives than in nonendogenous depressives. In contrast, endogenous depressives were found to have significantly higher

Table 4. Brain regional PNMT activity, adrenaline content and α_1- and α_2-adrenoceptor number in F344 and Buf rats. (From U'Prichard 1984)

Region	F 344 (*n*)	Buf (*n*)
Medulla, pons		
PNMT	0.43 ± 0.39 (9)	0.15 ± 0.02 (9)*
Adrenaline	8.8 ± 0.8 (5)	3.0 ± 0.4 (7)*
[^{3}H]-RAUW	20.9 ± 3.1 (6)	76.7 ± 14.3 (5)*
[^{3}H]-PAC	33.7 ± 2.8 (6)	84.1 ± 16.8 (5)*
[^{3}H]-Prazosin	36.1 ± 3.2 (4)	97.6 ± 11.3 (2)*
Hypothalamus		
PNMT	0.89 ± 0.05 (9)	0.09 ± 0.03 (10)*
Adrenaline	74.0 ± 4.0 (9)	14.0 ± 2.0 (10)*
[^{3}H]-RAUW	86.6 ± 12.3 (2)	160.5 ± 20.9 (2)*
[^{3}H]-PAC	74.3 ± 0.8 (2)	148.3 ± 12.9 (2)*
[^{3}H]-Prazosin	31.3 ± 7.9 (2)	68.5 ± 5.9 (2)*
Cerebral cortex		
[^{3}H]-RAUW	179.9 ± 8.5 (2)	188.9 ± 10.7 (2)
[^{3}H]-PAC	141.4 ± 6.7 (2)	129.5 ± 12.1 (2)
[^{3}H]-Prazosin	882.00 ± 16.4 (2)	89.1 ± 30.7 (2)

Results expressed as mean ± SEM
Units: PNMT activity, units/g tissue; adrenaline, ng/g tissue; ligand Bmax values, fmol/mg protein. Ligand concentrations: [^{3}H]-RAUW, 0.1–3.0 n*M*; [^{3}H]-PAC, 0.1–3.0 n*M*; [^{3}H]-prazosin, 0.05–2.0 n*M*
* Significantly different from F 344 ($p < 0.05$)

diastolic blood pressure values than the nonendogenous depressive group (Nakagawara et al. 1987). Do these results give hints at the close relations between endogenous depression and hypertension, as does the blunted GH response to clonidine in endogenous depressives (Sect. 1.7.1) as well as in hypertensive patients (Barbieri et al. 1980) and the high mortality due to cardio-vascular diseases of manic depressive patients (Weeke 1979) and their family members (Slater 1938)?

At the moment, we do not exactly know whether the results of our preliminary study indicate that endogenous depressive patients may also more often suffer from essential hypertension. We are carrying out further investigations with regard to this problem, but we have to keep in mind, from animal studies with spontaneously hypertensive rats, "that the genes responsible for the hypertensive trait and those responsible for the hyperactivity trait were not tightly linked in hybrid population (between spontaneously hypertensive and Wistar-Kyoto rats) suggesting that different genetic factors were involved in the transmission of each of these traits" (Hendly et al. 1983). This might be the case in human beings too.

On the other hand, α_2-adrenoceptor numbers in the brain (only in medulla pons and hypothalamus), but also in adrenal medulla, are genetically determined in two more strains of rats, i.e., Fisher 344 and Buffalo rats (U'Prichard 1984; U'Prichard et al. 1988) (Table 4). Neuroendocrinological and behavioral studies are being carried out by the group of J. Maj in Krakow, in collaboration with us, for the purpose of finding out whether Fisher 344 rats, with low α_2-adrenoceptor density, have a blunted GH response to clonidine compared to Buffalo rats, with high α_2-adrenoceptor density. Up to now, it was believed that Fisher rats behave differently than Buffalo or Wistar rats as regards motor activity, giving the impression of being "anxious" animals that are irritated and vocalize when touched. Also, the clonidine effect on motor activity is significantly different in Fisher 344 rats than in other rat strains (Maj, personal communication) An earlier study demonstrated a significantly lower spontaneous behavioural activity in F 344 as compared to Buffalo rats (Segal et al. 1972). Furthermore, the basal firing rate in locus ceruleus in F 344 rats was significantly lower than in Buffalo rats, and 3-week DMI treatment in Fisher 344 rats led to a much stronger β-downregulation than in Buffalo rats (U'Prichard et al. 1988). Perhaps Fisher rats can be used as a good animal model for "endogenous depression," if the α_2-adrenoceptor sensitivity is definitely involved in the vulnerability and pathophysiology of endogenous depression and mood regulation. However, I do not believe that a phasic or bipolar course, as in endogenous depression, will be found.

Another interesting point that may be important for the clinical studies, is that the number of α_2-adrenoceptors in tissue is highly correlated to PNMT activity, the enzyme transforming NA to adrenaline (U'Prichard 1984) (Table 4). The adrenal PNMT activity in these rat strains is "inherited at a single major autosomal locus, superimposed upon a small polygenic component, co-inherited to brain enzyme activity" (Loftus et al. 1988). In spontaneous hypertensive rats, which show a higher GH response to clonidine than Wistar-Kyoto strains (Eriksson et al. 1980), the PNMT activity in specific brain areas (A1 and A2 of the brain stem) is elevated compared to Wistar-Kyoto rats (Saavedra et al. 1976). This means that there are perhaps interesting relationships between GH response to clonidine, α_2-adrenoceptors, PNMT activity, adrenaline and perhaps also NPY (see above) in depressed and hypertensive patients. With the help of a highly specific and more sensitive method for PNMT activity in blood than that used by Vogel et al. (1976), we are at the moment searching for a correlation between GH response to clonidine, α_2-adrenoceptor sites in platelets and PNMT activity in blood in depressives and in patients with essential hypertension, as well as in mono- and dizygotic twins. PNMT activity and Bmax of platelet α_2-adrenoceptor binding are thus far more concordant in monozygotic twins than in dizygotic twins (to be published).

6 Summary and Conclusion

The different catecholamine challenge tests with cortisol and GH stimulation showed, in most of the studies, a significant difference between endogenous depressives and patients with major depressive disorders on the one hand, and healthy subjects on the other. The amphetamine and perhaps also the DMI cortisol stimulation tests seem to be a state-dependent marker for endogenous depression. Further studies with clonidine both in the free interval and in families and twins with affective disorders are needed to see whether a blunted GH response to clonidine is a trait marker or a vulnerability factor for depression. Similar investigations ought to be done with DMI. It appears to me most important to clear up in future studies the relation of the blunted GH response in endogenous depressive patients in the different stimulation tests to the opioid system, to CRF and NPY in particular, and whether CSF NPY is a trait marker for major depression and is related to the blunted GH response to clonidine. Moreover, it has to be ascertained why guanfacine, a more selective α_2-adrenoceptor agonist than clonidine, does not show a blunted GH response in endogenous depression.

In regard to most pharmacological and biochemical effects, clonidine and guanfacine are qualitatively similar, but in most experiments clonidine was found to be about ten times more potent than guanfacine (Scholtysik et al. 1980). Furthermore, guanfacine does not block DA turnover and does not lead to a fall in mean arterial blood pressure after topical application to the ventral surface of the medulla oblongata in the cat, as does clonidine (Scholtysik and Jerie 1976). Above (Sect. 1.7), it has been stated briefly that clonidine, at least in the same area, namely the caudal medulla oblongata, reduces the adrenaline turnover acting as adrenoceptor agonist. This effect is related to the vasodepressor activity of clonidine (Fuxe et al. 1979 b). Perhaps the different influences of guanfacine and clonidine on adrenaline metabolism in other brain areas are also responsible for the difference in GH stimulation in endogenous depression.

Realizing our preliminary results showing a significant correlation between endogenous depression and hypertension, and the epidemiologic studies carried out in regard to cardio-vascular diseases as cause of death of manic depressive patients and their family members, it appears worthwhile to search for neurobiologically and genetically determined abnormal relations in both diseases, in particular in the regulation between α_2-adrenoceptors and NPY.

We have seen that so far, with the exception of the clonidine GH stimulation test, the results obtained in the various catecholamine challenge studies with endogenous and major depressive disorder patients compared to healthy controls are not uniform. Various factors may be responsible here.

Firstly, estrogen status and age are important variables that influence GH stimulation. They ought to be taken into account more accurately in future studies. Whether these variables might be responsible also for the divergent results obtained in the two latest ITT studies, as compared to the many older ones, remains open. As for the ITT, insulin insensitivity, and not low sensitivity of the α-adrenoglucoreceptor, appears to be the reason for the blunted GH response in a subgroup of depressive patients.

Secondly, our present diagnostic systems – ICD, RDC and DSM-III – are not concordant and thus may produce different results. Therefore, the participants of the Dahlem Workshop on the origins of depression (Angst 1983) recommended, at the present state of biological research, special methodological and patient variables with regard to diagnosis as well as very careful description of the patients (Checkley and Rush 1983). Up to now, however, most studies have not considered these points.

In my view, the significant differences between endogenous and nonendogenous depression found in many neuroendocrine studies cannot be explained in terms of age or sex divergencies alone. These differences have to be further examined so as to decide whether there is really a division between lower output (presynaptic) and lower sensitivity (postsynaptic) depression. Solution of this problem would have an important impact on all animal models of depression, like learned helplessness, bulbectomy, and others. For, in my view, all of these represent only the nonendogenous type of depression, a presynaptic one with an amine deficit, showing no phases with free intervals and no bipolarity. These points, however, are the most important characteristics of the endogenous type of the disease. This means that the course of the illness is even more essential for the nosological diagnosis than the symptomatology alone. In today's biological research on depression the different types of depression ought not to be mixed up.

I believe that the results of the different challenge tests indicate that disturbances in α_2-adrenoceptor regulation may be involved in pathophysiology of endogenous depression. Some studies investigating blood cells in depression point in the same direction (Wood et al. 1985). I have shown that various other transmitter, cotransmitter, and hormonal systems have a strong influence on the regulation of adrenoceptor function, some of them, like cortisol, being more a state-dependent marker. GH and cortisol secretion have some common reciprocally working mechanisms. Are these involved in the neurobiological disturbances of affective disorders? But animal and human twin studies have demonstrated that α_2-adrenoceptor binding sites are also strongly genetically controlled, probably linked to PNMT activity. In this connection interaction between α_2-adrenoceptors and NPY receptors also appears fascinating, as NPY may be a marker for vulnerability of major depression – perhaps related to the blunted GH response in some catecholamine challenge tests? There is evidence today that the adrenaline system in particular is strongly involved in these processes, not only as regards GH secretion, but probably also in mood regulation, recently postulated by Svensson et al. (1986) on the basis of pharmacological and physiological studies in animals. Furthermore, the clinical results of Rafaelsen et al. (1986) and Berger et al. (1984) showed significantly lower CSF adrenaline concentrations in depressive patients with hypochondriasis and somatic anxiety. On the other hand, Ackenheil et al. (1979) found higher plasma and Stokes et al. (1984) higher urinary adrenaline levels in depressive patients. In the study by Stokes et al. a high, significantly positive correlation was demonstrated between HPA hyperactivity and adrenaline values. For the moment, I do not believe it possible to come to clear conclusions from these studies as regards specific disturbances in the adrenaline system. A hypoactivity of the dopaminergic system in depression should also be further investigated.

On the basis of our present knowledge of the neurobiological mechanisms in the brain, gained in clinical and genetic studies, we have to assume that many neurotransmitter, cotransmitter, and hormonal systems are disturbed, when a depressive phase or reaction begins. Genetically determined vulnerability and environmental influences are both responsible. It may be expected that the rapidly developing neurobiological basic research and the new gene technology will contribute an answer to the question of vulnerability in the near future. Perhaps neuroendocrinological challenge tests will help to identify special genes that are related to vulnerability to the different types of depression. It remains to be seen whether adrenoceptor genes are involved – β-adrenoceptor gene, for instance, was recently cloned and cDNA nucleotide and amino acid sequence established (Dixon et al. 1986).

*Acknowledgements.*The author wishes to thank Mrs. I. v. Denffer and Mr. K. DeWitt for their assistance in translating and completing the manuscript.

References

Ackenheil M, Albus M, Müller F, Müller TH, Welter D (1979) Catecholamine response to short-time stress in schizophrenic and depressive patients. In: Usdin E, Kopin IJ, Barchas J (eds) Catecholamines: basic and clinical frontiers. Pergamon, New York, pp 1937–1939

Ackenheil M, Albus M, Bondy B, Müller-Spahn F, Münch U, Naber D (1985) Neuroendocrine and receptor binding studies in schizophrenia. In: Pichot P, Berner P, Wolf R, Than K (eds) Psychiatry, vol 2. Plenum, New York, pp 215–220

Aghajanian GK (1985) The neurobiology of opiate withdrawal: receptors, second messengers, and ion channels. Pair Oaks Hosp Psychiatry Lett III (10):57–60

Agnati LF, Fuxe K, Yu Z-Y, Härfstrand A, Okret S, Wikström A-C, Goldstein M, Zoli M, Vale W, Gustafsson JÅ (1985) Morphometrical analysis of the distribution of corticotropin releasing factor glucocorticoid receptor and phenylethanolamine-methyltransferase immunoreactive structures in the paraventricular hypothalamic nucleus of the rat. Neurosci Lett 54:147–152

Ågren H, Lundqvist G (1984) Low levels of somatostatin in human CSF mark depressive episodes. Psychoneuroendocrinology 9:233–248

Allen JM, Gibson SJ, Adrian TE, Polak JM, Bloo SR (1984) Neuropeptide Y in human spinal cord. Brain Res 308:145–148

Angst J (1983) The origins of depression: current concepts and approaches (Dahlem Konferenzen). Life sciences research report 26. Springer, Berlin Heidelberg New York

Ansseau M, Scheyvaerts M, Doumont A, Poirrier R, Legros JJ, Franck G (1984) Concurrent use of REM latency, dexamethasone suppression, clonidine, and apomorphine tests as biological markers of endogenous depression: a pilot study. Psychiatry Res 12:261–272

Ansseau M, v Frenckell R, Cerfontaine J-L, Papart P, Franck G, Timsit-Berthier M, Geenen V, Legros J-J (1987) Neuroendocrine evaluation of catecholaminergic neurotransmission in mania. Psychiatry Res 22:193–206

Arató M, Rihmer Z, Banki CM, Grof P (1983) The relationships of neuroendocrine tests in endogenous depression. Prog Neuro-Pharmacol Biol Psychiatry 7:715–718

Ashcroft GW, Eccleston D, Murray LG, Glen AJ, Crawford TBB, Pullar JA, Shields PJ, Walter DS, Blackburn JM, Chonnechan J, Lonergan M (1972) Modified amine hypothesis for the ethiology of affective illness. Lancet II:573–577

Asnis GM, Lemus CZ, Halbreich U (1986) The desipramine cortisol test – a selective noradrenergic challenge (relationship to other cortisol tests in depressives and normals). Psychopharmacol Bull 22(3):571–578

Barbieri C, Ferrari C, Caldera R, Curtarelli G (1980) Growth hormone secretion in hypertensive patients: evidence for a derangement in central adrenergic function. Clinical Science 58:135–138

Berger M, Doerr P, Lund R, Bronisch T, Zerssen vD (1982) Neuroendocrinological and neurophysiological studies in major depressive disorders: are there biological markers for the endogenous subtype? Biol Psychiatry 17:1217–1241

Berger PA, King R, Lemoine P, Mefford IN, Barchas JD (1984) Cerebrospinal fluid epinephrine concentrations: discrimination of subtypes of depression and schizophrenia. Psychopharmacol Bull 20(3):412–415

Berrettini WH, Nurnberger JI, Dimaggio DA (1986) Neuropeptide Y immunoreactivity in human cerebrospinal fluid. Peptides 7:455–458

Berrettini WH, Nurnberger JI, Simmons-Alling S (1987) Growth hormone releasing factor in human cerebrospinal fluid. Psychiatry Res 22:141–147

Bleuler M (1979) Endokrinologische Psychiatrie. In: Kisker KP, Meyer JE, Müller C, Strömgren E (eds) Psychatrie der Gegenwart, vol I/1. Springer, Berlin Heidelberg New York, pp 257–342

Boyd AE, Lebovitz HE, Pfeiffer JB (1970) Stimulation of human growth hormone secretion by L-DOPA. N Engl J Med 283:1425
Boyer P, Schaub C, Pichot P (1982) Growth hormone response to clonidine test in depressive states. Neuroendocrinol Lett 4:178
Brambilla F, Lampertico M, Sali L, Cavagnini CI, Invitti C, Maggioni M, Candolfi C, Panerai AE, Müller EE (1987) Clonidine stimulation in anorexia nervosa: growth hormone, cortisol, and beta-endorphin responses. Psychiatry Research 20:19–31
Brodde OE, Anlauf M, Graben N, Bock KD (1982) Age-dependent decrease of α_2-adrenergic receptor number in human platelets. Eur J Pharmacol 81:345–347
Brown WA, Corriveau DP, Ebert MH (1978) Acute psychologic and neuroendocrine effects of dextroamphetamine and methylphenidate. Psychopharmacology 58:189–195
Brown GM, Friend WC, Chambers JW (1979) Neuropharmacology of hypothalamic-pituitary regulation. In: Tolis G, Labrie F, Martin JB, Naftolin F (eds) Clinical neuroendocrinology. A pathophysiological approach. Raven, New York, pp 47–81
Bunney WE Jr, Davis JM (1965) Norepinephrine in depressive reactions. Arch Gen Psychiatry 13:483
Calil HM, Lesieur P, Gold PW, Brown GM, Potter WZ (1984) Hormonal responses to zimelidine and desipramine in depressed patients. Psychiatry Res 13:231–242
Cantor EH, Abraham S, Spector S (1981) Central neurotransmitter receptors in hypertensive rats. Life Sci 28:519–526
Casper RC, Davis JM, Pandey GN, Garver DL, Dekirmenjian H (1977) Neuroendocrine and amine studies in affective illness. Psychoneuroendocrinology 2:105–114
Catalano M, Bellodi L, Lucca A, Brambilla F (1984) Lithium and alpha-2-adrenergic receptors: effects of lithium ions on clonidine-induced growth hormone release. Neuroendocrinol Lett 6:61–66
Cella SG, Picotti GB, Morgese M, Mantegazza P, Müller EE (1984) Presynaptic stimulation leads to growth hormone release in the dog. Life Sci 34:447–454
Chalmers JP, Baldessarini RJ, Wurtman RJ (1971) Effects of L-DOPA on norepinephrine metabolism in the brain. Proc Natl Acad Sci 68:662–666
Charney DS, Heninger GR, Sternberg DE, Hafstadt KM, Giddings S, Landis H (1982a) Adrenergic receptor sensitivity in depression. Effects of clonidine in depressed patients and healthy subjects. Arch Gen Psychiatry 39:290–294
Charney DS, Heninger GR, Sternberg DE (1982b) Failure of chronic antidepressant treatment to alter growth hormone response to clonidine. Psychiatry Res 7:135–138
Charney DS, Heninger GR, Sternberg DE (1984) The effect of mianserin on alpha-2 adrenergic receptor function in depressed patients. Br J Psychiatry 144:407–416
Checkley SA (1979) Corticosteroid and growth hormone responses to methylamphetamine in depressive illness. Psychol Med 9:107–115
Checkley SA, Arendt J (1984) Pharmacoendocrine studies of GH, PRL, and melatonin in patients with affective illness. In: Brown GM, Koslow SH, Reichlin S (eds) Neuroendocrinology and psychiatric disorder. Raven, New York, pp 165–190
Checkley SA, Corn TH (1986) Factors responsible for reduced growth hormone (GH) responses to clonidine in patients with endogenous depression. In: Shagass C, Josiassen RC, Bridger WH, Weiss KJ, Stoff D, Simpson GM (eds) Biological psychiatry 1985. Developments in psychiatry, vol 7. Proceedings of IVth World Congress of Biological Psychiatry. Elsevier, Amsterdam, pp 799–801
Checkley SA, Crammer JL (1977) Hormone responses to methylamphetamine in depression: a new approach to the noradrenaline depletion hypothesis. Br J Psychiatry 131:582–586
Checkley SA, Slade AP, Shur E (1981) Growth hormone and other responses to clonidine in patients with endogenous depression. Br J Psychiatry 138:51–55
Checkley SA, Rush AJ (1983) Functional indices of biological disturbance. In: Angst J (ed) The origins of depression: current concepts and approaches (Dahlem Konferenzen). Springer, Berlin Heidelberg New York Tokyo, pp 425–445
Checkley SA, Glass IB, Thompson C, Corn T, Robinson P (1984) The GH response to clonidine in endogenous as compared to reactive depression. Psychol Med 14:773–777

Checkley SA, Corn TH, Glass IB, Burton SW, Burke CA (1986) The responsiveness of central alpha-2 adrenoceptors in depression. In: Deakin JFW (ed) The biology of depression. Gaskell, London, pp 101–120
Corn TH, Honig A, Thompson C, Bridges PK, Bartlett JR, Checkley SA (1984a) A neuroendocrine study of stereotactic sub-caudate tractotomy. Br J Psychiatry 144:417–420
Corn TH, Hale AS, Thompson C, Bridges PK, Checkley SA (1984b) A comparison of the growth hormone responses to clonidine and apomorphine in the same patients with endogenous depression. Br J Psychiatry 144:636–639
Corn TH, Thompson C, Checkley SA (1984c) Effects of desipramine treatment upon central adrenoreceptor function in normal subjects. Br J Psychiatry 145:139–145
Czernik A (1982) Zur Psychophysiologie und Neuroendokrinologie von Depressionen. Springer, Berlin Heidelberg New York
Czernik A, Kleesiek K, Steinmeyer EM (1980) Änderungen neuroendokrinologischer Parameter im Verlauf von Depressionen. Nervenarzt 51:662–667
Delini-Stula A, Hunn C (1985) Neophobia in spontaneous hypertensive (SHR) and normotensive control (WKY) rats. Behav Neural Biol 43:206–211
Delitala G, Maioli M, Pacificio A, Brianda S, Palermo M, Manelli H (1983) Cholinergic receptor control mechanisms for L-DOPA, apomorphine, and clonidine-induced growth hormone secretion in man. J Clin Endocrinol Metab 57:1145–1149
Denef C, Baes M, Schamme C (1986) Paracrine interactions in the anterior pituitary: role in the regulation of prolactin and growth hormone secretion. In: Ganong WF, Martini L (eds) Frontiers in neuroendocrinology, vol 9. Raven, New York, pp 115–148
Dixon RA, Kobilka BK, Strader DJ, Benovic JL, Dohlman HG, Frielle T, Bolanowski MA, Bennett CD, Rands E, Diehl RE, Mumford RA, Slater EE, Sigal IS, Caron MG, Lefkowitz RJ, Strader CD (1986) Cloning of the gene and cDNA for mammalian β-adrenergic receptor and homology with rhodopsin. Nature 321:75–79
Dolan RJ, Calloway SP (1986) The human growth hormone response to clonidine: relationship to clinical and neuroendocrine profile in depression. Am J Psychiatry 143:772–774
Endo M (1970) Plasma growth hormone levels during insulin hypoglycemia in atypical psychosis. Folia Endocrinol Jpn 45:1295–1296
Endo M, Endo J, Nishikubo M, Yamaguchi T, Hatotani N (1974) Endocrine studies in depression. In: Mieken (ed) Psychoneuroendocrinology. Karger, Basel
Eriksson E (1985) Experimental psycho-neuro-endocrinology: brain α2-adrenoceptor function and growth hormone release. Medi, Göteborg
Eriksson E, Modigh K (1984) Depression, α2-receptors and sex hormones: neuroendocrine studies in rat. In: Usdin E, Åsberg M, Bertilsson L, Sjöqvist F (eds) Frontiers in biochemical and pharmacological research in depression. Raven, New York, pp 161–178
Eriksson E, Eden S, Modigh K (1980) Enhanced growth hormone response to clonidine in the spontaneously hypertensive rat. Clin Exp Hypertens 2:341
Eriksson E, Eden S, Modigh K (1981) Importance of norepinephrine alpha-2-receptor activation for morphine-induced rat growth hormone secretion. Neuroendocrinology 33:91–96
Everett GM, Toman JEP (1959) Mode of action of *Rauwolfia* alkaloids and motor activity. Biological psychiatry, vol 1. Grune and Stratton, New York
Feinberg M, Greden JF, Carroll BJ (1981) The effect of amphetamine on plasma cortisol in patients with endogenous and nonendogenous depression. Psychoneuroendocrinology 6(4):355–357
Fichter MM, Pirke K-M (1986) Effect of experimental and pathological weight loss upon the hypothalamo-pituitary-adrenal axis. Psychoneuroendocrinology 11:295–305
Fichter MM, Weyerer S, Meller I, Eiberger T, Witzke W, Rehm J, Dilling H, Hippius H (1987) Ergebnisse der oberbayerischen Verlaufsuntersuchung. In: Schmidt MH (ed) Psychiatrische Epidemiologie. Abschlußband des SFB 116. VHC, Heidelberg
Flier JS, Kahn CR, Roth J (1979) Receptors, antireceptors, antibodies, and mechanisms of insulin resistance. N Engl J Med 300:413

Frantz AG, Rabkin MT (1965) Effects of estrogen and sex differences on secretion of human growth hormone. J Clin Endocrinol Metab 25:1470–1480
Frazer A (1975) Adrenergic responses to depression: implications for a receptor defect. In: Mendels J (ed) The psychobiology of depression. Spectrum, New York
Fuxe K, Agnati LF (1985) Receptor-receptor interactions in the central nervous system. A new integrative mechanism in synapses. Med Res Rev 5(4):441–482
Fuxe K, Hökfeld T, Johansson O, Ganten D, Goldstein M, Perez de la Mora M, Possani L, Tapia R, Teran L, Palacios R, Said S, Mutt V (1976) Monoamine neuron systems in the hypothalamus and their relation to the GABA and peptide containing neurons. In: Colloque de synthèse. Série action thématique rapport no 7. INSERM, Paris, pp 17–40
Fuxe K, Andersson K, Löfström A, Hökfelt T, Ferland L, Agnati LF, Perez de la Mora M, Schwarcz R, Eneroth P Gustafsson JÅ, Skett P (1979a) Neurotransmitter mechanisms in the control of the secretion of hormones from the anterior pituitary. In: Fuxe K, Hökfelt T, Luft R (eds) Central regulation of the endocrine system. Plenum, New York
Fuxe K, Jonsson G, Bolme P, Andersson K, Agnati LF, Goldstein M, Hökfelt T (1979b) Reduction of adrenaline turnover in cardiovascular areas of rat medulla oblongata by clonidine. Acta Physiol Scand 107:177–179
Fuxe K, Andersson K, Härfstrand A, Agnati LF, Eneroth P, Janson AM, Vale W, Thorner M, Goldstein M (1986a) Medianosomes as integrative units in the external layer of the median eminence. Studies on GRF/catecholamine and somatostatin/catecholamine interactions in the hypothalamus of the male rat. Neurochem Int 9(1):155–170
Fuxe K, Agnati LF, Härfstrand A, Andersson K, Mascagni F, Zoli M, Kalia M, Battistini N, Benfenati F, Hökfelt T, Goldstein M (1986b) Studies on peptide comodulator transmission. New perspective on the treatment of disorders of the central nervous system. In: Emson PC, Rossor MN, Tohyama M (eds) Progress in brain research, vol 66. Elsevier, Amsterdam, pp 341–368
Fuxe K, Härfstrand A, Agnati LF, Kalia M, Neumeyer A, Cintra A, Zini I, Zoli M, Terenius L, Goldstein M (1986c) Evidence for a vasodepressor role of neuropeptide Y and adrenaline mechanisms in the medulla oblongata of the rat. In: Nakamura K (ed) Brain and blood pressure. Elsevier, Amsterdam, pp 99–112
Genazzani AR, Lemarchand-Beraudth TH, Aubert ML, Felber JP (1975) Pattern of plasma ACTH, HGH and cortisol during menstrual cycle. J Clin Endocrinol Metab 41:431–437
Gil-Ad I, Gurewitz R, Marcovici O, Rosenfeld J, Laron Z (1984) Effect of aging on human plasma growth hormone response to clonidine. Mechan Ageing Dev 27:97–100
Gibson SJ, Polak JM, Allen JM, Adrian TE, Kelly JS, Bloom SR (1984) Distribution and origin of a novel brain peptide, neuropeptide Y, in the spinal cord of several mammals. J Comp Neurol 227:78–91
Glass IB, Checkley SA, Shur E, Dawling S (1982) The effect of desipramine upon central adrenergic function in depressed patients. Br J Psychiatry 141:372–376
Gold PW, Goodwin FK (1977) Neuroendocrine response to levodopa in affective illness. Lancet I:1007
Gregoire F, Brauman H, de Buck R, Corvilain J (1977) Hormone release in depressed patients before and after recovery. Psychoneuroendocrinology 2:303–312
Grof E, Brown GM, Grof P (1983) Neuroendocrine strategies in affective disorders. Prog Neuro-Psychopharmacol Biol Psychiatry 7:557–562
Gruen PH, Sachar EJ, Altman N, Sassin J (1975) Growth hormone responses to hypoglycemia in postmenopausal depressed women. Arch Gen Psychiatry 32:31–33
Halbreich U, Goldstein S (1987) Aging and hormones. In: Nemeroff CB, Loosen PT (eds) Handbook of clinical psychoneuroendocrinology. Guilford, New York
Halbreich U, Sachar EJ, Asnis GM, Quitkin F, Nathan RS, Halpern FS, Klein DF (1982) Growth hormone response to dextroamphetamine in depressed patients and normal subjects. Arch Gen Psychiatry 39:189–192
Halbreich U, Asnis GM, Zumoff B, Nathan RW, Shindledecker R (1984) Effect of age and sex on cortisol secretion in depressives and normals. Psychiatry Res 13:221–229

Halbreich U, Asnis GM, Shindledecker R, Zumoff B, Nathan RS (1985a) Cortisol secretion in endogenous depression. I. Basal plasma levels. Arch Gen Psychiatry 42:904–908
Halbreich U, Asnis AG, Shindledecker R, Zumoff B, Nathan RS (1985b) Cortisol secretion in endogenous depression. II. Time-related functions. Arch Gen Psychiatry 42:909–914
Härfstrand A, Fuxe K, Agnati LF, Ganten D, Eneroth P, Tatemoto K, Mutt V (1984) Studies on neuropeptide-Y catecholamine interactions in central cardiovascular regulation in the α-chloralose anaesthetized rat. Evidence for a possible way of activating the α_2-adrenergic transmission line. Clin Exp Hypertens [A] 6(10–11):1947–1950
Härfstrand A, Fuxe K, Agnati LF, Eneroth P, Zini I, Zoli M, Andersson K, v. Euler G, Terenius L, Muti V, Goldstein M (1986) Studies on neuropeptide Y –catecholamine interactions in the hypothalamus and in the forebrain of the male rat. Relationship to neuroendocrine function. Neurochem In 8(3):355–376
Härfstrand A, Eneroth P, Agnati LF, Fuxe K (1987) Further studies on the effects of central administration of neuropeptide Y on neuroendocrine function in the male rat: relationship to hypothalamic catecholamines. Regul Pept 17:167–179
Hendley ED, Atwater DG, Myers MM, Whitehorn D (1983) Dissociation of genetic hyperactivity and hypertension in SHR. Hypertension 5(2):211–217
Höhe M (1986) Wirkung von Opiaten auf neuroendokrine und psychische Parameter. Untersuchungen an Probanden als Grundlage für neuroendokrinologische Depressionsstudien. Thesis. Psychiatric Hospital, University of Munich, Munich
Höhe M, Duka TH, Doenicke A, Matussek N (1984) Dose-dependent influence of fentanyl on prolactin, growth hormone and mood. Neuropeptides 5:269–272
Höhe M, Valido G, Matussek N (1986) Growth hormone response to clonidine in endogenous depressive patients: evidence for a trait marker in depression. In: Shagass C, Josiassen RC, Bridger WH, Weiss KJ, Stoff D, Simpson GM (eds) Biological psychiatry 1985. Developments in psychiatry, vol 7. Proceedings of IVth World Congress of Biological Psychiatry. Elsevier, Amsterdam, pp 862–864
Honer WG, Glas IB, Corn T, Checkley SA (1984) Measurement of the GH and other responses to clonidine at different times of the day in normal subjects. Psychoneuroendocrinology 9:279–284
Hunt GE, O'Sullivan BT, Johnson GFS, Smythe GA (1986) Growth hormone and cortisol secretion after oral clonidine in healthy adults. Psychoneuroendocrinology 11(3):317–325
Insel TR, Mueller III EA, Gillin JC, Siever LJ, Murphy DL (1984) Biological markers in obsessive-compulsive and affective disorders. J Psychiatry Res 18(4):407–423
Janowsky DS, El-Yousef MK, Davis JM, Sekerke HJ (1972) A cholinergic hypothesis of mania and depression. Lancet II:6732
Jhanwar-Uniyal M, Leibowitz SF (1986) Impact of circulating corticosterone on α_1- and α_2-noradrenergic receptors in discrete brain areas. Brain Research 368:404–408
Jimerson DC, Cutler NR, Post RM, Rey A, Gold PW, Brown GM, Bunney WE Jr (1984) Neuroendocrine responses to apomorphine in depressed patients and healthy control subjects. Psychiatry Res 13:1–12
Johansson O (1986) A detailed account of NPY-immunoreactive nerves and cells of the human skin. Comparison with VIP-, substance P- and PHI-containing structures. Acta Physiol Scand 128:147–152
Kafka MS, Siever LJ, Nurnberger JI, Uhde TW, Targum S, Cooper DMJ, van Kammen DP, Tokola NS (1985) Studies of peripheral monoamine receptors in affective illness. Psychopharmacol Bull 21:599–602
Kahn CR, Flier JS, Bahr RS, Archer JA, Gordon P, Martin MM, Roth J (1976) The syndrome of insulin resistance and acanthosis nigricans: insulin-receptor disorder in man. N Engl J Med 294–739
Katona CLE, Theodorou AE, Davies SL, Yamaguchi Y, Tunnicliffe CA, Hale AS, Horton RW, Kelly JS, Paykel ES (1986) Platelet binding and neuroendocrine responses in depression. In: Deakin JFW (ed) The biology of depression. Gaskell, London, pp 121–136

Kaye WH, Gwirtsman HE, Lake CR, Liever LJ, Jimerson DC, Ebert MH, Murphy DL (1985) Disturbances of norepinephrine metabolism and α-2 adrenergic receptor activity in anorexia nervosa: relationship to nutritional state. Psychopharmacology Bull 21(3):419–423

Kendler KS, Davis KL (1977) Elevated corticosteriods as a possible cause of abnormal neuroendocrine function in depressive illness. In: Usdin E (ed) Communications in psychopharmacology. Pergamon, Oxford

Kleinberg DL, Noel GL, Frantz AG (1971) Chlorpromazine stimulation and L-DOPA suppression of plasma prolactin in man. J Clin Endocrinol Metab 33:873–876

Klerman GL (1986) Evidence for increase in rates of depression in North America and Western Europe in recent decades. In: Hippius H, Klerman GL, Matussek N (eds) New results in depression research. Springer, Berlin Heidelberg New York, pp 7–15

Kobinger W (1979) α-Adrenoreceptors and the action of clonidine-like drugs. Wenner-Gren-Center Intern Symp Ser 33:143–150

Koslow SH, Stokes PE, Mendels J, Ramsey A, Casper R (1982) Insulin tolerance test: human growth hormone response and insulin resistance in primary unipolar depressed, bipolar depressed and control subjects. Psychol Med 12:45–55

Kräuchi K, Wirz-Justice A, Willener R, Campbell IC, Feer H (1983) Spontaneous hypertensive rats: behavioral and corticosterone response depend on circadian phase. Physiol Behav 30:35–40

Krishnan KRR, Manepalli A, Rayasam ML, Melville G, Daughtry G, Rivier J, Vale W, Thorner MO, Nemeroff CB (1986) Somatotroph response to GHRF in depression. 15th CINP congress, San Juan, Puerto Rico (Book of abstracts, poster sessions)

Krulich L, Mayfield MA, Steele MK, McMillen BA, McCann SM, Koenig JI (1982) Differential effects of pharmacological manipulations of central α_1- and α_2-adrenergic receptors on the secretion of thyrotropin and growth hormone in male rats. Endocrinology 110(3):796–804

Kunos G (1980) Reciprocal changes in α- and β-adrenoreceptor reactivity – myth or reality? Trends Pharmacol Sci 7:282–284

Kunos G (1981) Modulation of adrenergic reactivity and adrenoreceptors by thyroid hormones. In: Kunos G (ed) Neurotransmitter receptors. Adrenoreceptors and catecholamine action, part A. Wiley, New York, pp 297–333

Kunos G, Tchakarov L, Ishac JN, Kan WH (1985) Hepatic α_1-adrenergic receptors: structure and heterologous regulation. In: Lefkowitz J, Lindenlaub E (eds) Adrenergic receptors: molecular properties and therapeutic implications. Symposia medica Hoechst 19. Schattauer, Stuttgart, pp 105–131

Laakmann G (1980) Beeinflussung der Hypophysenvorderlappen-Hormonsekretion durch Antidepressiva bei gesunden Probanden, neurotisch und endogen depressiven Patienten. Nervenarzt 51:725–732

Laakmann G, Schumacher G, Benkert O, v Werder K (1977) Stimulation of growth hormone secretion by desimipramine and chlorimipramine in man. J Clin Endocrinol Metab 44:101

Laakmann G, Zygan K, Schoen HW, Weiss A, Wittmann M, Meissner R, Blaschke D (1986 a) Effects of receptor blockers (methysergide, propranolol, phentolamine, yohimbine, and prazosin) on the desimipramine-induced pituitary hormone stimulation in humans. Part I: growth hormone. Psychoneuroendocrinology 11:447–461

Laakmann G, Hinz A, Neulinger E (1986 b) Pharmacoendocrine studies in healthy subjects and depressed patients. In: Hippius H, Klerman GL, Matussek N (eds) New results in depression research. Springer, Berlin Heidelberg New York

Laakmann G, Neuhauser H, Hinz A, Dick G (1986 c) Stimulation of growth hormone, prolactin and cortisol by GRFH and DMI in healthy subjects and depressed patients. 15th CINP congress, San Juan, Puerto Rico (Book of abstracts, poster sessions)

Lake CR, Sternberg DE, van Kammen DP, Ballenger JC, Ziegler MG, Post RM, Kopin IJ, Bunney WE (1980) Schizophrenia: elevated cerebrospinal fluid norepinephrine. Science 207, 18:331–333

Lake CR, Gullner HG, Polinsky RA (1981) Essential hypertension: central and peripheral norepinephrine. Science 211:955–957

Lal D, Tolis G, Martin JB, Brown GM, Guyda H (1975) Effect of clonidine on growth hormone, prolactin, luteinizing hormone, follicle-stimulating hormone, and thyroid-stimulating hormone in the serum of normal men. J Clin Endocrinol Metab 41:827–832
Langer G, Heinze G, Reim B, Matussek N (1975) Growth hormone response to *d*-amphetamine in normal controls and in depressive patients. Neurosci Lett 1:185–189
Langer G, Heinze G, Reim B, Matussek N (1976) Reduced growth hormone responses to amphetamine in endogenous depressive patients. Arch Gen Psychiatry 33:1471–1475
Leaton RN, Cassella JV, Whitehorn (1983) Locomotor activity, auditory startle and shock thresholds in spontaneously hypertensive rats. Physiol Behav 31:103–109
Lechin F, van der Dijs B, Jakubowicz D, Camero R, Villa S, Lechin E, Gomez F (1985a) Effects of clonidine on blood pressure, noradrenaline, cortisol, growth hormone, and prolactin plasma levels in high and low intestinal tone subjects. Neuroendocrinology 40:253–261
Lechin F, van der Dijs B, Jakubowicz D, Camero RE, Villa S, Arochia L, Lechina AE (1985b) Effects of clonidine on blood pressure, noradrenaline, cortisol, growth hormone, and prolactin plasma levels in high and low intestinal tone depressed patients. Neuroendocrinology 41:156–162
Linkowsky P, Brauman H, Mendlewicz J (1983) Prolactin and growth hormone response to levodopa in affective illness. Neuropsychobiology 9:108–112
Loftus DJ, Guchhait RB, U'Prichard DC, Stolk JM (1988) Inheritence of epinephrine formation rates and phenylethanolamine *N*-methyltransferase structural heterology in inbred rat strains. In: Stolk G, U'Prichard D, Fuxe K (eds) Epinephrine in the central nervous system. Oxford University Press, New York
Maany I, Mendels J, Frazer A, Brunswick D (1979) A study of growth hormone release in depression. Neuropsychobiology 5:282
Maj J, Przegalinski E, Mogilnicka E (1984) Hypotheses concerning the mechanism of action of antidepressant drugs. Rev Physiol Biochem Pharmacol, vol 100. Springer, Berlin Heidelberg New York Tokyo
Martin JB (1973) Neural regulation of growth hormone secretion. Engl J Med 228:1381–1392
Matussek N (1966) Neurobiologie und Depression. Med Wochenschr 20:109
Matussek N (1980) Stoffwechselpathologie der Zyklothymie und Schizophrenie. In: Kisker KP, Meyer JE, Müller C, Strömgren E (eds) Psychiatrie der Gegenwart, vol I/2. Springer, Berlin Heidelberg New York, pp 65–113
Matussek N (1982) Erweiterung und Einschränkung der Dopamin-Hypothese der Schizophrenie. In: Huber G (ed) Endogene Psychosen: Diagnostik, Basissymptome und biologische Parameter. Schattauer, Stuttgart, pp 315–318
Matussek N, Holsboer F (1987) Ätiologie und Pathogenese: Biologischer Hintergrund. In: Müller C, Kisker KP, Lauter H, Meyer JE, Strömgren E (eds) Psychiatrie der Gegenwart, vol V. Affektive Psychosen. Springer, Berlin Heidelberg New York Tokyo, pp 204–240
Matussek N, Ackenheil M, Hippius H, Müller F, Schröder H-T, Schultes H, Wasilewski B (1980) Effect of clonidine on growth hormone release in psychiatric patients and controls. Psychiatry Res 2:25–36
Matussek N, Ackenheil M, Herz M (1984) The dependence of the clonidine growth hormone test on alcohol drinking habits and the menstrual cycle. Psychoneuroendocrinology 9:173–177
Matussek N, Ackenheil M, Höhe M, Müller-Spahn F (1986) Clonidine growth hormone stimulation test in depressive patients before and after treatment. In: Shagass C, Josiassen RC, Bridger WH, Weiss KJ, Stoff D, Simpson GM (eds) Biological psychiatry 1985. Developments in psychiatry, vol 7. Proceedings of the IVth World congress of biological psychiatry. Elsevier, New York, pp 788–790
McCarty R, Chiueh CC, Kopin IJ (1978) Spontaneously hypertensive rats: adrenergic hyperresponsivity to anticipation of electric shock. Behav Biol 23:180–188
McWilliam JR, Meldrum BS (1983) Noradrenergic regulation of growth hormone secretion in the baboon. Endocrinology 112(1):254–259

Meesters P, Kerkhofs M, Charles G, Decoster C, Vanderelst M, Mendlewicz J (1985) Growth hormone release after desipramine in depressive illness. Eur Arch Psychiatry Neurol Sci 235:140–142
Meltzer HY, Kolakowska T, Fang VS, Fogg L, Robertson A, Lewine R, Strahilevitz M, Busch D (1984) Growth hormone and prolactin response to apomorphine in schizophrenia and the major affective disorders. Relation to duration of illness and depressive symptoms. Arch Gen Psychiatry 41:512–519
Merimee TJ, Fineberg SE (1971) Studies of the sex based variation of human growth hormone secretion. J Clin Endocrinol Metab 33:896–902
Middeke M (1985) Adrenoceptors in hypertension. In: Middeke M, Holzgreve H (eds) New aspects in hypertension – adrenoceptors. Springer, Berlin Heidelberg New York
Misra CH, Shelat HA, Smith RC (1980) Effect of age on adrenergic and dopaminergic receptor binding in rat brain. Life Sci 27:521–526
Möller HJ, Kissling W, Bottermann P (1984) Serial application of clonidine tests during antidepressive treatment with chlorimipramine. Pharmacopsychiat 17:184–187
Motulskly HJ, Insel PA (1982) Adrenergic receptors in man. Direct identification, physiologic regulation, and clinical alterations. N Engl J Med 307:18–29
Mueller PS, Heninger GR, McDonald RK (1969 a) Insulin tolerance test in depression. Arch Gen Psychiatry 21:587–594
Mueller PS, Heninger GR, McDonald RK (1969 b) Intravenous glucose tolerance test in depression. Arch Gen Psychiatry 21:470–477
Müller-Spahn F, Bondy B, Ackenheil M (1985) Apomorphine and clonidine induced growth hormone secretion in different states of schizophrenic illness: a predictor of relapse? IV. World congress of biological psychiatry, Philadelphia. In: Book of abstracts. p 139
Nagata M, Franco-Cereceda A, Svensson TH, Lundberg JML (1986) Clonidine treatment elevates content of neuropeptide Y in cardiac nerves. Acta Physiol Scand 128:312–322
Nakagawara N, Witzke W, Matussek N (1987) Hypertension in depression. Psychiatry Res 21:85–86
Nathan RS, Sachar EJ, Asnis GM, Halbreich U, Halpern FS (1981) Relative insulin insensitivity and cortisol secretion in depressed patients. Psychiatry Res 4:291–300
Nurnberger JI Jr, Gershon ES, Jimerson DC, Buchsbaum M, Gold P, Brown GM, Ebert M (1981) Pharmacogenetics of *d*-amphetamine response in man. In: Gershon ES, Matthysse S, Breakefield XO, Ciaranello RD (eds) Genetic strategies in psychobiology and psychiatry. Boxwood, Pacific Grove, pp 257–268
Nurnberger JI Jr, Gershon ES, Simmons S, Ebert M, Kessler LR, Dibble ED, Jimerson SS, Brown GM, Gold P, Jimerson DC, Guroff JJ, Storch FI (1982) Behavioral, biochemical and neuroendocrine responses to amphetamine in normal twins and "well-state" bipolar patients. Psychoneuroendocrinology 7(2/3):163–176
Price LH, Charney DS, Heninger GR (1986) Effects of trazodone treatment on alpha-2 adrenoceptor function in depressed patients. Psychopharmacology 89:38–44
Propping P, Friedl W (1983) Genetic control of adrenergic receptors on human platelets. A twin study. Hum Genet 64:105–109
Rafaelsen OJ, Christensen NJ, Gjerris A (1986) Cerebrospinal fluid (CSF) adrenaline in depression. Clinical Neuropharmacology 9(4):170–171
Randrup A, Braestrup C (1977) Uptake of biogenic amines by newer antidepressant drugs: relevance to the dopamine hypothesis of depression. Psychopharmacology 53:309
Randrup A, Munkvad I, Fog R, Gerlach J, Molander L, Kjellberg B, Scheel-Kruger J (1975) Mania, depression and brain dopamine. In: Essman WB, Valzelli L (eds) Current developments in psychopharmacology, vol 2. Plenum, New York, pp 206–248
Rees L, Butler PWP, Gosling C, Besser GM (1970) Adrenergic blockade and corticosteroid and growth hormone responses to methylamphetamine. Nature 228:565–566
Risch S, Janowsky D, Judd L, Gillin J, Ehlers C (1986) Attenuated growth hormone response to human growth hormone releasing factor in depressed subjects vs. matched controls. 15th CINP congress, San Juan, Puerto Rico. (Book of abstracts, poster session)

Rivier C, Vale (1984) Corticoptropin-releasing factor (CRF) acts centrally to inhibit growth hormone secretion in rat. Endocrinology 114:2409–2411
Roth KA, Katz RJ, Sibel M, Mefford IN, Barchas JD, Carroll BJ (1981) Central epinephrinergic inhibition of corticosterone release in rat. Life Sci 28:2389–2394
Rubinow DR, Gold PW, Post RM, Ballenger JC, Cowdry R, Bollinger J, Reichlin S (1983) CSF somatostatin in affective illness. Arch Gen Psychiatry 40:409–412
Saavedra JM, Grobecker H, Axelrod J (1976) Adrenaline-forming enzyme in brainstem: elevation in genetic and experimental hypertension. Science 191:483–484
Sachar EJ, Finkelstein J, Hellmann L (1971) Growth hormone responses in depressive illness. I. Response to insulin tolerance test. Arch Gen Psychiatry 25:263–269
Sachar EJ, Mushrush G, Perlow M, Weitzman ED, Sassin J (1972) Growth hormone response to L-DOPA in depressed patients. Science 178:1304–1305
Sachar EJ, Altman N, Gruen PH, Glassman A, Halpern FS, Sassin J (1975) Human growth hormone response to levodopa. Arch Gen Psychiatry 32:502–503
Sachar EJ, Asnis G, Nathan RS, Halbreich U, Tabrizi MA, Halpern FS (1980) Dextroamphetamine and cortisol in depression. Arch Gen Psychiatry 37:755–757
Sachar EJ, Puig-Antich J, Ryan ND, Asnis GM, Rabinovich H, Halpern FS, Novacenko H, Ostrow L, Nelson B, Davies M, Goetz D (1985) Three tests of cortisol secretion in adult endogenous depression. Acta Psychiatry Scand 71:1–8
Sawa Y, Odo S, Nakasawa T (1982) Growth hormone secretion by tricyclic and nontricyclic antidepressants in healthy volunteers and depressives. In: Langer SZ, Takahashi R, Segawa T, Briley M (eds) New vistas in depression. Pergamon, New York, pp 309–315
Schildkraut JJ (1965) The catecholamine hypothesis of affective disorders: a review of supporting evidence. Am J Psychiatry 122:509
Schittecatte M, Charles G, Defauw Y, Meesters P, Wilmotte J (1988) Growth hormone response to clonidine in panic disorder patients. Psychiatry Res (to be published)
Scholtysik G, Jerie P (1976) Pharmacological and clinical effects of BS 100-141, a new antihypertensive agent. In: Scriabine A, Sweet CS (eds) New antihypertensive drugs. Spectrum, New York, pp 259–368
Segal DS, Kuczenski RT, Mandell AJ (1972) Strain differences in behavior and brain tyrosine hydroxylase activity. Behav Biol 7:75–81
Seidel UP (1977) Hypertension and paranoid states. Lancet I:906
Siever LS, Davis KL (1985) Overview: toward a dysregulation hypothesis of depression. Am J Psychiatry 142:1017–1031
Siever LJ, Risch SC, Murphy DL (1981) Central cholinergic-adrenergic imbalance in the regulation of affective state. Psychiatry Res 5:108–109
Siever LJ, Uhde TW, Silberman EK, Jimerson DC, Aloi JA, Post RM, Murphy DL (1982 a) The growth hormone response to clonidine as a probe of noradrenergic receptor responsiveness in affective disorder patients and controls. Psychiatry Res 6:171–183
Siever LJ, Uhde TW, Insel TR, Roy BF, Murphy DL (1982 b) Growth hormone response to clonidine unchanged by chronic chlorgyline treatment. Psychiatry Res 7:139–144
Siever LJ, Insel TR, Jimerson DC, Lake CR, Uhde TW, Aloi J, Murphy DL (1983) Growth hormone response to clonidine in obsessive-compulsive patients. Br J Psychiatry 142:184–187
Siever LJ, Uhde TW, Jimerson DC, Post RM, Lake CR, Murphy DL (1984) Plasma cortisol responses to clonidine in depressed patients and controls. Arch Gen Psychiatry 41:63–68
Siever LJ, Insel TR, Hamilton JA, Aloi J, Murphy DL (1985) A comparison between the growth hormone responses to amphetamine and clonidine. Psychiatry Res 16:79–82
Siever LJ, Coccaro EF, Benjamin E, Rubinstein K, Davis KL (1986) Adrenergic and serotonergic receptor responsiveness in depression. In: Porter R, Bock G, Clark S (eds) Antidepressant and receptor function. Ciba foundation symposium 123. Wiley, Chichester, pp 148–158
Slade AP, Checkley SA (1980) Neuroendocrine study of the mechanism of action of ECT. Br J Psychiatry 137:217–221

Slater E (1938) Zur Erbpathologie des manisch-depressiven Irreseins. Die Eltern und Kinder von Manisch-Depressiven. Z Gesamt Neurol Psychiatry 163:1–45
Sowers JR, Nyby M, Jasberg K (1982) Dopaminergic controls of prolactin and blood pressure: altered control in essential hypertension. Hypertension 4:431–438
Stokes PE, Koslow S, Swann A, Stoll P, Maas J, Kocsis J, Jawaid J (1984) Neurochemical/neuroendocrine interactions in depression and mania. Clin Neuropharmacol 7(1):836–837
Sulser F (1982) Antidepressant drug research: its impact on neurobiology and psychobiology. In: Costa E, Racagni G (eds) Typical and atypical antidepressants: molecular mechanisms. Raven, New York
Sulser F (1986) The "serotonin-noradrenaline link" and transmembrane signalling: physiological and pharmacotherapeutic implications. In: Hippius H, Matussek N (eds) Advances in pharmacotherapy, vol 2. Karger, Basel, pp 175–188
Svensson TH, Fuxe K, Härfstrand A (1986) Pharmacological and physiological studies on brain adrenaline and noradrenaline neurons. Clin Neuropharmacol 9(4):180
Terry LC (1984) Catecholamine regulation of growth hormone and thyrotropin in mood disorders. In: Brown GM et al. (eds) Neuroendocrinology and psychiatric disorder. Raven, New York, pp 237–254
Terry LC (1986) Neuroendocrine regulation by brain epinephrine. Clin Neuropharmacol 9(4):172–174
Terry LC (1988) Role of epinephrine in regulation of anterior pituitary hormone secretion. In: Stolk G, U'Prichard D, Fuxe K (eds) Epinephrine in the central nervous system. Oxford University Press, New York
Tilson HA, Chamberlain JH, Gylys JA, Buyniski JP (1977) Behavioral suppressant effects of clonidine in strains of normotensive and hypertensive rats. Eur J Pharmacol 43:99–105
Tuomisto J, Männistö P (1985) Neurotransmitter regulation of anterior pituitary hormones. Pharmacol Rev 37:249–332
Uhde TW, Vittone BJ, Siever LJ, Kaye WH, Post RM (1986) Blunted growth hormone response to clonidine in panic disorder patients. Biol Psychiatry 21:1077–1081
U'Prichard DC (1984) Biochemical characteristics and regulation of brain α2-adrenoceptors. In: Salama AI (ed) Presynaptic modulation of postsynaptic receptors in mental diseases. Ann NY Acad Sci 430:55–75
U'Prichard D, Loftus D, Stolk G (1988) Functional association between epinephrine and α-adrenoceptor in discrete brain areas: analysis in bred rat strains. In: Stolk G, U'Prichard D, Fuxe K (eds) Epinephrine in the central nervous system. Oxford University Press, New York
Van Valkenburg C, Winokur G (1984) Hypertension and paranoia. Am J Psychiatry 141:999–1000
Vizi S (1983) Releasing-modulating adrenoceptors. In: Kunos G (ed) Neurotransmitter receptors. Adrenoreceptors and catecholamine action, part B. Wiley, New York, pp 65–107
Vogel WH, Lewis LE, Boehme DH (1976) Phenylethanolamine-*N*-methyltransferase activity in various areas of human brain, tissues and fluids. Brain Res 115:357–359
v Zerssen (1981) Befindlichkeits-Skala. In: CIPS (ed) Internationale Skalen für Psychiatrie. Beltz, Weinheim
v Zerssen D, Berger H, Doerr P (1984) Neuroendocrine dysfunction in subtypes of depression. In: Shah NK, Donald AG (eds) Psychoneuroendocrine dysfunction. Plenum, New York
Weeke A (1979) Causes of death in manic-depressives. In: Schou M, Strömgren E (eds) Origin, prevention and treatment of affective disorders. Academic Press, London New York, pp 289–299
Weissman MM, Myers JK, Leaf PJ, Tischler GL, Holzer CE (1986) The affective disorders: results from the epidemiologic catchment area study (ECA). In: Hippius H, Klerman GL, Matussek N (eds) New results in depression research. Springer, Berlin Heidelberg New York, pp 16–26
Weizman A, Gil-Ad I, Weizman R, Hering R, Bechar M, Tyano S, Laron Z (1984) Basal plasma HGH and cortisol levels and the effect of clonidine administration in female migrainous patients. Neuropsychobiology 12:106–111

Widerlöv E, Wahlestedt, Hakanson R, Ekman R (1986) Altered brain neuropeptide function in psychiatric illnesses – with special emphasis on NPY and CRF in major depression. Clin Neuropharmacol 9(4):572–574

Winokur G (1986) Familial classification of depressive illness. In: Hippius H, Klerman GL, Matussek N (eds) New results in depression research. Springer, Berlin Heidelberg New York

Wirz-Justice A, Kräuchi K, Campbell IC, Feer H (1983) Adrenoceptor changes in spontaneous hypertensive rats: a circadian approach. Brain Res 262:233–242

Wood K, Swade C, Coppen A (1985) Platelet α-adrenergic receptors in depression: ligand binding and aggregation studies. Acta Pharmacol Toxicol 56[suppl 1]:203–211

Wright JH, Jacisin JJ, Radin NS, Bell RA (1978) Glucose metabolism in unipolar depression. Br J Psychiatry 132:386–393

Serotonin and Mood: Neuroendocrine Aspects

H. Y. Meltzer and J. F. Nash Jr.

Contents

1 Introduction

The role of serotonin (5-HT) in the pathophysiology of mood disorders has been intensively investigated during the past 25 years (Meltzer and Lowy 1987). If 5-HT is involved in the etiology of affective disorders and the mechanism of action of thymoleptic drugs, it must have a significant role, direct or indirect, in the regulation of somatic processes that are disturbed in these mood disorders. In fact, 5-HT has been shown to have an important influence on mood itself, as well as on sleep, sexual activity, appetite, circadian rhythms, neuroendocrine function, anxiety, motor activity, and cognitive function, all of which are disturbed in affective disorders (Meltzer and Lowy 1987).

There are multiple lines of evidence on which the 5-HT hypothesis of affective disorders is based (Murphy et al. 1978; Meltzer and Lowy 1987). These include the following: (1) low cerebrospinal fluid (CSF) levels of 5-hydroxyindoleacetic acid (5-HIAA), the major metabolite of 5-HT (van Praag et al. 1970; Agren 1980); (2) low levels of total and free plasma tryptophan (Moller et al. 1976; Moller 1985); (3) decreased platelet 5-HT uptake (Tuomisto et al. 1979; Meltzer et al.

Department of Psychiatry, School of Medicine, Case Western Reserve University, Cleveland, OH 44106, USA

Current Topics in Neuroendocrinology, Vol. 8

1981 a); (4) decreased platelet imipramine binding sites (Paul and Rehavi 1981; Meltzer et al. 1984 a; Briley and Langer 1986); and (5) altered hormone responses to serotonergic probes in depressed or manic patients (Meltzer and Lowy 1987). Furthermore, the 5-HT precursors L-tryptophan and 5-hydroxytryptophan (5-HTP) have been reported to possess limited antidepressant effects, which are enhanced when combined with classical antidepressants such as monoamine oxidase (MAO) inhibitors or tricyclic antidepressants (van Praag 1981, 1983; Thompson and Rankin 1982; Moller et al. 1986).

There is also evidence for an association between low 5-HT turnover and suicide (Asberg et al. 1976; Brown et al. 1982). Post-mortem brain samples obtained from suicide completers have generally been found to have reduced levels of 5-HT and/or 5-HIAA as well as having an increased number of 5-HT_2 binding sites (Beskow et al. 1976; Mann et al. 1986). Suicide attempters have a greater cortisol response to 5-HTP administration than those who did not make any attempts (Meltzer et al. 1984 b). Thus, a central 5-HT deficiency has also been suggested to play an important role in impulsive behavior which is associated with psychopathological conditions such as depression, suicidal behavior, and thrill-seeking behavior, as well as alcoholism and drug abuse (Virkkunen et al. 1987).

Neuroendocrine studies have provided a key means to further investigate the role of 5-HT in mood disorders. For example, 5-HT has been found to modulate the secretion of prolactin (PRL), growth hormone (GH), adrenocorticotropin (ACTH), thyroid-stimulating hormone (TSH), luteinizing releasing hormone (LH-RH) and β-endorphin (Tuomisto and Mannisto 1985). The discovery of multiple 5-HT receptor subtypes and the development of recaptor-specific 5-HT agonists and antagonists has provided fresh impetus for research into the role of 5-HT in regulating hormone secretion. Many other neurotransmitters and neuromodulators also affect the secretion of hormones (Tuomisto and Mannisto 1985). Thus, it is essential to understand the interplay of these neurotransmitters with each other. There is also extensive evidence that cortisol can affect 5-HT turnover and other aspects of its metabolism.

The purposes of this article are to review and give a critique of the clinical studies on the effects of 5-HT challenge agents on hormone secretion in normal volunteers, and in depressed and other psychiatric patients. The possible importance of 5-HT for the variety of disturbances of the hypothalamo-pituitary-adrenal (HPA) axis in depression as well as the effect of glucocorticoids on 5-HT metabolism will also be briefly discussed.

2 Regulation of Hormone Secretion

2.1 General Endocrine Mechanisms

Neuroendocrinology is, in part, the study of the physiological processes involved in the neural regulation of hormone secretion. A brief description of the neuroendocrine system will be included for those readers who may be unfamiliar with its general mechanism. Cortical and limbic system nuclei, especially those located in the hippocampus and amygdala, play a crucial role in regulating the hypothalamic-pituitary axis which in turn controls the target endocrine organs. Nuclei within the hypothalamus secrete releasing factors, usually peptides such as corticotropin releasing factor (CRF), thyrotropin releasing hormone (TRH), or biogenic amines such as dopamine, at the median eminence, into the pituitary portal circulation. This hypothalamo-hypophysial system includes the hypothalamo-adenohypophysial and the hypothalamo-neurohypophysial elements, both of which terminate on the primary plexus of the hypophysial portal vessels. Neurotransmitters and neuropeptides released from median eminence nerve terminals enter the primary portal capillaries and are conveyed by the hypophysial portal veins to the highly vascularized anterior pituitary where they interact with specific target cells to stimulate or inhibit pituitary hormone secretion. The pituitary hormones enter the general circulation which conveys them to the specific target organs. Increased plasma concentrations of the pituitary hormones, as well as target organ secretions such as cortisol, provide inhibitory feedback which limits the further secretion of hypothalamic releasing factors and pituitary hormones. Thus, the neuroendocrine system represents a dynamic, highly regulated physiological system which responds to internal or external perturbations in such a way as to maintain homeostasis.

2.2 Serotoninergic Influence on Hormone Secretion

Numerous neurotransmitters originating from anatomically distinct areas of the brain innervate the hypothalamus (Halàsz 1978). Although serotoninergic neurons comprise only a very small percentage of the total neurons in the brain, serotoninergic axons innervate many forebrain structures. There are serotoninergic influences on the neuroendocrine system via pathways from the medial and dorsal raphé nuclei to hypothalamic nuclei and the median eminence (Feldman and Quenzer 1984). Dorsal raphé nucleus 5-HT neurons which project to the hypothalamus appear to be involved in the regulation of PRL secretion (Fessler et al. 1984). Both animal and human studies suggest that pharmacological stimulation of 5-HT mechanisms will activate the HPA axis, resulting in elevated plasma levels of ACTH and corticosteroids (Fuller 1981; Holmes et al. 1982). Animal studies suggest that the serotoninergic mechanisms which activate the HPA axis and PRL secretion are independent of each other (Van de Kar et al. 1985). TRH secretion appears to be controlled by a dual mechanism. There is evidence for 5-HT dependent stimulation of TRH based in the dorsal or central raphé nuclei to the hypothalamus (Jordan et al. 1979) and an inhibitory effect either from the median raphé nucleus to the hypothalamus (Di Renzo et al. 1979) or an intrinsic hypothalamic 5-HT neuronal pathway. The nuclei which mediate the effects of 5-HT on growth hormone and ACTH secretion are much less clear (Tuomisto and Mannisto 1985).

Multiple 5-HT receptor subtypes have been known to exist for many years. Gaddum and Picarelli (1957) were the first to report that two distinct 5-HT receptors, "M" and "D", were present in peripheral smooth muscle. The development of specific 5-HT radioligands has resulted in great advances in the characterization of multiple 5-HT receptor subtypes in both peripheral and central locations. 5-HT receptors in brain defined by radioligand binding studies include the 5-HT_{1A}, 5-HT_{1B}, 5-HT_{1C}, and 5-HT_2 subtypes (Hoyer et al. 1986a, b; Conn and Sanders-Bush 1987). Peroutka et al. (1981) demonstrated that the physiological effects of 5-HT_1 and 5-HT_2 specific agonists may be distinct. For example, it has been recognized for many years that activation of central 5-HT systems results in a behavioral syndrome characterized by limb abduction, splayed hind limbs, forepaw treading, straub tail, side-to-side head weaving, and head twitching (Jacobs 1976). The head-twitch response to 5-HT agonists is blocked by specific 5-HT_2 antagonists such as ketanserin (Peroutka et al. 1981); however, these antagonists have little or no effect on the other behavioral components of the 5-HT syndrome (Lucki et al. 1984). Similarly, antagonistic hypothermic and hyperthermic responses have been reported following 5-HT_{1A} and 5-HT_2 receptor activation respectively (Gudelsky et al. 1986). In other instances, however, synergistic effects may be expected. As the following section will clearly illustrate, various 5-HT challenge agents elicit similar or diverse neuroendocrine responses, suggesting that different 5-HT receptor subtypes may modulate different responses.

3 The Effect of 5-HT Challenge Drugs on Neuroendocrine Responses in Humans

3.1 5-HT Precursors

3.1.1 Tryptophan

5-HT is synthesized from its amino acid precursor L-tryptophan. Following the absorption of L-tryptophan in the diet or after pharmacological doses, most of it is bound to serum albumin and transported throughout the body. There is also a low concentration of free tryptophan in plasma. At the blood-brain barrier, a transport mechanism responsible for the uptake of large neutral amino acids (LNAA), including tyrosine, phenylalanine, leucine, isoleucine and valine as well as tryptophan, selectively transports these amino acids into the brain. The amino acids compete with each other so that the amount of tryptophan which enters the brain is a function of the ratio of its concentration and the concentration of other LNAAs (Fernstrom 1983). The conversion of tryptophan to 5-HT occurs via tryptophan hydroxylase which is located exclusively in 5-HT neurons. Under normal physiological conditions, tryptophan hydroxylase is believed to be unsaturated. Both dietary and pharmacological manipulations of tryptophan have been reported to alter the concentration of brain 5-HT (Fernstrom 1983), For example, serum tryptophan concentrations correlate significantly with both lumbar and ventricular CSF tryptophan (Curzon et al. 1980) and a strong correlation was demonstrated between plasma free tryptophan and cerebral cortical tissue tryptophan following i.v. administration of L-tryptophan prior to subcaudate tracheotomy (Gillman et al. 1981). These data suggest that L-tryptophan would selectively activate serotoninergic mechanisms and provide a useful challenge drug to assess 5-HT function.

This rationale has lead to many neuroendocrine studies using L-tryptophan as a specific 5-HT stimulus. These studies are summarized in Table 1. The results suggest that in normal controls, rapid i.v. infusion of L-tryptophan in doses equal to or greater than 100 mg/kg reliably increases the secretion of PRL and GH; furthermore, Winokur et al. (1986) demonstrated that the i.v. infusion of L-tryptophan failed to elevate serum cortisol and thyrotropin. In contrast to these results, oral administration of L-tryptophan appears to have very little effect on the secretion of PRL or, in most cases, GH. Modlinger et al. (1979, 1980) reported that high oral doses (10 g) of L-tryptophan elicit increased ACTH and cortisol secretion. However, these results are somewhat difficult to interpret since no placebo comparison group was included to control for the administration of L-tryptophan which was accomplished via fluid loading (1.5 l over 4 h). Both Woolf and Lee (1977) and Hyyppä et al. (1979) reported a decrease in serum cortisol levels following oral L-tryptophan administration, but the data are not convincing. We have found no consistent cortisol, PRL, or GH response to L-tryptophan (100 mg/kg, orally) in 10 unmedicated depressed patients (Meltzer and Nash, unpublished data). This is, in part, because slow, erratic absorption of oral L-tryptophan produces much lower plasma levels of L-tryptophan than the i.v. route.

Although the i.v. administration of L-tryptophan appears to increase PRL and GH secretion in most normal volunteers, the assumption that the effects of L-tryptophan are mediated exclusively via serotoninergic mechanisms has recently been questioned (Koyama and Meltzer 1986; van Praag et al. 1987; Meltzer and Koyama, unpublished data). Van Praag et al. (1987) reported that both the oral and intravenous routes produced comparable increases in CSF 5-HIAA accumulation 8 h after 59 of L-tryptophan was administered (post probenecid). However, CSF levels of homovanillic acid (HVA), the main dopamine (DA) metabolite, and 3-methoxy-4-hydroxyphenylglycol (MHPG), the main CNS metabolite of norepinephrine (NE), were found to be decreased or unaffected following i.v. or oral tryptophan load, respectively. Similarly, Meltzer and Koyama found that following the i.v. administration of L-tryptophan, serum HVA levels decreased significantly, suggesting interference with DA synthesis, release, uptake, or metabolism. The bolus administration of L-tryptophan may competitively inhibit the uptake of tyrosine into the CNS or into DA neurons and thus alter the synthesis of DA and/or NE or lead to a decrease in the firing of DA neurons via an inhibitory effect of 5-HT. These data suggest that L-tryptophan can affect serotoninergic and catecholaminergic mechanisms and that the hormone response to large doses of L-tryptophan might be due, in part, to its influence on catecholamine synthesis and release.

The effects of L-tryptophan on hormone secretion have been compared in normal controls and depressed patients in order to determine whether serotoninergic mechanisms are impaired in depression. Table 1 presents a summary of the results obtained in these studies. Westenberg et al. (1982) administered 5 g L-tryptophan, orally, to normal controls and depressed patients and failed to detect any differences in hormone secretion between these two groups. However, as previously mentioned, oral administration of L-tryptophan does not reliably stimulate the secretion of any pituitary hormone. On the other hand, both Heninger et al. (1984) and Meltzer and Koyama (unpublished data) found that the PRL response

Table 1. The effects of L-tryptophan administration on neuroendocrine hormone secretion in man

Authors	Year	Subjects	Dose	Route	Results			
					PRL	GH	Cortisol	ACTH
MacIndoe and Turkington	1973	8 N.C.	5–10 g	i.v.	↑			
Müller et al.	1974	32 N.C.	70 mg/kg	p.o.		↑		
Woolf and Lee	1977	11 N.C.	10 g	p.o.	↑*	↑*	↓	↓
Wiebe et al.	1977	8 N.C.	90 mg/kg	p.o.	↔			
Hyyppä et al.	1979	5 N.C.	2 g	p.o.	↔	↑	↓	
Fraser et al.	1979	20 N.C.	5 g	p.o.	↑*	↑*		
Modlinger et al.	1979	9 N.C.	2 and 10 g	p.o.			↑	
Modlinger et al.	1980	11 N.C.	10 g	p.o.	↔		↑	↑
Glass et al.	1980	8 N.C.	5 g	p.o.	↔			
Charney et al.	1982	10 N.C.	7 g	i.v.	↑	↑		
Cowan et al.	1986	10 N.C.	7.5 g	i.v.	↑	↑		
Winokur et al.	1986	11 N.C.	5–10 g	i.v.	↑	↑	↔	
Westenberg et al.	1982	5 N.C.	5 g	p.o.	↔	↔	↔	
		9 dep.			↔	↔	↔	
Heninger et al.	1984	19 N.C.	7 g	i.v.	↑			
		25 dep.			↑**			
Meltzer et al.	1987	13 N.C.	100 mg/kg	i.v.	↑	↑		
		13 dep			↑**	↑**		

N.C., normal controls; dep., depressed patients; p.o., oral; i.v., intravenous; * not significantly elevated; ** significantly smaller increase than in corresponding control subjects to the i.v. administration of L-tryptophan was blunted in depressed patients as compared to age- and sex-matched normal controls. Furthermore, Meltzer and Koyama found that the GH response to L-tryptophan was also blunted in depressed patients as compared to normal controls. More importantly, these authors also found that the serum levels of L-tryptophan differed with respect to peak and "total" (area under the curve) amount, i.e., the area under the curve following the i.v. infusion of L-tryptophan was significantly less in depressed patients than in normal controls. After excluding these differences in serum L-tryptophan concentration, the PRL responses were no longer significantly different whereas the GH responses remained significantly blunted in the depressed patients compared to normal controls.

These data suggest that the availability of L-tryptophan determines the hormone response and that the disposition of the L-tryptophan load was significantly different in the depressed patients compared to normal controls. After treatment with antidepressants, these depressed patients still had significantly lower levels of L-tryptophan (Ohmori et al., unpublished data) and GH following a similar load (Meltzer et al., unpublished data). We have also found evidence for a dysregulation of tryptophan hydroxylase in depressed patients in that no correlation between L-tryptophan and 5-HIAA levels in serum was present whereas there was a highly significant correlation in normal controls. In both depressed patients and controls there were highly significant correlations between serum L-tryptophan and kynurenine, the major metabolite of the tryptophan pyrrolase pathway (Ohmori et al., unpublished data). The blunted GH but normal PRL response to an i.v. tryptophan load after adjusting for differences in serum L-tryptophan levels in depressed patients suggests that the dysregulation of tryptophan hydroxylase may not be present in all 5-HT neurons.

Subacute administration of amitriptyline or desipramine to depressed patients enhanced the PRL response to i.v. L-tryptophan (Charney et al. 1984) as did tranylcypromine, an MAO inhibitor (Price et al. 1985). Chlorimipramine treatment enhanced the PRL and GH response following infusion of L-tryptophan (50 mg/kg) in six normal volunteers, suggesting a 5-HT mechanism was involved in these hormone responses (Anderson and Cowen 1986). However, chlorimipramine might have made more L-tryptophan available for nonspecific effects via an effect on L-tryptophan metabolism or uptake into 5-HT neurons. These authors also found no inhibitory effect of two 5-HT antagonists, cyproheptadine or ketanserin (Cowen et al. 1986), on PRL or GH secretion following i.v. L-tryptophan. They concluded from these studies that 5-HT_1 rather than 5-HT_2 receptors may mediate these hormone responses but it is also possible that a nonserotoninergic mechanism is involved.

There are a few studies of the effect of i.v. L-tryptophan on hormone secretion in patients with nonaffective psychiatric disorders. The PRL response to i.v. L-tryptophan did not differ in patients with panic disorder or agoraphobia and normal controls. No effect of prior treatment with alprazolam, a benzodiazepine anxiolytic, on the PRL response to i.v. L-tryptophan in panic patients was noted (Charney and Heninger 1986). Schizophrenics treated with neuroleptics were reported to have an enhanced PRL response to i.v. L-tryptophan (Cowen et al. 1985).

In summary, although differences between normal controls and depressed patients exist with respect to hormone secretion in response to L-tryptophan loading, the 5-HT specificity of this response is questionable. Whereas i.v. infusion of L-tryptophan stimulates 5-HT function, it is still not clear to what extent the hormone responses to L-tryptophan infusion are serotoninergic in origin. Furthermore, it is difficult to determine whether differences in pre- or postsynaptic mechanisms are responsible for the blunted GH response observed in depressed patients following L-tryptophan loading.

3.1.2 5-Hydroxytryptophan (5-HTP)

Following conversion of tryptophan to 5-HTP by tryptophan hydroxylase, the next step in the synthesis of 5-HT is the conversion of 5-HTP to 5-HT via the enzyme, L-aromatic amino acid decarboxylase. Since this enzyme is nonspecific and located in many monoaminergic neurons, the specificity of 5-HTP as a 5-HT neuroendocrine challenge agent has been seriously questioned (Fuxe et al. 1971; Ng et al. 1972; van Praag et al. 1987). Nevertheless, as shown in Table 2, numerous clinical neuroendocrine studies utilizing 5-HTP have been carried out. As is the case with L-tryptophan, the effect of 5-HTP on hormone secretion depends on the dose and route of drug administration. For example, i.v. administration of 5-HTP produces a more consistent increase in PRL secretion than does oral administration (Table 2). In two of the studies in which i.v. administration of 5-HTP failed to stimulate the secretion of PRL, the subjects had been pretreated with a peripheral decarboxylase inhibitor. The decarboxylase inhibitor carbidopa has been reported to increase PRL secretion (Lancranjan et al. 1977; Garfinkel et al. 1979), possibly by decreasing the release of DA in the median eminence. It also suggests the possibility that 5-HTP may stimulate hormone release via a mechanism which is outside the blood-brain barrier. Similarly, GH was increased following i.v. 5-HTP administration in three studies, whereas the oral administration of 5-HTP fails consistently to stimulate GH secretion (Meltzer et al. 1982, 1984c).

The effects of 5-HTP on ACTH and cortisol secretion have been studied infrequently. In normal controls, two studies have reported that 5-HT administration stimulated cortisol secretion (Imura et al. 1973; Mashchak et al. 1983), whereas two studies found that cortisol was unchanged following 5-HTP administration (Westenberg et al. 1982; Meltzer et al. 1984c) in normal volunteers. However, we have found that the oral administration of 200 mg L-5-HTP produced a significant increase in the secretion of ACTH and cortisol in normal volunteers (Meltzer and Nash, unpublished observation). In previous studies (Meltzer et al 1984b–d), 200 mg DL-5-HTP failed to significantly elevate serum cortisol in normal volunteers, as compared to placebo administration. Our data suggest that, in sufficient doses, the oral administration of L-5-HTP stimulates cortisol secretion in normal volunteers.

There have been three controlled neuroendocrine challenge studies which have compared the hormone response in depressed patients with that obtained in normal controls following 5-HTP administration. Takahashi et al. (1973) found no differences in GH secretion between normal controls and manic patients following oral administration 200 mg L-5-HTP. However, in depressed patients, the

Table 2. The effects of 5-hydroxytryptophan administration on neuroendocrine hormone secretion in man

Authors	Year	Subjects	Isomer	Dose	Route	Results			
						PRL	GH	Cortisol	ACTH
Imura et al.	1973	8 N. C.	–	150 mg	p. o.		↑	↑	↑
Nakai et al.	1974	8 N. C.	–	150 mg	p. o.		↑		
Kato et al.	1974	21 N. C.	–	200 mg	p. o.	↑			
Handwerger et al.	1975	6 N. C.	L-5-HTP	200 mg	p. o.	↔	↔		
Pühringer et al.	1976	16 N. C.	L-5-HTP*	200 mg	i. v.	↔	↑		
Beck-Peccoz et al.	1976	10 N. C.	–	200 mg	p. o.	↔			
Lancranjan et al.	1977	11 N. C.	L-5-HTP*	200 mg	i. v.	↔	↑		
Mashchak et al.	1983	15 N. C.	L-5-HTP*	0.8 mg/kg	i. v.	↑	↑	↑	
Ansseau et al.	1983	4 N. C.	–	200 mg	p. o.		↔		
Sueldo et al.	1986	9 N. C.	–*	0.2–0.8 mg/kg	i. v.	↑			
Takahashi et al.	1974	7 N. C.	L-5-HTP	200 mg	p. o.		↑		
		4 manic					↑		
		6 dep.					↔***		
Westerburg et al.	1982	5 N. C.	L-5-HTP*	200 mg	p. o.	↔	↔	↔	
		9 dep.				↔	↔	↔	
Meltzer et al.	1984c	15 N. C.	DL-5-HTP	200 mg	p. o.			↔	
		31 dep.						↑**	
		16 manic						↑**	

N. C., normal controls; dep., depressed patients; p. o., oral; i. v., intravenous; * a peripheral decarboxylase inhibitor was administered prior to 5-HTP; ** significantly greater increase than in corresponding controls subjects; *** significantly smaller effect than in corresponding control subjects

GH response to 5-HTP was blunted as compared to normal controls. Similarly, the GH response in depressed patients to the insulin tolerance test has also been reported to be blunted (Amsterdam et al. 1987).

Westenberg et al. (1982) found that the administration of 200 mg L-5-HTP (enteric coated tablets), following 3 days' pretreatment with 50 mg carbidopa three times a day, failed to stimulate the secretion of PRL, GH or cortisol in either normal controls or depressed patients. The lack of an effect of 5-HTP on the secretion of any of these hormones is in contrast with several of the earlier studies (Table 2) and is not easily explained. However, the peak plasma 5-HTP levels they observed came considerably later (130 min after administration) than those previously reported by Magnussen and Engbaek (1978), 30–60 min after administration of 5-HTP. The time course for the peak 5-HTP serum concentrations in this latter study corresponds more closely to the peak change in serum cortisol reported by Meltzer et al. (1984c) as well as to the peak change in GH reported by Takahashi et al. (1974).

More recently, Meltzer et al. (1984c) reported that the cortisol response to 200 mg orally administered DL-5-HTP in depressed and manic patients was greater than that in normal controls. Moreover, the cortisol response was positively correlated with the severity of depression and suicidality. Furthermore, in a separate study CSF 5-HIAA levels were found to be negatively correlated with the cortisol response to 5-HTP in a group of depressed patients (Koyama et al. 1987). The enhanced cortisol response in the depressed patients was proposed to be due to a 5-HT postsynaptic receptor supersensitivity as a consequence, in part, of reduced concentrations of central 5-HT. This hypothesis is based on animal studies in which central deficiencies of brain 5-HT following lesions or pretreatment with parachlorphenylalanine (PCPA), a tryptophan hydroxylase inhibitor, resulted in enhanced neuroendocrine responses to serotoninergic challenge drugs (Kuhn et al. 1981; Quattrone et al. 1981; Meltzer et al. 1982). Administration of tricyclic antidepressants normalized the cortisol responses to 5-HTP challenge in depressed patients (Meltzer 1984d), possibly by down-regulating 5-HT receptor numbers (Fuxe et al. 1985). Conversely, chronic lithium administration enhanced the 5-HTP-induced increase in cortisol in manic patients (Meltzer et al. 1984d).

3.2 Fenfluramine

Fenfluramine, an anorexiant drug widely used in the treatment of obesity, is the only 5-HT-releasing drug whose neuroendocrine effects have been extensively tested in humans. In addition to increasing the release of 5-HT, fenfluramine also inhibits its reuptake into 5-HT neurons (Fuxe et al. 1975; Garrantini et al. 1975; Quattrone et al. 1978) so it does not provide a discrete measure of 5-HT stores. Obviously, postsynaptic responses to 5-HT will also influence the secretory response to fenfluramine and its active metabolites. There is also evidence that the PRL response to fenfluramine is inhibited by pretreatment with naloxone (Foresta et al. 1985) suggesting an opiate-dependent influence on serotoninergic neurons which must also be considered in evaluating the PRL response to fenfluramine. The L-isomer of fenfluramine may also increase DA availability

Table 3. The effects of fenfluramine on neuroendocrine hormone secretion in man

Authors	Year	Subjects	Dose	Route	Results			
					PRL	GH	Cortisol	ACTH
Quattrone et al.	1983	9 N. C.	20–100 mg	p. o.	↑			
Lewis and Sherman	1984	9 N. C.	0.5–1.5 mg/kg	p. o.			↑	↑
Casanueva et al.	1984	6 N. C.	20 mg	i. v.	↑	↔		
Scarduelli et al.	1985	16 N. C.	40 and 80 mg	p. o.	↑			
Foresta et al.	1985	7 N. C.	60 mg	p. o.	↑			
Siever et al.	1984	10 N. C. 18 dep.	60 mg	p. o.	↑ ↑**			
Mühlbauer and Müller-Oerlinghausen	1985	12 N. C. 8 manic-dep.	60 mg	p. o.			↑* ↑*	

N. C., normal controls; dep., depressed patients; p. o. oral; i. v., intravenous; * not significantly elevated; ** significantly smaller increase than in corresponding control subjects

(Invernizzi et al. 1986), which would tend to diminish the PRL response to the racemic mixture of fenfluramine which is ordinarily used in clinical studies. However, it is most likely that the major effect of fenfluramine on PRL secretion is mediated through a 5-HT dependent release of hypothalamic PRL releasing factor(s) or by inhibiting the release of DA from tuberoinfundibular DA neurons (Quattrone et al. 1978).

Table 3 summarizes the neuroendocrine effects of fenfluramine in humans. There are four studies demonstrating a PRL response in normal controls. However, not all subjects will give a PRL response to fenfluramine, which has led van Praag et al. (1987) to question its usefulness as a neuroendocrine probe.

Siever et al. (1984) reported the effects of fenfluramine administration on PRL secretion in normal controls and depressed patients. The PRL response was blunted in the depressed patients, consistent with the hypothesis of diminished serotoninergic activity in this group. However, had there been a compensatory upregulation of 5-HT receptors relevant to the PRL response or decreased uptake of 5-HT released by fenfluramine, no blunting of the PRL response to fenfluramine might have been predicted unless the 5-HT deficiency were of major proportions. We have found no evidence of a blunted PRL response in unmedicated depressed patients following fenfluramine (60 mg orally) compared to normal controls (Meltzer, unpublished data).

Lewis and Sherman (1984) reported that oral administration of fenfluramine in doses of 0.5–1.5 mg/kg increased cortisol and ACTH secretion; these responses were blocked by the 5-HT antagonist cyproheptadine in a dose-dependent manner. However, in a study by Muhlbauer and Müller-Oerlinghausen (1985), fenfluramine administration was reported to block the morning decline of serum cortisol in eight bipolar depressive drug-free patients in remission and normal controls but it did not produce an increase in serum cortisol levels over baseline in either the normal controls or the drug-free patients studied. We have also found no consistent effect of fenfluramine on serum cortisol levels in comparison with placebo in unmedicated depressed patients (Meltzer et al., unpublished data). Muhlbauer and Müller-Oerlinghausen did find an increase in cortisol in lithium-treated, remitted bipolar depressive patients given fenfluramine.

Further studies with fenfluramine as a challenge agent appear warranted in patients with mood disorders to clarify some of the inconsistencies in the literature. Adequate attention to issues such as subtype of depression, type and severity of symptoms, previous and current medication status, sex, response to placebo, and appropriate control and comparison groups will be needed.

3.3 Direct 5-HT Agonists

Because of the possibility that there are alterations in postsynaptic 5-HT receptors in patients with mood disorders (Meltzer et al. 1984c; Mann et al. 1986), there is considerable interest in the effect on hormone secretion of specific, direct-acting 5-HT agonists. The hormone response to such agents is not directly influenced by presynaptic serotoninergic activity other than what might be expected from a slight synergism with pharmacologic doses of the direct-acting agonists.

Until recently, the only available direct-acting 5-HT agonists for neuroendocrine studies were indole hallucinogens (Meltzer et al. 1983) which cannot be used in patients due to their adverse psychic effects. However, recent advances in the characterization of 5-HT receptor subtypes has led to the development of 5-HT agonists which lack hallucinogenic effects.

Table 4 presents a summary of the neuroendocrine response to direct-acting 5-HT agonists. For the most part, these data suggest that regardless of the receptor specificity, the secretion of PRL and, when measured, cortisol appears to be enhanced following the administration of different 5-HT agonists.

Mescaline (3,4,5-trimethoxy-β-phenylethylamine) and other phenethylamines have been reported to display a low affinity for 5-HT receptors in ligand binding assays; nevertheless, analogues of mescaline appear to interact preferentially with 5-HT_2 receptors, although the data are not as convincing for mescaline (Titeler et al. 1985; Penington and Reiffenstein 1986). Mescaline acts mainly via 5-HT-dependent mechanisms (Meltzer et al. 1978). Demisch and Neubauer (1979) reported that the oral administration of mescaline increased the secretion of prolactin and, although not significantly, GH. The hormone response to mescaline may be confounded with its stressful psychic effects. Similarly, Meltzer et al. (1981 b), found that *N,N*-dimethyltryptamine (*N,N*-DMT), an indole hallucinogen, stimulated the secretion of both PRL and GH in a small sample of experienced drug users. *N,N*-DMT has been reported to have a greater affinity for 5-HT_{1A} receptors, based on radioligand binding studies (Glennon 1987). Meltzer et al. (1981 b) also found that phencyclidine (PCP), when administered to the same group of subjects, failed to augment the secretion of these hormones, suggesting that the stimulation of 5-HT mechanisms rather than the psychotomimetic effects of *N,N*-DMT were responsible for the elevation in hormones. Interestingly, administration of lysergic acid diethylamide (LSD), had no effect on PRL secretion but increased GH secretion in this small group of subjects. The psychotomimetic effects of LSD were more potent than those of mescaline or *N,N*-DMT, yet administration did not elicit a PRL response. LSD has equal binding affinity for both 5-HT_1 and 5-HT_2 receptors (Glennon 1987), either of which might be expected to stimulate PRL secretion (see below). A plausible explanation for the lack of effect of LSD on PRL secretion and its enhancement of the GH secretion may be that the DA agonist properties of LSD would overcome the enhancement of PRL secretion via 5-HT-dependent mechanisms (Meltzer et al. 1977). Regardless of these results, these three agents have serious practical limitations in their use as neuroendocrine challenge drugs in psychiatric patients, as a result of their potent hallucinogenic effects.

Parati et al. (1980) reported that the arylpiperazine quipazine has no effect on PRL or GH secretion but does significantly elevate cortisol secretion. Quipazine is a nonspecific 5-HT agonist which has roughly equal affinity for both 5-HT_1 and 5-HT_2 binding sites (Glennon 1987). More recently, two other arylpiperazines, *m*-chlorophenyl piperazine (mCPP) and 1-piperazinyl-6-chloropyrazine (MK-212) have been studied in humans. mCPP has a high binding affinity for 5-HT_{1B} receptors (Glennon et al. 1986). Mueller et al. (1985, 1986) reported that mCPP stimulated the secretion of PRL and cortisol but had no effect on growth hormone secretion. The effects of mCPP were blocked by the nonspecific 5-HT antagonist

Table 4. The effect of direct-acting 5-HT agonists on neuroendocrine hormone secretion in man

Authors	Year	Subjects	Drug	Dose	Route	Results			
						PRL	GH	Cortisol	ACTH
			Receptor Subtype						
			$5\text{-}HT_1$						
Meltzer et al.	1981	3 drug users	*N,N* dimethyl-tryptamine	0.7 mg/kg	i. m.	↑	↑		
Meltzer et al.	1983	8 N. C.	Buspirone	30–90 mg	p. o.	↑	↑		
Mueller et al.	1985	15 N. C.	mCPP	0.5 mg/kg	p. o.	↑		↑	
Mueller et al.	1986	10 N. C.	mCPP	0.5 mg/kg	p. o.	↑	↔	↑	↑
			$5\text{-}HT_2$						
Demisch and Neubauer	1979	5 N. C.	Mescaline	5 mg/kg	p. o.	↑	↑*		
Lowy et al.	1988	8 N. C.	MK-212	10–40 mg	p. o.	↑	↔	↑	↑
			Mixed $5\text{-}HT_1/5\text{-}HT_2$						
Parati et al.	1980	8 N. C.	Quipazine	50 mg	p. o.	↔		↑	
Meltzer et al.	1981	3 drugs users	LSD	100 ug	p. o.	↔	↑		

N. C., normal controls; p. o., oral; i. m., intramuscular; * not significantly elevated metergoline. Our group (Lowy and Meltzer 1988) showed that the reputed 5-HT_2 agonist MK-212 stimulates the secretion of PRL and cortisol in a dose-dependent manner. Preliminary results with MK-212 (20 mg orally) suggest the cortisol response is enhanced in patients with unipolar depression (Meltzer et al., unpublished data). MK-212 has little effect on GH secretion. MK-212 displays a low binding affinity for either 5-HT_1 or 5-HT_2 receptors; however, it has high efficacy in vivo at 5-HT receptors (Clineschmidt et al. 1985). More importantly, Koenig et al. (1987) reported that the stimulatory effects of MK-212 on corticosterone and β-endorphin secretion were blocked by ketanserin, the prototypic 5-HT_2 antagonist, in rats. Thus, it is conceivable that the low binding affinity of MK-212 at 5-HT_2 receptors accounts for the lack of hallucinogenic effects, while at the same time it can stimulate the secretion of various hormones.

Finally, the novel anxiolytic buspirone has been reported to possess significant ligand binding affinity for 5-HT_{1A} receptors (Peroutka 1985). Meltzer et al. (1987) reported that buspirone stimulated the secretion of PRL, cortisol, and GH in normal volunteers. It is probable that some or all of these effects are due to a 5-HT_{1A} mechanism but specific evidence, in humans is lacking. We are currently studying the effects of buspirone on PRL, cortisol, and GH secretion in unmedicated depressed patients.

3.4 5-HT Uptake Inhibitors

Tricyclic antidepressants are believed to enhance serotoninergic transmission by inhibiting the reuptake process which terminates 5-HT activation of postsynaptic receptors (Willner 1985). However, until recently, most tricyclics were pharmacologically nonspecific in that they not only inhibited 5-HT but also blocked the reuptake of NE either directly or via active metabolites. More recently, newer, selective 5-HT uptake inhibitors such as fluoxetine, zimelidine, citalopram, alaproclate, and fluvoxamine have been developed which have little effect on the uptake of other monoamines. More importantly, in the majority of controlled clinical trials, these newer antidepressants have been shown to be clinically effective in the treatment of depression (Meltzer and Lowy 1987). Thus, the neuroendocrine effects of these agents are of interest both theoretically and clinically not only in patients with mood disorders but also in normal controls.

A number of studies examining the acute neuroendocrine effects of specific 5-HT uptake inhibitors in normal controls have been performed and are summarized in Table 5. In general, these data suggest that the specific 5-HT uptake inhibitors have little effect by themselves on PRL, cortisol, or ACTH secretion in humans. In the majority of the studies in which GH secretion was measured, the results were quite variable and appeared to be sex related. That is, the 5-HT uptake inhibitors more consistently increased GH in females than males. Trazodone tended to decrease the basal levels of PRL in normal controls. The mechanism of action of trazodone is unclear; however, there is evidence that it antagonizes the effect of 5-HT agonists (Meltzer et al. 1981). Therefore, the reduction in basal PRL secretion may be the result of blockage of 5-HT stimulatory input. We have found that trazodone treatment will block the effect of 5-HTP or MK-212 on the

Table 5. The effect of serotoninergic antidepressant drugs on neuroendocrine hormone secretion in man

Authors	Year	Subjects	Drug	Dose	Route	Results			
						PRL	GH	Cortisol	ACTH
Meltzer et al.	1977b	2/2 dep.	Imipramine/ amitriptyline	75 mg	p. o.	↔			
Laakmann et al.	1977	6 N. C.	Chlorimipramine	75 mg	i. m.		↑*		
		6 N. C.	Chlorimipramine	100 mg	p. o.		↑*		
Jones et al.	1977	4 N. C.	Chlorimipramine	50 mg	p. o.	↔			
Syvälahti et al.	1979a	14 N. C.	Alaprolclate	200 mg	p. o.	↔	↔		
		14 N. C.	Zimelidine			↔	↔	↑*	
Syvälahti et al.	1979b	5 N. C.	Zimelidine	100 mg	p. o.	↔			
Masala et al.	1979	5 N. C.	Fluoxetine	30 mg	p. o.	↔			
Schulz et al.	1982	6 N. C.	Amitriptyline	20–60 mg	i. v.		↑*		
		6 N. C.	Amitriptyline	40–100 mg	p. o.		↑*		
Laackmann et al.	1984a	12 N. C.	Chlorimipramine	25 mg	i. v.	↑	↑*		
Laackmann et al.	1984b	6 N. C.	Chlorimipramine	25 mg	i. v.			↑	
Holsboer et al.	1985	6 N. C.	Zimelidine	100 mg	p. o.				↑**
Calil et al.	1984	11 dep.	Zimelidine	100 mg	p. o.	↔	↔	↔	
Roccatagliata et al.	1979	10 psych.	Trazodone	100 mg	i. v.	↓*			
Holl et al.	1986	15 N. C.	Chlorimipramine	50 mg	p. o.			←	↔
Rolandi et al.	1981	10 N. C.	Trazodone	50 mg	p. o.	↓			

N. C., normal controls; dep., depressed patients; psych., psychotic patients; p. o., oral; i. m., intramuscular; i. v., intravenous; * not significantly elevated; ** potentiated response to metyrapone pretreatment cortisol or PRL and cortisol responses, respectively (Meltzer et al., unpublished data). The route of administration of 5-HT uptake blockers, unlike 5-HT precursors, does not overtly influence their neuroendocrine effects. We have observed a marked potentiation by fluoxetine and sertraline of the cortisol and PRL responses to 5-HTP in man (Meltzer, unpublished data). This enhancement is not observed with the less selective uptake blockers such as amitriptyline and desipramine. This suggests that in vivo there may be a functional difference in 5-HT uptake blockade between these classes of drugs or that other mechanisms intervene to produce different responses to the specific or nonspecific 5-HT uptake blockers. Studies are needed to determine whether the hormone response to 5-HTP in patients receiving specific 5-HT uptake blockers differs from that of normal controls, and whether this is related to clinical response or other measures of 5-HT function. The data suggest that 5-HT uptake inhibitors are not sufficient neuroendocrine challenge agents by themselves but they may be useful when combined with 5-HTP.

The ability of 5-HT uptake blockers to sustain an effective concentration of 5-HT in the synapse without needing pharmacological doses of 5-HTP to promote 5-HT synthesis may be limited due to the stimulation of presynaptic autoreceptors which suppress the release of 5-HT. It has been shown that the behavioral and endocrine effects of 5-HT uptake inhibitors in rodents are absent when these agents are administered alone but they do potentiate the effects of the 5-HT precursors (Fuller 1981, 1986). Therefore, the limited neuroendocrine response to the administration of 5-HT uptake inhibitors alone is not surprising.

3.5 Antagonism of 5-HT-Induced Neuroendocrine Responses

A serotoninergic neuroendocrine challenge agent must meet at least three basic requirements: (1) the challenge drug should stimulate or inhibit hormone secretion through 5-HT mechanisms with a high degree of specificity; (2) a dose-dependent relationship between the drug and hormone response should be demonstrable; and (3) if mechanisms other than 5-HT-related ones influence the hormone response to the challenge agent these should be known and independently assessed. To data, 5-HTP, MK-212, mCPP, and perhaps buspirone come closest to meeting these criteria. The use of antagonists helps to clarify the role of 5-HT in response to challenge drugs. MacIndoe and Turkington (1973) found that the nonspecific 5-HT antagonist methysergide partially attenuated the secretion of PRL following 10 g i.v. L-tryptophan in two female subjects. Similarly, Fraser et al. (1979) reported that 4 g cyproheptadine completely blocked the GH response to 5 g orally administered L-tryptophan. Conversely, Cowen et al. (1985) failed to obtain a significant attenuation of either the PRL or growth hormone response to 7.5 g i.v. L-tryptophan following pretreatment with 4 g cyproheptadine. In two separate studies, the effects of the oral administration of 200 mg 5-HTP on PRL and GH secretion were significantly blunted by pretreatment with cyproheptadine (Kato et al. 1974; Nakai et al. 1974). Furthermore, Lewis and Sherman (1984) reported that, in four normal controls pretreated every 6 h for 2.5 days with cyproheptadine (0.6 mg/kg), the cortisol and ACTH responses to fenfluramine (1.5 mg/kg) were significantly blunted. Most recently, Mueller et al. (1986) found that metergoline pretreatment in 10 normal controls completely blocked the stimulatory effects of mCPP on PRL, cortisol, and ACTH. However, as previously mentioned the interpretation of the results obtained in these studies is questionable since nonselective 5-HT antagonists were used.

Perhaps, with the introduction of highly selective 5-HT_2 antagonists such as ketanserin, the precise role of 5-HT in hormone secretion may be more accurately investigated. To date, the neuroendocrine studies in which ketanserin has been used have involved physiological challenge tests such as exercise- and insulin-induced hypoglycemia. For example, the insulin tolerance test has been shown to increase the secretion of PRL, GH and ACTH, effects which have been reported to be mediated via serotoninergic mechanisms (Kletzky et al. 1980). Thus, Prescott et al. (1985) reported that ketanserin blocked PRL and ACTH release during insulin-induced hypoglycemia. Similarly, Meirleir et al. (1985) found that ketanserin attenuated exercise-induced PRL secretion. Although very few studies have been performed, the results obtained in these studies are encouraging.

3.6 5-HT Antagonists

To date, very few studies have been performed on the neuroendocrine effects of 5-HT antagonists. In the majority of studies, 5-HT antagonists have been used to block the neuroendocrine effects of 5-HT challenge drugs. Nevertheless, there are rare instances in which the effects of 5-HT antagonists themselves have been of primary interest. For example, Mendelson et al. (1975) reported that the nonspecific 5-HT antagonist methysergide blocked sleep-induced PRL secretion in 10 normal controls. However, methysergide is also a DA agonist and this probably accounts for its ability to suppress PRL secretion. Similarly, Chihara et al. (1976) found that the early morning rise in plasma cortisol was blocked by another nonselective 5-HT antagonist, cyproheptadine, in five normal controls. However, the former study reported an enhancement of sleep-related GH secretion, whereas the latter study found attenuation of sleep-induced growth hormone secretion. Ferrari et al. (1978) reported a suppression of basal PRL secretion by metergoline and methysergide, drugs which are both DA agonists and 5-HT antagonists. Both nocturnal secretion as well as basal levels of cortisol were reported to be inhibited by cyproheptadine (Delitala et al. 1975); however, Lewis and Sherman (1984) found that cyproheptadine did not significantly alter basal ACTH or cortisol levels. These data provided limited evidence that 5-HT is involved in the secretion of PRL, GH and cortisol under normal physiological conditions. However, these data must be interpreted cautiously since the antagonists used are nonspecific.

Recently, a group of quinazoline derivatives which selectively block 5-HT_2 receptors and are devoid of any agonist action have become available for preclinical and clinical investigations. The prototypic 5-HT_2 antagonist ketanserin has been reported to have no effect on basal or TRH-stimulated TSH and PRL levels in 10 normal volunteers (O'Malley et al. 1984). Similarly, Rocco et al. (1983) and Gordin et al. (1985) failed to observe any influence of ketanserin on PRL secretion. Further studies of the effect of ketanserin on the basal and stimulated cortisol secretion will be of interest since there is extensive evidence that, at least in the rat, some corticosteroid responses due to 5-HT stimulation are related to 5-HT_2 receptors (Koenig et al. 1987).

3.7 HPA Axis and 5-HT

There are a number of well-established abnormalities of the HPA axis in major depression: (a) increased 24 h urinary free cortisol; (b) increased plasma cortisol concentrations; (c) disturbances in the diurnal rhythm of cortisol secretion; (d) increased CSF cortisol concentrations; and (e) failure to suppress serum cortisol levels adequately after a low dose of the high-potency corticosteroid dexamethasone. The evidence for these abnormalities has been reviewed elsewhere (Meltzer and Lowy 1986).

There is considerable evidence that abnormalities in the serotoninergic system may mediate many aspects of the hypersecretion of cortisol and the failure of the glucocorticoid-sensitive feedback system to function properly. We have reviewed some of this evidence elsewhere (Meltzer et al. 1984d). Thus, diminished serotoninergic activitiy has been reported to increase basal levels of corticosterone (Scapagnini et al. 1971; Vernilios-Danellis et al. 1973) as well as to interrupt circadian rhythms of corticosterone secretion in the rat (Scapagnini et al. 1971; Szafarczyk et al. 1979; Honma et al. 1979). Träskman et al. (1980) found a positive correlation between CSF 5-HIAA and cortisol levels ($r = 0.46$, $n = 19$, $p < 0.05$) which, however, is more consistent with a stimulatory effect of 5-HT on cortisol secretion.

There is preclinical and clinical evidence that failure to suppress cortisol levels after dexamethasone (the DST) may be due to diminished serotoninergic activity. McIntyre et al. (1984) found that lesioning of the 5-HT system with 5,7-dihydroxytryptamine blocked the ability of dexamethasone to suppress corticosterone secretion in the rat. Nuller and Ostroumova (1980) found that administration of tryptophan (3.5–7 g/day for 7–14 days) significantly increased the suppression of cortisol by dexamethasone in depressed patients, whereas L-DOPA treatmant did not. Similar results have been reported by Lopez-Ibor et al. (1985). In this study an association between low CSF 5-HIAA levels, nonsuppression of cortisol in the DST, response to 5-HTP treatment, and suicidal thoughts or attempts in 21 depressed patients was also reported. CSF levels of 5-HIAA following probenecid did not predict the response to dexamethasone in depressed patients in one study (Davis et al. 1981) but did correlate positively with serum cortisol levels 8 and 15 h after dexamethasone in the study of Banki and Arato (1983). Roy et al. (1986) did not find any association between CSF 5-HIAA and dexamethasone nonsuppression in 27 depressed patients. Finally, Meltzer et al. (1984c) found no relationship between dexamethasone nonsuppression and the 5-HTP-induced increase in cortisol in depressed patients. In sum, the evidence for a relationship between 5-HT abnormalities and HPA axis abnormalities is mixed but there are enough positive indications to warrant continued study.

It is also noteworthy that adrenocorticoids have been reported to influence brain 5-HT metabolism in a number of important ways. Although there are considerable discrepancies between studies and some evidence for regional differences in glucocorticoid effects, the overall effect of glucocorticoids appears to be an increase in 5-HT synthesis and turnover (Azmita and McEwen 1969; Azmita et al. 1970; Telegdy and Vermes 1975; Neckers and Sze 1975; Van Loon et al. 1981; De Kloet et al. 1982). Glucocorticoids ma also decrease 5-HT release (Vermes et al. 1976) and increase its neuronal uptake (Tukianen 1981). The latter two effects might decrease availability of 5-HT but could also contribute to an upregulation of 5-HT receptors. There is evidence that glucocorticoids may inhibit (Nausieda et al. 1982) or facilitate (Nausieda et al. 1982; Hall 1980) 5-HT-dependent behavior. Thus, the apparent decrease in serotoninergic activity might be secondary to excessive adrenocorticoid activity. However, insofar as stress enhances serotoninergic activity, and the effects of stress on 5-HT turnover are dependent upon an intact HPA axis, then glucocorticoids may help to enhance the serotoninergic activity (McEwen 1987). In this regard, adrenalectomy has been found to enhance 5-HT_1 receptor densitiy in the dorsal raphé and hippocampus (De Kloet et al. 1986).

Although the precise relationships are unclear, it appears that there are interactions between 5-HT and the HPA axis that could play an important role in the initial development of mood disorders and then their progress as increased glucocorticoid production modulates serotoninergic activity and vice versa.

4 Conclusions

There is solid evidence that 5-HT can modulate the secretion of cortisol, PRL, and GH in man. Animal studies suggest effects on renin and vasoactive intestinal polypeptide as well. Because of this, it has been possible to asssess some aspects of serotoninergic activity in patients with mood disorders by administering the precursors of 5-HT, the 5-HT releaser fenfluramine, 5-HT uptake blockers, and direct-acting 5-HT agonists. As no drug has a single mechanism of action nor is the secretion of any hormone controlled only by 5-HT, interpretation of the effects of administering these agents on hormone secretion is very difficult. Treatment with thymoleptic drugs has variable effects on the endocrine responses. Current data suggest diminished availability of 5-HT with a possible enhanced response to exogenous 5-HT. However, these studies are still fairly crude. Factors such as pharmacokinetic differences, subtype of depression, severity of depression, effects of previous drug treatment, sex, and character traits such as impulsivity and violence must be considered along with suicidality in interpreting hormone responses. The specificity of any findings must be demonstrated before they can have value as diagnostic markers.

5-HT may contribute to the excessive activity of the HPA axis frequently found in depressed or manic patients. The abnormal DST response sometimes found in major depression may be related to diminished 5-HT. Glucocorticoids may affect 5-HT synthesis, turnover, and uptake, and the density of 5-HT receptors and thus contribute to or diminish the apparent serotoninergic deficit in mood disorders.

It appears likely that the endocrine effects of 5-HT may facilitate further study of the role of 5-HT in the etiology of these disorders and the mechanism of action of thymoleptic drugs.

Acknowledgements. Supported, in part, by USPHS MH 41684, by a USPHS Research Scientist Award to HYM, by NARSAD fellowship to JFN, by grants from the Cleveland, Bingham, and Sawyer Foundations and GCRC MO IRR-00080-25. The authors thank Lee Mason for her excellent secretarial assistance and Dr. Martin Lowy for his thoughtful comments.

References

Ågren H (1980) Symptom patterns in unipolar and bipolar depression correlating with monoamine metabolites in the cerebrospinal fluid. II. Suicide. Psychiatry Res 3:225–236

Amsterdam JD, Schweizer E, Winokur A (1987) Multiple hormone responses to insulin-induced hypoglycemia in depressed patients and normal volunteers. Am J Psychiatry 144:170–175

Anderson IM, Cowen PJ (1986) Clomipramine enhances prolactin and growth hormone responses to L-tryptophan. Psychopharmacol 89:131–133

Asberg M, Traskman L, Thoren P (1976) 5-HIAA in the cerebrospinal fluid: a biochemical suicide predictor? Arch Gen Psychiatry 33:1193–1197

Azmita EC, McEwen BS (1969) Corticosterone regulation of tryptophan hydroxylase in midbrain of the rat. Science 166:1274–1276

Azmita EC, Algeri S, Costa E (1970) In vivo conversion of ^{3}H-L-tryptophan into ^{3}H-serotonin in brain areas of adrenalectomized rats. Science 169:201–203

Banki CM, Arató M (1983) Amine metabolites, neuroendocrine findings, and personality dimensions as correlates of suicidal behavior. Psychiatry Res 19:253–261

Beck-Peccoz P, Ferrari C, Rondena M, Paracuhi A, Faglia G (1976) Failure of oral 5-hydroxytryptophan administration to affect prolactin secretion in man. Horm Res 7:303–307

Beskow J, Gottfries OG, Roos BE, Winblad B (1976) Determination of monoamine and monoamine metabolites in the human brain: postmortem studies in a group of suicides and in a control group. Acta Psychiatr Scand 53:7–20

Briley MS, Langer SZ (1986) Platelet (^{3}H) imipramine binding in the affective disorders: trait versus state characteristics. Am J Psychiatry 143:711–717

Brown GL, Ebert MH, Goyer PF, Jimerson CD, Klein WJ, Bunney WE, Goodwin FK (1982) Aggression, suicide and serotonin: relationships to CSF amine metabolites. Am J Psychiatry 139:741–746

Calil HM, Lesieur P, Gold PW, Brown GM, Zavadil III AP, Potter WZ (1984) Hormonal responses to zimelidine and desipramine in depressed patients. Psychiatry Res 13:231–242

Casanueva FF, Villanueva L, PeNalva H, Cabezas-Cerrato J (1984) Depending on the stimulus, central serotonergic activation by fenfluramine blocks or does not alter growth hormone secretion in man. Neuroendocrinol 38:302–308

Charney DS, Heninger GR (1986) Serotonin function in panic disorders: the effect of intravenous tryptophan in healthy subjects and patients with panic disorder before and during alprazolam treatment. Arch Gen Psychiatry 43:1059–1065

Charney DS, Heninger GR, Reinhard JF, Sternberg DE, Hafstead KM (1982) The effect of intravenous L-tryptophan on prolactin and growth hormone and mood in healthy subjects. Psychopharmacology 77:217–222

Charney DS, Heninger GR, Sternberg DE (1984) Serotonin function and mechanism of action of antidepressant treatment: effects of amitryptyline and desipramine. Arch Gen Psychiatry 41:359–365

Chihara K, Kato Y, Maeda K, Matsukara S, Imura H (1976) Suppression by cyproheptadine of human growth hormone and cortisol secretion during sleep. J Clin Invest 57:1393–1402
Clineschmidt BV, Reiss DR, Pettibone DJ, Robinson JL (1985) Characterization of 5-hydroxytryptamine receptors in rat stomach fundus. J Pharmacol Exp Ther 235:696–708
Conn PJ, Sanders-Bush E (1987) Central serotonin receptors: effector systems, physiology, roles and regulation. Psychopharmacology 92:267–277
Cowen PJ, Gadhvi H, Gosden B, Kolakowska T (1985) Responses of prolactin and growth hormone to L-tryptophan infusion: effects in normal subjects and schizophrenic patients receiving neuroleptics. Psychopharmacology 86:164–169
Cowen PJ, Geaney DP, Schächter M, Green AR, Elliott JM (1986) Desipramine treatment in normal subjects effects on neuroendocrine responses to tryptophan and on platelet serotonin (5-HT) related receptors. Arch Gen Psychiatry 43:61–67
Curzon G, Kantamaneni BD, Van Boxel P, Gillman PK, Bartlett JR, Bridges PK (1980) Substances related to 5-hydroxytryptamine in plasma and in lumbar and ventricular fluids of psychiatric patients. Acta Psychiatr Scand 61 (Suppl 280):3–19
Davis KL, Hollister LE, Mathé AA, Davis BM, Rothpearl AB, Faull KF, Hsieh JYK, Barchas JD, Berger PA (1981) Neuroendocrine and neurochemical measurements in depression. Am J Psychiatry 138:1555–1562
De Kloet ER, Kovacs GL, Szabo G, Telegdy G, Bohus B, Versteeg DHG (1982) Decreased serotonin turnover in the dorsal hippocampus of rat brain shortly after adrenalectomy: selective normalization after corticosterone administration. Brain Res 239:659–663
De Kloet ER, Sybesma H, Reul H (1986) Selective control by corticosterone of serotonin$_1$ receptor capacity in raphé-hippocampal system. Neuroendocrinology 42:513–522
De Meirleir K, L'Hermite-Balériaux M, L'Hermite M, Rost R, Hollyman W (1985) Evidence for serotonergic control of exercise-induced prolactin secretion. Horm Metab Res 17:380–381
Delitala G, Masala A, Alagna S, Devilla L (1975) Effect of cyproheptadine on the spontaneous diurnal variations of plasma ACTH-cortisol, and ACTH-GH secretion induced by L-dopa. Biomed Expr 23:406–409
Demisch L, Neubauer M (1979) Stimulation of human prolactin secretion by mescaline. Psychopharmacology 64:361–363
Di Renzo GF, Quattrone A, Schettini G, Preziosi P (1979) Effect of selective lesioning of serotonin-containing neurons on the TSH-inhibiting actions of D-fenfluramine in male rats. Life Sci 24:489–494
Feldman RS, Quenzer LF (1984) Fundamentals of neuropsychopharmacology. Sinauer, Sunderland, MA
Fernstrom JD (1983) Role of precursor availability in control of monoamine biosynthesis in brain. Physiol Rev 63:484–546
Ferrari C, Caldara R, Romussi M, Rampini G, Tellolo P, Zaatar S, Curtarelli G (1978) Prolactin suppression by serotonin antagonists in man; further evidence for serotonergic control of prolactin secretion. Neuroendocrinology 25:319–328
Fessler RG, Deyo SN, Meltzer HY, Miller RJ (1984) Evidence that the medial and dorsal raphé nuclei mediate serotonergically-induced increases in prolactin release from the pituitary. Brain Res 299:231–237
Foresta C, Scanelli G, Indino M, Federspil G, Scandellari C (1985) Naloxone reduces the fenfluramine-induced prolactin release in man. Clin Endocrinol 22:539–543
Fraser WM, Tucker HS, Grubb SR, Wigand JP, Blackard WG (1979) Effect of L-tryptophan on growth hormone and prolactin release in normal volunteers and patients with secretory pituitary tumors. Horm Metab Res 11:149–155
Fuller RW (1981) Serotonergic stimulation of pituitary-adrenocortical function in rats. Neuroendocrinology 32:118–127
Fuller RW (1986) Pharmacological modification of serotonergic function: drugs for the study and treatment of psychiatric and other disorders. J Clin Psychiatry 47(suppl 4):4–7
Fuxe K, Butcher LL, Engel J (1971) DL-5-Hydroxytryptophan-induced changes in central monoamine neurons after peripheral decarboxylase inhibition. J Pharm Pharmacol 23:420–424

Fuxe K, Farnebo L-O, Hamberger B, Ögren S-O (1975) On the "in vivo" and "in vitro" actions of fenfluramine and its derivates on central monoamine neurons, especially 5-hydroxytryptamine neurons, and their relation to the anorectic activity fenfluramine. Postgrad Med J 51(suppl 1):35–45

Fuxe K, Ögren S-O, Agnati LF, Benfenati F, Ferdholm B, Andersson K, Zini I, Eneroth P (1985) Chronic antidepressant treatment and central 5-HT synapses. Neuropharmacology 22:389–400

Gaddum JH, Picarelli ZP (1957) Two kinds of tryptamine receptor. Br J Pharmacol 12:323–328

Garfinkel PE, Brown GM, Warsh JJ, Stancer HC (1979) Neuroendocrine responses to carbidopa in primary affective disorders. Psychoneuroendocrinology 4:13–20

Garratini S, Buczo W, Jori A, Samanin R (1975) On the mechanism of action of fenfluramine. Postgrad Med J 51(suppl 1):27–35

Gillman PK, Bartlett JR, Bridges PK, Hunt A, Patel AJ, Kantamaneni BD, Curzon G (1981) Indolic substances in plasma, cerebrospinal fluid and frontal cortex of human subjects infused with saline or tryptophan. J Neurochem 37:410–417

Glass AR, Smallridge RC, Schaff M, Diamond RC (1980) Absent prolactin response to L-tryptophan in normal and acromegalic subjects. Psychoneuroendocrinology 5:261–265

Glennon RA (1987) Central serotonin receptors as targets for drug research. J Med Chem 30:1–12

Glennon RA, Slusher RM, Lyon RA, Titeler M, McKenney JD (1986) 5-HT_1 and 5-HT_2 binding characteristics of some quipazine analogues. J Med Chem 29:2375–2380

Gordin A, Mustajoki P, Pelhoren R (1985) Ketanserin without effects on basal anterior pituitary hormone secretion in healthy subjects. J Endocrinol Invest 8:73–75

Gudelsky GA, Koenig JI, Meltzer HY (1986) Thermoregulatory responses to serotonin (5-HT) receptor stimulation in the rat. Evidence for opposing roles of 5-HT_2 and 5-HT_{1a} receptors. Neuropharmacology 25:1307–1313

Halàsz B (1978) Functional anatomy of the hypothalamus. In: Cox B, Morris ID, Weston AH (eds) Pharmacology of the hypothalamus. University Park Press, Baltimore MD, pp 5–28

Hall ED (1980) Glucocorticoid enhancement of serotonergic facilitation of cat spinal monosynaptic motor neuron excitation. Exp Neurol 68:589–594

Handwerger S, Plonk JW, Lebovits HE, Bivens CH, Feldman JM (1975) Failure of 5-hydroxytryptophan to stimulate prolactin and growth hormone secretion in man. Horm Metab Res 7:214–216

Heninger GR, Charney DJ, Sternberg DE (1984) Serotonergic function in depression. Prolactin response to intravenous tryptophan in depressed patients and healthy subjects. Arch Gen Psychiatry 41:398–402

Holl WR, Fehm HL, Voigt KH, Steiner K, Teller W (1986) Acute effects of desipramine and clomipramine on pituitary-adrenal axis in man. Horm Metab Res 18:134–137

Holmes MC, DiRenzo GD, Beckford B, Gillham B, Jones MT (1982) Role of serotonin in the control of secretion of corticotropin releasing factor. J Endocrinol 93:151–160

Holsboer F, Müller OA, Winter K, Doerr HG, Sippell WG (1983) Effect of serotonin uptake inhibition by zimelidine on hypothalamic-pituitary-adrenal activity. Psychpharmacology 80:85–87

Honma K-I, Watanabe K, Kiroshige T (1979) Effects of parachlorophenylalanine and 5,6-D-hydroxytryptamine on the free-running rhythms of locomotor activity and plasma corticosterone in the rat exposed to continous light. Brain Res 169:531–544

Hoyer D, Pazos A, Probst A, Palacios JM (1986a) Serotonin receptors in the human brain. I. Characterization and autoradiographic localization of 5-HT_{1A} recognition sites. Apparent absence of 5-HT_{1B} recognition sites. Brain Res 376:85–96

Hoyer D, Pazos A, Probst A, Palacios JM (1986b) Serotonin receptors in the human brain. II. Characterization and autoradiographic localization of 5-HT_{1C} and 5-HT_2 recognition sites. Brain Res 376:97–107

Hyyppä MT, Jolma T, Liira V-A, Kytömäki O (1979) L-Tryptophan treatment and the episodic secretion of pituitary hormones and cortisol. Psychoneuroendocrinology 4:29–35

Imura H, Nakai Y, Yoshimi T (1973) Effect of 5-HTP on growth hormone and ACTH release in man. J Clin Endocrinol Metab 36:204–206

Invernizzi R, Berettera C, Garattini S, Samanin R (1986) D- and L-isomers of fenfluramine differ markedly in their interaction with brain serotonin and catecholamines in the rat. Eur J Pharmacol 20:9–15

Jacobs BL (1976) An animal behavior model for studying central serotonergic synapses. Life Sci 19:777–786

Jones RB, Luscombe DK, Groom GV (1977) Plasma prolactin concentrations in normal subjects and depressive patients following oral clomipramine. Postgrad Med J 53 (suppl 4):166–171

Jordan D, Pigeon P, McRae-De Gueurce A, Pujol JF, Mornex R (1979) Participation of serotonin in thyrotropin release. II. Evidence for the action of serotonin on the phasic release of thyrotropin. Endocrinology 105:975–979

Kato Y, Nakai Y, Imura H, Chihara K, Ohgo S (1974) Effects of 5-hydroxytryptophan on plasma prolactin levels in man. J Clin Endocrinol 38:695–697

Kletzky OA, Marrs RP, Nicoloff JT (1980) Effects of cyproheptadine on insulin induced hypoglycermia secretion of PRL, GH and cortisol. Clin Endocrinol 13:231–234

Koenig JI, Gudelsky GA, Meltzer HY (1987) Stimulation of corticosterone and β-endorphin secretion in the rat by selective 5-HT receptor subtype activation. Eur J Pharmacol 137:1–8

Koyama T, Lowy MT, Meltzer HY (1987) 5-Hydroxytryptophan-induced cortisol response and CSF 5-HIAA. Am J Psychiatry 144:334–347

Kuhn CM, Vogel RA, Mailman RB, Mueller RA, Schanberg SM, Breese GR (1981) Effect of 5,7-dihydroxytryptamine on serotonergic control of prolactin secretion and behavior in rats. Psychopharmacology 73:188–193

Laakmann G, Schumacher G, Benkert O (1977) Stimulation of growth hormone secretion by desipramine and chlorimipramine in man. J Clin Endocrinol Metab 44:1010–1013

Laakmann G, Gugath M, Kuss H-J, Zygan K (1984a) Comparison of growth hormone and prolactin stimulation induced by chlorimipramine and desipramine in man in connection with chlorimipramine metabolism. Psychopharmacology 82:62–67

Laakmann G, Wittman M, Gugath M, Mueller OA, Treusch J, Wahlster U, Stalla GK (1984b) Effects of psychotropic drugs (desipramine, chlorimipramine, sulpiride and diazepam) on the human HPA axis. Psychopharmacology 84:66–70

Lancranjan I, Wirz-Justice A, Pühringer W, Del Pozo E (1977) Effect of L-5-hydroxytryptophan infusion on growth hormone and prolactin secretion in man. J Clin Endocrinol Metab 45:588–593

Lewis DA, Sherman BM (1984) Serotonergic stimulation of adrenocorticotropin secretion in man. J Clin Endocrinol Metab 58:458–462

López-Ibor JJ Jr, Saiz-Ruiz J, Pérez de los Cobos JC (1985) Biological correlations of suicide and aggressivity in major depressions (with melancholia): 5-hydroxyindoleacetic acid and cortisol in cerebral spinal fluid, dexamethasone suppression test and therapeutic response to 5-hydroxytryptophan. Neuropsychobiology 14:67–74

Lowy MT, Meltzer HY (1988) Stimulation of serum cortisol and prolactin secretion in man by MK-212, a centrally active serotonin agonist. Biol Psychiatry (in press)

Lucki I, Nobler MS, Frazer A (1984) Differential effects of serotonin antagonists on two behavioral models of serotonin receptor activation in the rat. J Pharmacol Exp Ther 228:133–139

MacIndoe JH, Turkington RW (1973) Stimulation of human prolactin secretion by intravenous infusion of L-tryptophan. J Clin Invest 52:1972–1978

Magnussen I, Engbaek F (1978) The effects of aromatic amino acid decarboxylase inhibitors on plasma concentrations of 5-hydroxytryptophan in man. Acta Pharmacol Toxicol 43:36–42

Mann JJ, Stanley M, McBride A, McEwen BS (1986) Increased $serotonin_2$ and β-adrenergic receptor binding in the frontal cortices of suicide victims. Arch Gen Psychiatry 43:954–959

Masala A, Delitala G, Devilla L, Alagna S, Rovasio PP (1979) Enhancement of insulin-induced prolactin secretion by fluoxetine in man. J Clin Endocrinol Metab 49:350–352

Mashchak CA, Kletzky OA, Spencer C, Artal R (1983) Transient efect of L-5-hydroxytryptophan on pituitary function in men and women. J Clin Endocrinol Metab 56:170–176
McEwen BS (1987) Glucocorticoid-biogenic amine interactions in relation to mood and behavior. Biochem Pharmacol 36:1755–1763
Mclntrye IM, Oxenkrug GF, Stanley M, Gershon S (1984) The effect of 5,7-dihydroxytryptamine on the serum corticosterone resistance to suppression by dexamethasone. Brain Res 309:156–158
Meltzer HY, Lowy MT (1986) Neuroendocrine function in psychiatric disorders and behavior. In: Berger PA, Brodie KH (eds) American handbook of psychiatry, vol 7. Basic Books, New York, pp 111–150
Meltzer HY, Lowy MT (1987) The serotonin hypothesis at depression. In: Meltzer HY (ed) Psychopharmacology: the third generation of progress. Raven, New York, pp 513–526
Meltzer HY, Fessler RG, Simonovic M, Doherty J, Fang VS (1977 a) Lysergic acid diethylamide: evidence for stimulation of pituitary dopamine receptors. Psychopharmacology 54:39–44
Meltzer HY, Piyakalmala S, Schyve P, Fang VS (1977 b) Lack of effect of tricyclic antidepressants on serum prolactin levels. Psychopharmacology 51:185–187
Meltzer HY, Fessler RG, Simonovic M, Fang VS (1978) The effect of mescaline, 3,4-dimethoxylphenethylamine and 2,5-dimethoxy-4-methylamphetamine on rat plasma prolactin: evidence for serotonergic mediation. Life Sci 23:1185–1192
Meltzer HY, Arora RC, Baber R, Tricou BJ (1981 a) Serotonin uptake in blood platelets of psychiatric patients. Arch Gen Psychiatry 38:1322–1326
Meltzer HY, Simonovic M, Fang VS, Goode DJ (1981 b) Neuroendocrine effects of psychotomimetic drugs. J McLean Hosp 6:115–137
Meltzer HY, Simonovic M, Sturgeon RD, Fang VS (1981 c) Effects of antidepressants, lithium and electroconvulsive treatment on rat serum prolactin levels. Acta Psychiatr Scand 63(suppl 290):100–121
Meltzer HY, Wiita B, Tricou BJ, Simonovic M, Fang V, Manov G (1982) Effect of serotonin precursors and serotonin agonists on plasma hormone levels. In: Serotonin in biological psychiatry. Raven, New York, pp 117–138
Meltzer HY, Boutros NN, Simonovic M, Gudelsky GA, Fang VS (1983) Hallucinogenic drugs and neuroendocrine secretion. In: Endroczi E, de Weid D, Angelucci L, Scapagnini U (eds) Integrative neurohumoral mechanisms. Elsevier, Amsterdam, pp 463–477
Meltzer HY, Arora RC, Robertson A, Lowy M (1984 a) Platelet ^{3}H-imipramine binding and platelet 5-HT uptake in affective disorders and schizophrenia. Clin Neuropharmacol 7(suppl 1):320–321
Meltzer HY, Perline R, Tricou BJ, Lowy M, Robertson A (1984 b) Effect of 5-hydroxytryptophan on serum cortisol levels in major affective disorders. II. Relation to suicide, psychosis and depressive symptoms. Arch Gen Psychiatry 41:379–387
Meltzer HY, Umberköman-Wiita B, Robertson A, Tricou BJ, Lowy M, Perline R (1984 c) Effect of 5-hydroxytryptophan on serum cortisol levels in major affective disorders. I. Enhanced response in depression and mania. Arch Gen Psychiatry 41:366–374
Meltzer HY, Lowy M, Robertson A, Goodnick P, Perline R (1984 d) Effect of 5-hydroxytryptophan on serum cortisol levels in major affective disorders. III. Effect of antidepressants and lithium carbonate. Arch Gen Psychiatry 41:391–397
Meltzer HY, Gudelsky GA, Koenig JI, Lowy MT (1987) Neuroendocrine effects of buspirone: mediation by dopaminergic and serotonergic mechanisms. In: Tunnicliff G, Eisen A, Taylor D (eds) Buspirone: mechanisms and clinical aspects. Academic (in press)
Mendelson WB, Jacobs LE, Reichman JD, Othmer E, Cryer PE, Trivedi B, Daughaday WH (1975) Methysergide: suppression of sleep-related prolactin secretion and enhancement of sleep-related growth hormone secretion. J Clin Invest 56:690–697
Modlinger RS, Schonmuller JM, Arora SP (1979) Stimulation of aldosterone, renin and cortisol by tryptophan. J Clin Endocrinol Metab 50:599–603

Modlinger RS, Schonmuller JM, Arora SP (1980) Adrenocorticotropin release by tryptophan in man. J Clin Endocrinol Metab 50:360–363
Moller SE (1985) Tryptophan to competing amino acids ratio in depressive disorder: relation to efficacy of antidepressive treatments. Acta Psychiatr Scand 72:9–30
Moller SE, Kirk L, Fremming KH (1976) Plasma amino acids as an index for subgroups in manic depressive psychosis: correlation to effect of tryptophan. Psychopharmacology 49:205–213
Moller SE, de Beurs P, Timmerman L, Tan BK, Leijnse-Ybema HJ, Cohen S, Hopfner MH, Petersen HE (1986) Plasma tryptophan and tyrosine ratios to competing amino acids in relation to antidepressant response to citalopram and maprotiline. A preliminary study. Psychopharmacology 88:96–100
Mueller EA, Murphy DL, Sunderland T (1985) Neuroendocrine effects of m-chlorophenylpiperazine, a serotonin agonist, in humans. J Clin Endocrinol Metab 61:1179–1184
Mueller EA, Murphy DL, Sunderland T (1986) Further studies of the putative serotonin agonist, m-chlorophenylpiperazine: evidence for a serotonin receptor mediated mechanism of action in humans. Psychopharmacology 89:388–391
Muhlbauer HO, Muller-Oerlinghausen B (1985) Fenfluramine stimulation of serum cortisol in patients with major affective disorders and healthy controls: further evidence for a central serotonergic action of lithium in man. J Neural Trans 61:81–94
Müller EE, Brambilla F, Cavagnini F, Peracchi M, Panerai A (1974) Slight effect of L-tryptophan on growth hormone release in normal human subjects. J Clin Endocrinol Metab 39:1–5
Murphy DL, Campbell I, Costa J (1978) Current status of the indoleamine hypothesis of the affective disorders. In: Lipton MA, DiMascio A, Killiam KF (eds) Psychopharmacology: a generation of progress. Raven, New York, pp 1235–1248
Nakai Y, Imura H, Sakurai H, Kurahachi H, Yoshimi T (1974) Effect of cyproheptadine on human growth hormone secretion. J Clin Endocrinol Metab 38:446–449
Nausieda PA, Carvey PM, Weiner WJ (1982) Modification of central serotonergic and dopaminergic behaviors in the course of chronic corticosteroid administration. Eur J Pharmacol 78:335–343
Neckers L, Sze PY (1975) Regulation of 5-hydroxytryptamine metabolism in mouse brain by adrenal glucocorticoids. Brain Res 93:123–132
Ng LKY, Chase TN, Colburn RW, Kopin I (1972) Release of [^{3}H] dopamine by L-5-hydroxytryptophan. Brain Res 45:499–505
Nuller JL, Ostroumova MN (1980) Resistance to inhibiting effect of dexamethasone in patients with endogenous depression. Acta Psychiatr Scand 61:169–177
O'Malley BP, Jennings PE, Cook N, Barnett DB, Rosenthal FD (1984) The role of serotonin (5-HT) in the control of TSH and prolactin release in euthyroid subjects as assessed by the administration of ketanserin (5-HT_2 antagonist) and zimelidine (5-HT reuptake inhibitor). Psychoneuroendocrinology 9:13–19
Parati EA, Zanardi P, Cocchi D, Caraceni T, Müller EE (1980) Neuroendocrine effects of quipazine in man in healthy state or with neurological disorders. J Neural Trans 47:273–297
Paul SM, Rehavi M (1981) Depressed patients have decreased binding of tritiated imipramine to platelet serotonin "transporter". Arch Gen Psychiatry 38:1315–1317
Penington NJ, Reiffenstein RJ (1986) Direct comparison of hallucinogenic phenethylamines and D-amphetamine on dorsal raphé neurons. Eur J Pharmacol 122:373–377
Peroutka SJ (1985) Selective labelling of 5-HT_{1A} and 5-HT_{1B} binding sites in bovine brain. Brain Res 344:167–171
Peroutka SJ, Lebovitz RM, Snyder SH (1981) Two distinct central serotonin receptors with different physiological functions. Science 212:827–829
Prescott RWG, Kendall-Taylor P, Weightman DR, Watson MJ, Ratcliffe WA (1984) The effect of ketanserin, a specific serotonin antagonist on the PRL, GH, ACTH and cortisol responses to hypoglycemia in normal subjects. Clin Endocrinol 20:137–142
Price LH, Charney DS, Heninger GR (1985) Effects of tranylcypromine treatment of neuroendocrine, behavioral and autonomic responses to tryptophan in depressed patients. Life Sci 37:809–818

Pühringer W, Wirz-Justice A, Lancranjan (1976) Mood elevation and pituitary stimulation after i.v. L-5-HTP in normal subjects: evidence for a common serotonergic mechanism. Neurosci Lett 2:349

Quattrone A, Di Renzo G, Schettini G, Tedeschi G, Scopacaso F (1978) Increased plasma prolactin levels induced in rats by *d*-fenfluramine: relation to central serotonergic stimulation. Eur J Pharmacol 49:163–167

Quattrone A, Schettini G, Annunziato L, Di Renzo G (1981) Pharmacological evidence of supersensitivity of central serotonergic receptors involved in the control of prolactin secretion. Eur J Pharmacol 76:9–13

Quattrone A, Tedeschi G, Aguglia V, Scopacasa F, Di Renzo GF, Annunziato L (1983) Prolactin secretion in man: a useful tool to evaluate the activity of drugs on central 5-hydroxytryptaminergic neurons. Studies with fenfluramine. Br J Clin Pharmacol 16:471–475

Roccatagliata G, Murialdo G, Abano C, Giovale M, Zauli C, Polleri L (1982) Neuroendocrinological and clinical data on trazodone treatment in depressed patients. Neuropsychobiology 8:259–269

Rocco A, De Giorgio G, Buongiorno T, Proietti A, Lucchetti G, Falaschi P (1983) Prolactin response to R41468, a new specific 5-HT_2 antagonist. Neuroendocrinol Lett 5:47–53

Rolandi E, Magnani G, Milesi GM, Barreca T (1981) Effect of a pschoactive drug, trazodone, on prolactin secretion in man. Neuropsychobiology 7:17–19

Roy A, Agren H, Pickar D, Linnoila M, Doran AR, Cutler NR, Paul SM (1986) Reduced CSF concentrations of homovanillic acid and homovanillic acid to 5-hydroxyindoleacetic acid ratios in depressed patients: relationship to suicidal behavior and dexamethasone nonsuppression. Am J Psychiatry 143:1539–1545

Scapagnini U, Moberg GP, van Loon GR, Ganong WF (1971) Relation of brain 5-hydroxytryptamine content to the diurnal variation in plasma corticosterone in the rat. Neuroendocrinology 7:90–96

Scarduelli C, Mattei AM, Brambilla G, Zavaglia C, Adelasco P, Cavioni V, Ferrari C (1985) Effect of fenfluramine oral administration on serum prolactin levels in health and hyperprolactinemic women. Gynecol Obstet Invest 19:92–96

Schulz P, Reaven GM, Blaschke TF (1982) Growth hormone release after acute amitriptyline administration to normal human subjects. Psychopharmacology 76:299–301

Siever LJ, Murphy DL, Slater S, de la Vega E, Lipper S (1984) Plasma prolactin changes following fenfluramine in depressed patients compared to controls: an evaluation of central serotonergic responsivity in depression. Life Sci 34:1029–1039

Sueldo CE, Duda M, Kletzky (1986) Influence of sequential doses of 5-hydroxytryptophan on prolactin release. Am J Obstet Gynecol 154:424–427

Syvälahti E, Eneroth P, Ross SB (1979a) Acute effects of zimelidine and alaproclate, two inhibitors of serotonin uptake, on neuroendocrine function. Psychiatry Res 1:111–120

Syvälahti E, Nagy A, van Praag HM (1979b) Effects of zimelidine, a selective 5-HT uptake inhibitor, on serum prolactin levels in man. Psychopharmacology 64:251–253

Szafarczyk A, Ixart G, Malaval F, Nouguier-Soulé J, Assenmacher I (1979) Effects of lesions of the suprachiasmatic nuclei and of p-chlorophenylalanine on the circadian rhythms of adrenocorticotrophic hormone and corticosterone in the plasma, and on locomotor activity of rats. J Endocrinol 83:1–16

Takahashi S, Kondo H, Yoshimura M, Ochi Y (1974) Growth hormone response to administration of L-5-hydroxytryptophan (L-5-HTP) in manic-depressive psychoses. In: Psychoneuroendocrinology Workshop Conf. Int. Karger, Basel, pp 1235–1248

Telegdy G, Vermes G (1975) Effect of adrenocortical hormones on activity of the serotonergic system in limbic structures in rats. Neuroendocrinology 18:16–26

Thompson J, Rankin H (1982) The treatment of depression in general practice: a comparison of L-tryptophan, amitryptiline with placebo. Psychol Med 12:741–751

Titeler M, Herrick K, Lyon RA, McKenney JD, Glennon RA (1985) [^{3}H]DOB: a specific agonist radioligand for 5-HT_2 serotonin receptors. Eur J Pharmacol 117:145–146

Träskman L, Tybring G, Asberg M, Bertilsson L, Lantto O, Schalling D (1980) Cortisol in the CSF of depressed and suicidal patients. Arch Gen Psychiatry 37:761–767

Tukiainen E (1981) Effect of hypophysectomy and adrenalectomy on 5-hydroxytryptamine uptake by rat hypothalamic synaptosomes and blood platelets. Acta Pharmacol Toxicol 48:139–144

Tuomisto J, Tukiainen E, Ahlfors UG (1979) Decreased uptake of 5-hydroxytryptamine in blood platelets from patients with endogenous depression. Psychopharmacology 65:141–147

Tuomisto J, Mannisto P (1985) Neurotransmitter regulation of anterior pituitary hormones. Pharmacol Rev 37:251–332

Van de Kar LD, Urban JH, Richardson KD, Bethea CL (1985) Pharmacological studies on the serotonergic and nonserotonin-mediated stimulation of prolactin and corticosterone secretion by fenfluramine. Neuroendocrinology 41:283–288

van Loon GR, Shum A, Sole MJ (1981) Decreased brain serotonin turnover after short term adrenalectomy in rats: a comparison of four turnover methods. Endocrinology 108:1392–1402

van Praag HM (1981) Management of depression with serotonin precursors. Biol Psychiatry 16:291–310

van Praag HM (1983) In search of the mode of action of antidepressants. 5-HTP/tyrosine mixtures in depression. Neuropharmacology 22:433–440

van Praag HM, Korf J, Puite J (1970) 5-Hydroxyindoleacetic acid levels in the cerebrospinal fluid of depressive patients treated with probenecid. Nature 225:1259–1260

van Praag HM, Lemus C, Kahn R (1987) Hormonal probes of central serotonergic activity: do they really exist? Biol Psychiatry 22:86–98

Vermes I, Smelik PG, Mulder AH (1976) Effects of hypophysectomy: adrenalectomy and corticosterone treatment on uptake and release of putative central neurotransmitters by rat hypothalamic tissue in vitro. Life Sci 19:1719–1726

Vernikos-Danellis J, Berger P, Barchas JD (1973) Brain serotonin and pituitary adrenal function. Rec Prog Brain Res 39:301–308

Virkkunen M, Nuutila A, Goodwin FK, Linnoila M (1987) Cerebrospinal monoamine metabolite levels in male arsonists. Arch Gen Psychiatry 44:241–247

Westenberg HGM, van Praag HM, de Jong TVM, Thijssen JHH (1982) Post-synaptic activity in depressive serotonergic patients: evaluation of the neuroendocrine strategy. Psychiatry Res 7:361–371

Wiebe RH, Handwerger S, Hammond CB (1977) Failure of L-tryptophan to stimulate prolactin secretion in man. J Clin Endocrinol Metab 45:1310–1312

Willner P (1985) Antidepressants and serotonergic neurotransmission: an integrative review. Psychopharmacology 85:387–404

Winokur A, Lindberg ND, Lucki I, Phillips J, Amsterdam JD (1986) Hormonal and behavioral effects associated with intravenous L-tryptophan administration. Psychopharmacology 88:213–219

Woolf PD, Lee L (1977) Effect of the serotonin precursor, tryptophan, on pituitary hormone secretion. J Clin Endocrinol Metab 45:123–133

Cholinergic Mechanisms in Mood: Neuroendocrine Aspects

D. S. Janowsky [1], R. N. Golden [1], and S. C. Risch [2]

Contents

1 Introduction

Central cholinergic mechanisms are thought to play key roles in such critical and diverse brain functions as motor tone and coordination, sleep and dreaming, analgesia, cognition and memory, and neuroendocrine regulation (Risch and Janowsky 1984). In addition, acetylcholine has been implicated in the regulation of mood and in the pathogenesis of affective illness (Janowsky et al. 1972a). Data derived from both animal and human studies suggest that the relative balance between central cholinergic and adrenergic tone may play a role in the modulation of mood. Further, preliminary evidence suggests that patients suffering from affective disorders may be hypersensitive, both in terms of behavioral and neuroendocrine responses, to the pharmacologic effects of centrally acting cholinomimetic agents. Such hypersensitivity may be a reflection of central cholinergic receptor hypersensitivity.

1 Psychiatry Department, University of North Carolina, School of Medicine, Chapel Hill, NC 27514, USA
2 Psychiatry Department, School of Medicine, University of California at San Diego, La Jolla, CA 92093, USA

Current Topics in Neuroendocrinology, Vol. 8

In this paper, we will review the neuropharmacology of central cholinergic systems and the clinical and preclinical studies that have explored the role of acetylcholine in the regulation of mood. Further, we will summarize the available data regarding the neuroendocrine aspects of cholinergic function as they relate to affective illness.

2 Neuropharmacology of Central Cholinergic Systems

Although acetylcholine was the first neurotransmitter to be discovered, limitations in technology have, until recently, slowed the pace of investigations of its role in central nervous system function. Acetylcholine is synthesized by choline acetyltransferase, using choline and acetyl-coenzyme A as substrates. Brain acetylcholine formation is relatively dependent on bloodborne choline obtained from dietary intake, since choline cannot be synthesized by the brain in the requisite amounts. Choline may be taken into cholinergic neurons for neurotransmitter synthesis either by a low-affinity, passive diffusion process, or by a high-affinity, saturable sodium-adenosine triphosphate membrane potential-dependent transport system. This high-affinity choline transport system appears to be the rate-limiting step in acetylcholine synthesis, and the major regulatory factor in cholinergic neuron neurotransmitter content. In addition to this high-affinity transport system, the intracellular synthesis of acetylcholine is believed to be self-regulating, since diverse pharmacologic manipulations have been unable to increase brain content of acetylcholine by more than 100% (Cooper et al. 1982).

In the brain, there are nicotinic and muscarinic, pre- and postsynaptic receptors, and many cholinoceptic cells respond to both muscarinic and nicotinic agonists. In addition, some data suggest that cholinergic receptors may exist which have properties that are different from both the classic nicotinic and muscarinic types (Cooper et al. 1982). Unlike other central neurotransmitters, such as the monoamines, there is no high-affinity system for synaptic reuptake of acetylcholine. Instead, the primary process for synaptic inactivation of available acetylcholine appears to be through hydrolysis either by the neuronal acetylcholinesterase or via non-neuronal (glial) pseudocholinesterases.

Acetylcholine acts as an excitatory neurotransmitter in most areas of the brain. However, due to the nature of the brain areas excited by cholinergic neurotransmission, acetylcholine generally produces behavior-inhibiting or depressant effects. There is extensive cholinergic innervation of limbic and hypothalamic systems thought to be functionally important in the regulation of mood. Cholinergic fibers originating in the medial septal and diagonal band nuclei travel via the fornix to innervate the hippocampus and dentate gyrus. The hippocampus in turn may regulate cholinergic activity in other areas of the brain, such as the habenular nuclei, the anterior thalamic nuclei, the dorsal and deep tegmental nuclei, the subfornical organ, and the cerebellum. Lesions of the hippocampus have been shown to depress cortical levels of acetylcholine. The amygdalae also receive cholinergic innervation, and cholinomimetics applied to the amygdala produces rage responses. Cholinergic innervation of the locus coeruleus, the site of origin for the major noradrenergic systems, may be functionally important and especially relevant to the cholinergic-adrenergic balance hypothesis of affective illness discussed in Sect. 3.1 below (Risch and Janowsky 1984).

The cholinergic antagonists atropine and scopolamine decrease brain acetylcholine content, presumably by blocking presynaptic muscarinic receptors and thereby increasing acetylcholine release due to loss of presynaptic inhibitory feedback. Muscarinic agonists, on the other hand, inhibit the neuronal release of acetylcholine, presumably by activating presynaptic feedback receptors. To date, peripheral markers of central cholinergic activity have not been clearly identified, and neither plasma choline nor acetylcholinesterase activity appears to correlate well with central cholinergic function. The relationship between plasma acetylcholine and central acetylcholine is not clear. Nonetheless, the central neuroregulation of acetylcholine described above provides the potential for pharmacologic manipulation of central cholinergic activity by a variety of approaches, including precursor loading, receptor agonists and antagonists, and acetylcholinesterase inhibitors (Risch et al. 1987).

2.1 Acetylcholine and the Limbic System

Papez (1937) and later MacLean (1958) hypothesized that the limbic system comprises the major neuroanatomical substrate for the experience of emotion. The structural elements of the limbic system are believed to form an integrated circuit by which cortical centers of emotion (e.g., the hippocampus, amygdala, and septal nuclei) direct impulses to the hypothalamus via intervening structures, such as the hippocampal-fornix-mamillary-hypothalamic tract. In turn, hypothalamic stimuli may project via the mamillothalamic tract to the anterior nuclei of the thalamus, then via the cingulate cortex and isthmus to the parahippocampal cortex, and finally to the adjacent hippocampus. In addition, there are similar reciprocal circuits between the amygdalae and the septal nuclei in the hypothalamus. Thus, there are neuroanatomical pathways allowing reciprocal feedback between the limbic system and the hypothalamus.

Support for Pepez's and MacLean's theories of the limbic system as the major anatomical loci of emotional states comes from a variety of animal stimulation and lesion experiments, as well as from certain clinical neuropathological observations in man. For example, electrical stimulation of the hippocampal formation leads to a state of increased arousal and anxiety, and may produce a behavioral syndrome that resembles aspects of temporal lobe epilepsy. In man, bilateral hippocampal damage can lead to development of Korsakoff's syndrome, with apathy, loss of initiative, and anterograde amnesia, accompanied by confabulation. The amygdalae, key components of the limbic system, are also related to the production of emotional excitement. Bilateral destruction of these nuclei leads to a striking state of fearlessness and placidity which has been labeled the Kluver-Bucy syndrome (Kluver and Bucy 1939). In man, stereotactic lesions of the amygdalae in patients with severe behavioral disturbances have been reported to normalize social behavior and markedly reduce social lability (Narabayashi et al. 1963). On the other hand, electrical stimulation of the amygdalae is associated with increased alertness, fear, and rage, as well as sympathetic and parasympathetic autonomic activity. These behavioral changes are accompanied by hormonal responses, including release of ACTH and gonadotropins from the anterior pituitary gland, suggesting that reciprocal circuitry exists between the limbic structures and the hypothalamus (Carpenter 1972).

Other elements of the limbic system also appear to be involved in the regulation of emotional states. The septal nuclei seem to function as a pleasure center; animals with electrodes implanted in this brain region continuously self-stimulate themselves at very high rates (Olds 1960). In human subjects, the stimulation of the septal nuclei has been reported to produce feelings of euphoria (MacLean 1957). Stimulation of the cingulate cortex produces a so-called arrest reaction in which an animal ceases all activities and appears to be alert and responsive to external stimuli. When the posterior area of the cingulate cortex is stimulated, pleasurable affect and sexual behavior appear to be induced (MacLean 1958). Thus it appears that the limbic system and its components are involved in the modulation of affective state.

In summary, it appears that the limbic and hypothalamic systems play significant roles in the regulation of affective responses. These two systems are interrelated at both the neuroanatomic and functional levels. Further, they receive specific cholinergic innervation; thus cholinergic tone may influence the activity of these mood-modulating systems and thereby influence affective state.

3 Acetylcholine and Mood Regulation

3.1 Cholinergic-Adrenergic Balance Hypothesis

In 1972, Janowsky and coworkers proposed an adrenergic-cholinergic balance hypothesis of affective illness. This hypothesis postulated that affect may represent a balance between cholinergic and adrenergic neurotransmitter activity in those areas of the brain that regulate mood. Depressive illness may represent a disease of relative cholinergic predominance, while mania could result from relative adrenergic predominance (Janowsky et al. 1972a).

Previous theories on the pathogenesis of affective illness had emphasized dysregulation in central catecholamine or indoleamine neurotransmitter systems (Prange 1964; Bunney and Davis 1965; Schildkraut 1965; Coppen et al. 1965). The additional consideration of central cholinergic activity in the regulation of mood and in the etiology of affective disorders was based in part on interactions between catecholaminergic and cholinergic systems elsewhere in the human body, such as in the autonomic regulation of the cardiovascular and gastrointestinal systems, and has subsequently derived further support from data derived from animal experiments and from clinical observations and human studies in which pharmacologic manipulations of central cholinergic activity have resulted in behavioral and affective changes.

3.2 Animal Studies and Models

Animal studies suggest that there is a strong interaction between adrenergic and cholinergic systems in the regulation of behavior. Centrally active cholinomimetics often produce inhibitory-depressant behavioral effects, including decreased locomotion and lethargy (Janowsky et al. 1972a). In contrast, sympathomimetics produce behavioral arousal and hyperactivity (Pradam and Kamat 1972; Janowsky et al. 1972b). Animals may be trained to electrically self-stimulate putative pleasure-generating centers of the brain, such as the septal nuclei, and centrally acting acetylcholinesterase inhibitors can markedly decrease this self-stimulation while at the same time increasing brain acetylcholine content (Domino and Olds 1968). Amphetamine increases low rates of self-stimulation while reducing high rates, possibly secondary to its increasing acetylcholine release at high doses (Myers 1964). Injection of acetylcholine or cholinomimetics into the rostral hippocampus (Hernandez-Peon 1965), lateral hypothalamus (Leibowitz and Miller 1969), amygdalae (MacLean and Delgado 1953), and septum (Myers 1964) can produce rage reactions in cats. Adrenergic stimulation of the lateral hypothalamus in cats produces sedation and a relative absence of emotional behavior (Hernandez-Peon 1965). Adrenergic blocking agents, when injected into the hypothalamus of rats, can provoke the characteristic rage reactions that are seen with cholinomimetic injections (Leibowitz and Miller 1969). Taken together, these observations lend support to a dynamic interaction between central cholinergic and adrenergic systems in regard to mood regulation.

There are data to suggest that adrenergic and cholinergic influences can antagonize each other. Amphetamine-induced behavioral activation is antagonized by centrally-acting cholinomimetics (Janowsky et al. 1972b; Myers 1964). Anticholinergics, on the other hand, may produce behavioral activation and hyperactivity and may augment the effects of psychostimulants (Pradam and Kamat 1972). At high doses, psychostimulants such as amphetamine and methylphenidate can induce stereotypic behaviors which can be augmented by anticholinergic agents and antagonized by cholinomimetics (Janowsky et al. 1972b).

Recently, studies involving selectively bred strains of rats have provided interesting data regarding the role of cholinergic systems in depression. The Roman High Avoiding strain, which was selectively bred for the ability to learn an active avoidance response (Goldman and Erickson 1983) is more sensitive to the behavioral effects of physostigmine than the Roman Low Avoiding strain (Martin et al. 1981). The Flinders Sensitive Line (FSL) strain of rats was bred for sensitivity to the cholinomimetic anticholinesterase drug diisopropyl fluorophosphate (Overstreet and Russell 1982). Further, FSL rats have decreased body weight, decreased locomotor activity, and increased numbers of hippocampal and striatal muscarinic acetylcholine receptors. Interestingly this rat strain is relatively inactive and appears to be quite sensitive to the induction of behavioral immobility following a forced swim test, a behavior considered to be a model for depression by some researchers (Overstreet and Russell 1984). Overstreet et al. (1987a) have recently reported that when normal rats develop upregulated muscarinic receptors due to the withdrawal of chronic scopolamine treatment, they also show a greater degree of behavioral immobility in the forced swim test, compared to control animals, and have greater cholinomimetic-induced elevations in serum corticoids. Further, anticholinergic drugs can reduce the behavioral immobility seen in the forced swim test and have been reported to reduce the effects of inescapable shock, a procedure widely used to produce learned helplessness (Goldman and Erickson 1983).

3.3 Clinical Studies and Observations

Some of the most striking evidence in support of the role of central cholinergic influences in the regulation of mood comes from observations of the psychological effects of acetylcholine-enhancing cholinesterase inhibitors. In the 1950s and 1960s several groups of investigators studied healthy volunteers, depressed patients, manic patients, and normal subject accidentally exposed to insecticide poisoning, and observed that cholinesterase inhibitors exert anergic-inhibitory effects, as well as intensifying depressive symptoms while antagonizing manic symptoms (Gershon and Shaw 1961; Bowers et al. 1964). Consistent with an adrenergic-cholinergic balance hypothesis of affective disorders and with the observations of the effects of cholinesterase inhibitors made by Rowntree et al. (1950), Janowsky and associates applied an agonist strategy and found that the centrally active cholinesterase inhibitor physostigmine caused a dramatic but brief reduction in hypomanic and manic symptoms in bipolar patients. Placebo and neostigmine, a cholinesterase inhibitor that acts peripherally but does not readily cross the blood-brain barrier, produced no changes. Physostigmine also rapidly converted mania to a syndrome consistent with a psychomotorically retarded depression. Following the administration of physostigmine, manic patients became significantly less talkative, euphoric, active, friendly, or grandiose and showed a decrease in flight of ideas. Some patients also described inhibitory symptoms, such as feelings of fatigue, anergia, apathy, and the perception of "having no thoughts" (Janowsky et al. 1973, 1974).

Physostigmine has also been found to cause depressed mood, presumably by shifting adrenergic-cholinergic balance to a cholinergic predominance. Thus Janowsky et al. (1973) and later Davis et al. (1978) found that physostigmine induced depressive symptoms in a subgroup of actively ill manics, as well as in a larger group of psychiatric patients, including depressives and schizoaffectives. Furthermore, physostigmine induced depression in the majority of a group of euthymic bipolar patients maintained on lithium. More recently, Risch and colleagues (1981 a) reported that some normal volunteers became depressed after receiving physostigmine.

Researchers have also induced depressed mood in subjects receiving the acetylcholine precursors choline, lecithin, and deanol. For example, Davis and coworkers (1979) and Tamminga and associates (1976) found an increase in depressive symptoms in schizophrenics with tardive dyskinesia who were treated with the acetylcholine precursor choline. Others have noted in a minority of cases that depression is a side effect of choline and of lecithin when these compounds are given in efforts to improve the memory deficits of Alzheimer's disease. Further, Casey (1979) has reported the induction of depressed mood and in some cases hypomania in a minority of patients with tardive dyskinesia treated with the presumed acetylcholine precursor deanol. Interestingly, Cohen and colleagues have preliminary data suggesting that the administration of the precursor choline may have antimanic effects in manic subjects (Cohen et al. 1980).

A considerable quantity of data suggest that the mood-depressant effects of various cholinomimetic drugs may be specific to patients with affective disorder. In 1974, Janowsky and colleagues reported that while nearly all psychiatric patients who received physostigmine developed an inhibitory syndrome, patients with depression or mania or a schizoaffective diagnosis became significantly more sad and depressed compared to schizophrenic subjects (Janowsky et al. 1974). More recently, the response to intravenous physostigmine in psychiatric inpatients, with and without diagnoses of affective disorders, has been studied. Preliminary results show that as a group the patients with affective disorder diagnoses are significantly more reactive to physostigmine's dysphoric and behavioral inhibiting effects (Janowsky et al. 1981). Also Casey (1979) noted that those patients with a strong history of affective disorder selectively showed affective symptoms while receiving deanol for tardive dyskinesia, in contrast to those patients who did not have an affective disorder history.

In addition to the mood-lowering and antimanic effects of cholinomimetic drugs, there is some evidence that centrally acting anticholinergic drugs have mood-elevating properties. Anticholinergic-antiparkinsonian agents, given to treat drug-induced parkinsonian symptoms, may cause euphoria, as manifested by a sense of well-being, increased sociability, and even a reversal of depressed moods. Further, these euphoric effects can occur in patients who have not previously taken neuroleptics. In addition, several old reports suggest that high doses of atropine and other anticholinergics may cause alleviation of depression, and a recent report has been published indicating that a tricyclic antidepressant-induced central anticholinergic syndrome may alleviate depression. Finally, a number of effective tricyclic antidepressants have been observed to have central anticholinergic properties, although some of the "second generation" antidepressants, such as trazodone, do not have antimuscarinic properties (Janowsky et al. 1983a).

4 Acetylcholine and Neuroendocrine Function

Historically, clinical observations of the affective state and behavior of patients with endocrine diseases, such as hypothyroidism and Cushing's disease, have had important heuristic value in relation to some of the major psychiatric syndromes. More recently, a great deal of attention has been focused on more subtle forms of neuroendocrine dysregulation that appear to be biological correlates of affective disorders (Golden and Potter 1986). The hypothalamic-pituitary-adrenal and the hypothalamic-pituitary-thyroid axes have probably been studied most extensively in this regard. In addition, other hypothalamic-pituitary endocrine systems appear to be dysregulated in depressed patients (Amsterdam et al. 1983). Female steroidal hormones, such as estrogen and estradiol, should also be of interest, since women are at greater risk than men for developing certain types of depressive illness. There appears to be important interactions between acetylcholine and each of these depression-relevant hormonal systems.

4.1 Acetylcholine and the Hypothalamic-Pituitary-Adrenal Axis

For more than a century, ever since reliable techniques have been available for measuring the output of the hypothalamic-pituitary-adrenal (HPA) axis, investigators have noted a relationship between cortisol metabolism and affective illness. Initial studies examined basal output and repeatedly found elevations in plasma cortisol concentrations and urinary-free cortisol excretion (Board et al. 1957; Carroll et al. 1976; Gibbon and Meltaugh 1962; Rubinow et al. 1981). Subsequent studies uncovered disruptions in the usual circadian rhythm of cortisol, including increased secretion at night and loss or blunting of the diurnal variation of secretion (Sachar et al. 1973). More recently, "challenge" paradigms, primarily the dexamethasone suppression test, have been increasingly utilized in an effort to study the dynamic functioning of the system. It is clear that a significant percentage of depressed patients showed dysregulation of the HPA axis, as manifested by failure to suppress cortisol output following the administration of the synthetic glucocorticoid dexamethasone (Carroll et al. 1981; Brown and Shuey 1980; Evans et al. 1983). Preliminary findings from ongoing investigations of the HPA axis, utilizing corticotropin-releasing factor (CRF) and adrenocorticotropic hormone (ACTH) "challenges", suggest that there may be multiple sites of dysregulation.

A large number of studies suggest that acetylcholine plays an important role in the regulation of ACTH release and that its effects at least parallel those occurring naturally in depression. CRF is released from in vitro preparations of rat hypothalamus in a dose-dependent manner by acetylcholine (Edwardson and Bennett 1974). This action of acetylcholine may involve both nicotinic and muscarinic mechanisms, since the acetylcholine-induced release of CRF is completely antagonized by the antinicotinic agent, hexamethonium, and partially antagonized by the antimuscarinic agent, atropine (Jones et al. 1969). However, in another preparation, the muscarinic agent bethanechol did not change basal levels of CRF (Steiner 1968). Norepinephrine can inhibit the acetylcholine-induced release of hypothalamic CRF, and hypothalamic neurons activated by microiontophoretic application of acetylcholine are also inhibited by norepinephrine (Steiner 1968).

In vivo studies also support a role for acetylcholine in the regulation of the HPA axis. Carbachol, a cholinergic agonist with both nicotinic and muscarinic properties, induces the release of corticosteroids from the adrenal gland when implanted in the hypothalamus of a variety of animal species. Presumably this effect takes place via ACTH release (Krieger and Krieger 1970). Conversely, hypothalamic implantation of atropine prevents stress-related ACTH release (Kaplanski and Smelik 1973).

Intravenous administration of nicotinic or muscarinic cholinergic agonists in rats and dogs can lead to adrenocortical activation and the subsequent release of 17-OH-corticosteroids, and this effect can be blocked by hypophysectomy or by radiation-induced lesions of the anterior median eminence (Suzuki et al. 1975). Carbachol, when injected into the lateral ventricle of rats, can also lead to activation of the HPA axis (Abe and Hiroshige 1974).

Data from a variety of studies suggest that cholinomimetic drugs may elevate serum ACTH and cortisol levels in man and thus, in a way, recreate the hypercortisolemic state of depression (Risch et al. 1982). Several recent clinical studies have documented the effects of cholinomimetics on neuroendocrine activity in man (Janowsky et al. 1983 b; Risch et al. 1982; Carroll et al. 1980). Physostigmine and arecoline infusions significantly increase serum ACTH and cortisol levels in normals and in psychiatric patients, as well as increasing β-endorphin levels. Further, physostigmine can reverse the dexamethasone-induced suppression of cortisol in healthy subjects which is seen in a significant number of depressed patients (Carroll et al. 1980; Doerr and Berger 1983). Patients suffering with affective illness appear to show significantly greater increases in ACTH and β-endorphin levels after physostigmine infusion than healthy subjects (Risch et al. 1982). Thus it appears that physostigmine can induce increases in HPA activity which parallel certain phenomena seen in patients with endogenous depression, including increased cortisol secretion, cortisol resistance to suppression by dexamethasone, and elevated ACTH and β-endorphin levels. Significantly, a similar finding has been noted by Overstreet et al. (1987 b), who found corticosterone release exaggerated in acetylcholine-sensitive rats following chronic scopolamine administration withdrawal, presumably due to unmasked upregulated muscarinic receptors.

In addition, it appears that physostigmine-induced elevations in serum ACTH, cortisol, and β-endorphin are due to a central muscarinic mechanism, since such increases are blocked by the centrally acting antimuscarinic agent scopolamine and do not occur when non-centrally acting neostigmine is given in doses equally potent to that of physostigmine (Janowsky et al. 1986 a).

The relationship between the HPA axis and cholinergic systems may be dynamic. While changes in cholinergic function may lead to changes in HPA "tone," it is also possible that the HPA axis may affect cholinergic systems as well. Several investigators have reported that rats who are chronically treated with synthetic glucocorticoids developed increased cholinergic function (Marquardt et al. 1982; Riker et al. 1979; Sze et al. 1983). Some of these studies have indicated that a specific regional increase in choline uptake develops, indicative of an increased turnover of acetylcholine, while others have reported increased receptor sensitivity. These data are supportive of the hypothesis that hypercortisolism may be a predisposing factor to depression, perhaps via the mechanism of a glucocorticoid-related increase in cholinergic receptor sensitivity. In fact, recent studies have found significantly elevated hippocampal muscarinic acetylcholine receptor concentrations following chronic exposure to glucocorticoids (Crocker and Overstreet, unpublished observations).

4.2 Acetylcholine and Other Anterior Pituitary Hormone Systems

Acetylcholine may also play a role in the regulation of prolactin release, although the literature in this area is inconsistent and contradictory. Most animal studies suggest an inhibitory role for nicotinic and muscarinic agonists in regard to prolactin release. For example, basal prolactin secretion in estrogen-treated rats is reduced by the administration of acetylcholine into the lateral ventricle, or by administration of centrally acting muscarinic or nicotinic cholinomimetics. This effect can be prevented by the prior administration of atropine (Grandison and Meites 1976). Also, daily nocturnal prolactin surges and stress-induced prolactin secretion in the rat can be prevented by centrally acting cholinomimetic agents. These effects appear to be predominantly of muscarinic origin, since nicotine alone has no effect on prolactin release. However, these prolactin surges can also be prevented by high doses of atropine, an effect which is antagonized by physostigmine, which seems contradictory to the above results (Risch et al. 1984).

In contrast, other studies in rats have found no effect of cholinergic agonists or antagonists on basal prolactin levels (Lawson and Gala 1975). On the other hand, consistent with results demonstrating a release of prolactin by cholinomimetics, Davis and Davis (1980) and Janowsky et al. (1980) have reported that physostigmine increases prolactin release, a phenomenon which is blocked by scopolamine and which does not occur after neostigmine administration. Thus it is apparent that the role of acetylcholine on prolactin secretion has not been fully clarified.

There have been relatively few investigations exploring the role of cholinergic mechanisms in the regulation of growth hormone secretion. Acetylcholine, pilocarpine, and physostigmine have all been reported to increase growth hormone release in rats, an effect which can be reversed by the concurrent administration of atropine (Bruni and Meites 1978). Similarly, Mendelson and colleagues have demonstrated that piperidine, a nicotinic cholinergic receptor agonist, can enhance sleep-related growth hormone secretion in humans (Mendelson et al. 1981). On the other hand, one animal study has reported that atropine partially suppresses episodic growth hormone secretion in rats (Martin et al. 1978). Further, methscopolamine, a non-centrally acting peripheral anticholinergic agent, inhibits nocturnal growth hormone secretion (Mendelson et al. 1978). However, administration of nicotine can blunt the neuroleptic- and barbiturate-induced growth hormone secretion in rats (Kato et al. 1974), and growth hormone levels in patients with tardive dyskinesia are reported to be positively correlated with the dose of anticholinergic medication (Burnett et al. 1978). Yet in a study of normal volunteers, atropine failed to affect insulin-induced, hypoglycemia-related increases in growth hormone secretion (Blackard and Waddell 1969). Thus as with prolactin, the role of central cholinergic mechanisms in growth hormone regulation is not fully understood. However, it is noteworthy that Janowsky et al. (1986a) found that physostigmine caused small yet statistically significant increases in serum growth hormone levels; these effects were blocked by scopolamine.

There have been relatively few studies examining the role of central cholinergic mechanisms in anterior pituitary β-endorphin regulation. However, in animal stress induction paradigms, β-endorphin and ACTH have been found to be released together, putatively in equimolar amounts (Guillemin et al. 1977). Based on these preclinical observations, our group examined the role that cholinergic mechanisms might play in stimulating the release of anterior pituitary β-endorphin (Risch et al. 1981b). In healthy volunteers and in a group of mixed psychiatric inpatients, we found that physostigmine increased plasma β-endorphin and ACTH concentrations. At the same time as described above, physostigmine led to significant increases in negative affect. In healthy subjects, the physostigmine-related increases in plasma β-endorphin and ACTH, as well as increases in plasma prolactin, were significantly correlated with the occurrence of physostigmine-induced behavioral changes. Other neuroendocrine changes, including increases in plasma cortisol, did not significantly correlate with the physostigmine-induced changes in behavioral measures, suggesting that the β-endorphin, ACTH, and prolactin correlations may be relatively specific and may correlate with central neurochemical mechanisms relating to depressed affect (Janowsky et al. 1980; Risch et al. 1981 b). In these studies, psychiatric inpatients suffering from affective disorders were more sensitive not only to the negative affect-inducing effects of physostigmine, but also to the prolactin-releasing effects of physostigmine, than nonaffective disorder psychiatric patients. Thus these data support the possible existence of central cholinergic receptor hypersensitivity in affective disorder patients. Further, they suggest that affective disorder subjects may have abnormalities of anterior pituitary prolactin and ACTH-β-endorphin regulation, as elicted by central cholinergic stimulation (Risch et al. 1984).

4.3 Acetylcholine and the Hypothalamic-Pituitary-Thyroid Axis

The psychopathological consequences of hypothyroidism have been recognized for a long time. More recently, investigations have explored more subtle forms of hypothalamic-pituitary-thyroid (HPT) axis dysregulation in depressive disorder. For example, numerous studies have demonstrated that a substantial number of depressed patients show a blunted thyroid-stimulating hormone (TSH) response to the administration of thyrotropin-releasing hormone (TRH) (Loosen and Prange 1982), while a smaller subgroup of patients show an augmented response (Gold et al. 1977; Targum et al. 1984). Addition of relatively low doses of triiodothyronine to tricyclic treatment converts a substantial number of "treatment-resistant" depressives to responders (Goodwin et al. 1982). Persistently low TSH response to TRH following treatment with electroconvulsive therapy predicts early relapse (Krog-Meyer et al. 1984). Thus dysregulation of the HPT axis is associated with depression, and persistent dysregulation may have prognostic implications.

Some data suggest that there are functional connections between cholinergic systems and the HPT axis. Interestingly, studies have shown that both thyroidectomy and chronic thyroxine treatment can lead to the development of cholinergic supersensitivity in rats (Gardiner et al. 1974; Gibson and Pollock 1975). Following thyroidectomy, there is a change in the maximal response of muscle tissue; chronic thyroxine exposure, on the other hand, leads to a shift in the dose-response curve to the right. The increased sensitivity of hypothyroid rats to acetylcholine suggests that the well-known phenomenon of secondary depression in the presence of hypothyroid illness may conceivably be mediated through changes in cholinergic systems, perhaps through the development of supersensitive cholinergic receptors.

4.4 Acetylcholine and Estrogens

Another hormone that appears to interact with cholinergic systems in animals is the female hormone estrogen, or estradiol (McEwen and Pfaff 1985). Hypothalamic muscarinic acetylcholine receptors are increased when ovariectomized female rats are treated with estradiol (Rainbow et al. 1980). Similar brain regions in male rats appear to be unresponsive to estradiol treatment (Dohanich et al. 1984). One study has suggested that these increased concentrations in muscarinic receptors may be functionally related to activation of female sexual behavior (Rainbow et al. 1984). Also, there are sex-related differences in serum cholinesterase activity which are felt to be primarily the result of estrogen-dependent stimulation of enzyme synthesis (Sawyer and Everett 1946).

Estrogen-related alterations in cholinergic systems may be involved in differential drug effects. For example, female rats appear to be relatively resistant to the effects of DFP, an inhibitor of serum cholinesterase and acetylcholinesterase, perhaps because they have inherently higher levels of serum cholinesterase than male rats (Overstreet et al. 1981; Netherton and Overstreet 1981). On the other hand, female rats may be more sensitive to the hypothermic effects of oxotremorine, a muscarinic agonist (Marquardt et al. 1982), perhaps as a consequence of higher concentrations of hypothalamic muscarinic receptors secondary to estrogen. Netherton (unpublished doctoral dissertation) found that female rats exhibit greater hypothermia and drinking behavior following direct administration of oxotremorine into the preoptic and lateral areas of the hypothalamus respectively. Thus, local application of muscarinic agonists to areas of the hypothalamus in which estrogen increases muscarinic receptor concentrations in females leads to greater physiologic response (Overstreet et al. 1987b).

4.5 Acetylcholine and Catecholamines

In addition to effects on pituitary and adrenal cortical neurohormones, central cholinergic mechanisms appear to cause activation of the sympathetic nervous system and to cause adrenal-medullary catecholamine release. Central administration of acetylcholine, direct-acting cholinergic agonists, and cholinesterase inhibitors has been shown to consistently increase blood pressure in a variety of animal species, including dogs, cats, and rats (Suh et al. 1975; Lang and Rush 1973; Day and Roach 1977). Directly acting muscarinic agonists which increase blood pressure include acetylcholine itself, carbachol, arecoline, and oxotremorine (Bhawe 1958; Lang and Rush 1973; Day and Roach 1977). Indirectly acting cholinomimetic agents, including the cholinesterase inhibitors physostigmine, neostigmine, and sarin, also regularly and dramatically increase blood pressure levels in nonanesthetized animals (Buccafusco et al. 1980; Buccafusco and Spector 1980; Brezenoff 1973), and cause a variable response in pulse rate depending on the species studied. Conversely, there are brain areas, including the premamillary hypothalamic areas, in which administration of cholinergic agonists and cholinesterase inhibitors regularly induce hypotension (Brezenoff and Giuliano 1982) and central cholinergic mechanisms may be bidirectionally important in baroreceptor regulation (Brezenoff and Giuliano 1982).

It appears that central muscarinic mechanisms, rather than peripheral and/or nicotinic mechanisms, predominantly control the blood pressure and heart rate altering effects of directly and indirectly acting cholinergic agents. In support of a central mechanism, only centrally acting cholinesterase inhibitors (i.e., physostigmine but not neostigmine) increase blood pressure if given systemically. In support of a central muscarinic mechanism, cholinergic drug effects on blood pressure are effectively blocked by prior systemic administration of centrally acting antimuscarinic agents such as atropine, but not by systemic administration of noncentrally acting anticholinergics such as methylatropine.

It is likely that cholinomimetic agents, such as acetylcholine, physostigmine, and carbachol, cause their central pressor effects by activating central sympathetic neurons, leading to an increase in sympathetic outflow. In support of this possibility, sympathetic efferent neuronal activity has been shown to be enhanced following central muscarinic stimulation (Kaul and Grewal 1968; Varagic and Vojvadic 1957), and administration of drugs which enhance peripheral sympathetic activity enhance the pressor response to cholinomimetics (Varagic 1955). Also, the stress-sensitive adrenal hormones, epinephrine and norepinephrine, are consistently released following administration of centrally acting cholinomimetics – effects blocked by centrally acting anticholinergic agents (Stamenovic and Varagic 1970; Cass and Spriggs 1961; Brezenoff 1973).

In addition to preclinical studies, Janowsky et al. (1986a, b) have shown that physostigmine, but not non-centrally acting neostigmine, can cause dramatic increases in serum epinephrine levels and small increases in serum norepinephrine levels. This increase in serum catecholamine appears due to a central muscarinic phenomena, since it does not occur with non-centrally acting neostigmine, and is blocked by centrally acting scopolamine, but not by non-centrally acting propantheline. In addition, our very preliminary data suggest that a mixed group of patients with affective disorders may show a blunting of physostigmine-induced serum epinephrine increases, compared to a group of control subjects. A lack of definitive results may be due to the small size of the subject populations studied. If it occurs, epinephrine blunting in response to physostigmine seems to exist in bipolar subjects regardless of their mood state (Janowsky et al. 1986b). This observation is consistent with the findings of Sitaram et al. (1984), who noted that both euthymic and depressed bipolar patients showed supershortening of REM latency after an arecoline infusion.

Koslow et al. (1983) noted higher 24-h urinary epinephrine and norepinephrine excretion levels in patients with major affective disorder than in control subjects. It may be of theoretical significance that, like naturally occurring elevations of epinephrine exretion in depressed patients, physostigmine also causes increases in serum epinephrine and possibly in serum norepinephrine levels. Furthermore, if epinephrine release is indeed chronically high in depressed subjects, the possible blunting of physostigmine-induced epinephrine release in patients with affective disorder may be due to compensatory downregulation of those central or peripheral mechanisms that regulate sympathetic phenomena.

5 Conclusions

We have summarized some of the accumulated data that suggest that cholinergic systems play an important role in the regulation of mood and possibly in the pathogenesis of mood disorders. Considering the interactive connections between cholinergic systems in brain regions that appear to influence mood and behavior, this hypothesis seems quite plausible. Animal experiments, clinical observations, and clinical studies support it. Further, acetylcholine interacts with several of the neuroendocrine systems that are relevant to and at times dysregulated in affective illness.

A few caveats need to be made: It is unlikely that there exists a single determinant for such complex phenomena as "mood" or the regulation of neurohormone systems. Further, it is unlikely that affective illness represents a "single" disease. No neurotransmitter, including acetylcholine, should be considered in isolation. There are important interactions between acetylcholine and central neurotransmitters, as emphasized in the cholinergic-adrenergic balance hypothesis discussed in Sect. 3.1 above. Finally, we have not reviewed some of the data that do not support the hypotheses discussed above. To date, there is no direct evidence of cholinergic alterations in the brains of affective disorder patients, although definitive studies in this regard have yet to be performed. Nonetheless, we believe that the weight of evidence from neuroanatomical and neuropharmacologic studies suggests that cholinergic mechanisms play an important role in the modulation of mood and in the regulation of depression-relevant neurohormone systems.

References

Abe K, Hiroshige T (1974) Changes in plasma corticosterone and hypothalamic CRF levels following intraventricular injection or drug-induced changes of brain biogenic amines in the rat. Neuroendocrinology 14:195–211

Amsterdam JD, Winokur A, Luck I, Caroff S, Synder P, Rickels K (1983) A neuroendocrine test battery in bipolar patients and healthy subjects. Arch Gen Psychiatry 40:515–521

Bhawe WB (1958) Experiments on the fate of histamine and acetylcholine after their injection into the cerebral ventricles. J Physiol 140:169–189

Blackard WG, Waddell CC (1969) Cholinergic blockade and growth hormone responsiveness to insulin hypoglycemia. Proc Soc Exp Biol Med 131:192–196

Board F, Wadeson R, Persky H (1957) Depressive affect and endocrine functions. Arch Neurol Psychiatry 78:612–620

Bowers MB, Goodman E, Sim VM (1964) Some behavioral changes in man following anticholinesterase administration. J Nerv Ment Dis 138:383

Brezenoff HE (1973) Centrally induced pressor responses to intravenous and intraventricular physostigmine evoked via different pathways. Eur J Clin Pharmacol 23:290–292

Brezenoff HE, Giuliano R (1982) Cardiovascular control of cholinergic mechanisms in the central nervous system. Annu Rev Pharmacol Toxicol 22:341–381

Brown WA, Shuey I (1980) Response to dexamethasone and subtype of depression. Arch Gen Psychiatry 37:747–751

Bruni JF, Meites J (1978) Effects of cholinergic drugs on growth hormone release. Life Sci 23:1351–1358

Buccafusco JJ, Spector S (1980) Role of central cholinergic neurons in experimental hypertension. J Cardiovasc Pharmacol 2:347–355
Buccafusco JJ, Finberg JPM, Spector S (1980) Mechanisms of the antihypertensive action of clonidine on the pressor response by physostigmine. J Pharmacol Exp Ther 212:58–63
Bunney WE Jr, Davis JM (1965) Norepinephrine in depressive reactions: a review. Arch Gen Psychiatry 13:483–494
Burnett GR, Prange AJ, Wilson EC, Synder SH (1978) Neuroendocrine-drug relations in tardive dyskinesia. Abstract NR21 of the annual meeting of the American Psychiatric Association, Atlanta, Georgia
Carpenter MA (1972) Core text of neuroanatomy. Williams and Wilkins, Baltimore
Carroll BJ, Curtis GC, Davies BM et al. (1976) Urinary free cortisol excretion in depression. Psychol Med 6:43–50
Carroll BJ, Greden JF, Haskett R, Feinberg M, Albala A, Martin F, Rubin R, Heath B, Sharp P, McLeod W, McLeod M (1980) Neurotransmitter studies of neuroendocrine pathology in depression. Acta Psychiatr Scand Suppl 280:183–199
Carroll BJ, Feinberg M, Greden JF, Tarika J, Albala AA, Haskett RF, Jame NM, Kronfol Z, Lohr N, Steiner M, DeVigne JP, Young E (1981) A specific laboratory test for the diagnosis of melancholia. Arch Gen Psychiatry 38:15–22
Casey DE (1979) Mood alterations during deanol therapy. Psychopharmacology 62:187–191
Cass R, Spriggs TLB (1961) Tissue amine levels and sympathetic blockade after guanethidine and bretylium. Br J Pharmacol 17:442–450
Cohen BM, Miller AL, Lipinski JF, Pope HG (1980) Lecithin in mania: a preliminary report. Am J Psychiatry 137:242–243
Cooper JR, Bloom FE, Roth RH (1982) The biochemical basis of neuropharmacology, 4th edn. Oxford University Press, New York
Coppen A, Shaw DM, Malleson A, Eccleston E, Gundy G (1965) Changes in 5-hydroxytryptophan metabolism in depression. Br J Psychiatry 3:993–998
Davis BM, Davis KL (1980) Cholinergic mechanisms and anterior pituitary hormone secretion. Biol Psychiatry 15(2):303–310
Davis KL, Berger PA, Hollister LE, Defraites E (1978) Physostigmine in mania. Arch Gen Psychiatry 35(1):119–122
Davis K, Hollister LE, Berger PA (1979) Choline chloride in schizophrenia. Am J Psychiatry 136:1581–1584
Day MD, Roach AG (1977) Cardiovascular effects of carbachol and other cholinomimetics administered into the cerebral ventricles of conscious cats. Clin Exp Pharmacol Physiol 4:431–432
Doerr P, Berger M (1983) Physostigmine-induced escape from dexamethasone suppression in normal adults. Biol Psychiatry 18:261–268
Dohanich GP, Barr PJ, Witcher JA, Clemen LG (1984) Pharmacological and anatomical aspects of cholinergic activation of female sexual behavior. Physiol Behav 32:1021–1027
Domino EF, Olds ME (1968) Cholinergic inhibition of self-stimulation behavior. J Pharmacol Exp Ther 164:202–211
Edwardson JA, Bennett GW (1974) Modulation of corticotropin-releasing factor release from hypothalamic synaptosomes. Nature 251:425–427
Evans DL, Burnett GR, Nemeroff CB (1983) The dexamethasone suppression test in a clinical setting. Am J Psychiatry 140:586–589
Gardiner DC, Gibson A, Pollock D (1974) A comparison of the effects of morphine withdrawal, thyroxine or thyroidectomy on the sensitivity of anococcygeus muscle to agonists and on serum thyroxine levels in the rat. Lif Sci 15:339–349
Gerson S, Shaw FH (1961) Psychiatric sequelae of chronic exposure to organophosphorus insecticides. Lancet 1:1371–1374
Gibbon JL, Meltaugh PR (1962) Plasma cortisol in depressive illness. Psychiatr Res 1:162–176

Gibson A, Pollock D (1975) Involvement of corticosteroids in supersensitivity produced in rat anococcygeus muscle by morphine withdrawal, thyroidectomy or a single injection of reserpine. J Pharmacol Exp Ther 192:390–398
Gold PW, Goldwin FK, Wehr T, Rebar R (1977) Pituitary thyrotropin response to thyrotropin-releasing hormone in affective illness: relationship to spinal fluid amine metabolites. Am J Psychiatry 134:1028–1031
Golden RN, Potter WZ (1986) Neurochemical and neuroendocrine dysregulation in affective disorders. Psychiatr Clin North Am 9:313–327
Goldman ME, Erickson CK (1983) Effects of acute and chronic administration of antidepressant drugs on the central cholinergic system: comparison with anticholinergic drugs. Neuropharmacology 22:1215–1222
Goodwin FK, Prange AJ Jr, Post RM, Muscettola G, Lipton MA (1982) Potentiation of antidepressant effects by l-triiodothyronine in tricyclic nonresponders. Am J Psychiatry 139:34–38
Grandison L, Meites J (1976) Evidence for adrenergic medication of cholinergic inhibition of prolactin release. Endocrinology 99:775–779
Guillemin R, Vargo T, Rossier J, Minick S, Ling N, Rivier C, Vale W, Bloom F (1977) Beta-endorphin and adrenocorticotropin are secreted concomitantly by the pituitary gland. Science 197:1367–1369
Hernandez-Peon R (1965) Central neuro-humoral transmission in sleep and wakefulness. Prog Brain Res 18:96–117
Janowsky DS, El-Yousef MK, Davis JM, Sekerke HJ (1972a) A cholinergic-adrenergic hypothesis of mania and depression. Lancet 2:6732–6735
Janowsky DS, El-Yousef M, Davis JM (1972b) Cholinergic antagonism of methylphenidate-induced stereotyped behavior. Psychopharmacologia 27:295–303
Janowsky DS, El-Yousef MK, Davis JM, Sekerle HJ (1973) Parasympathetic suppression of manic symptoms by physostigmine. Arch Gen Psychiatry 28:542–547
Janowsky DS, Davis JM, El-Yousef MK, Sekerke HJ (1974) Acetylcholine and depression. Psychosom Med 36:248–257
Janowsky DS, Risch SC, Parker D, Huey LY, Judd LL (1980) Increased vulnerability to cholinergic stimulation in affect disorder patients. Psychopharmacol Bull 16(4):29–31
Janowsky DS, Risch SC, Judd LL, Huey LY, Parker DC (1981) Cholinergic supersensitivity in affect disorder patients. Behavioral and neuroendocrine observations. Psychopharmacol Bull 17(3):129–132
Janowsky DS, Risch SC, Judd LL, Parker DC, Kalin NH, Huey LY (1983a) Behavioral and neuroendocrine effects of physostigmine in affect disorder patients. In: Clayton PJ, Barrett JE (eds) Treatment of depression: old controversies and new approaches. Proceedings of the American Psychopathological Association meeting. Raven, New York
Janowsky DS, Risch SC, Huey LY, Judd LL, Rausch J (1983b) Hypothalamic-pituitary-adrenal regulation, neurotransmitters, and affective disorders. Peptides 4:775–784
Janowsky DS, Risch SC, Kennedy B, Ziegler M, Huey L (1986a) Central muscarinic effects of physostigmine on mood, cardiovascular function, pituitary and adrenal neuroendocrine release. Psychopharmacology 89:150–154
Janowsky DS, Risch SC, Ziegler MG, Gillin JC, Huey L, Rausch J (1986b) Physostigmine-induced epinephrine release in patients with affective disorder. Am J Psychiatry 143:919–921
Jones M, Hillhouse E, Burden J (1969) The secretion of corticotropin-releasing hormone in vitro. In: Martini L, Ganong WF (eds) Frontiers of neuroendocrinology, vol 4. Raven, New York
Kaplanski J, Smelik PG (1973) Analysis of the inhibition of ACTH release by hypothalamic implants of atropine. Acta Endocrinol 73:651–659
Kato Y, Chirhara K, Ohgo S, Imura H (1974) Effect of nicotine on the secretion of growth hormone and prolactin in rats. Neuroendocrinology 16:237–242
Kaul CL, Grewal RS (1968) Effect of physostigmine upon the output of catecholamines from the adrenal gland of the rat. J Pharmacol Sci 57:1741
Kluver H, Bucy PC (1939) Preliminary analysis of functions of the temporal lobes in monkeys. Arch Neurol Psychiatry 42:949–1000

Koslow SH, Maas JW, Bowden CL et al. (1983) Urinary biogenic amines and metabolites in depression and mania: a controlled univariate analysis. Arch Gen Psychiatry 402:999–1010
Krieger HP, Krieger DT (1970) Chemical stimulation of the brain: effect on adrenal corticoid release. Am J Physiol 218:1632–1641
Krog-Meyer I, Kirkegaard C, Kijne B, Lumholtz B, Smith E, Lykke-Olesen L, Bjorum N (1984) Prediction of relapse with the TRH test and prophylactic amitriptyline in 39 patients with endogenous depression. Am J Psychiatry 141:945–948
Lang WJ, Rush ML (1973) Cardiovascular responses to injections of cholinomimetic drugs into the central ventricles of unanesthetized dogs. Br J Pharmacol 47:196–205
Lawson DM, Gala RR (1975) The influences of adrenergic, dopaminergic, cholinergic and serotonergic drugs on plasma prolactin levels in ovariectomized, estrogen-treated rats. Endocrinology 96:313–318
Leibowitz SF, Miller NE (1969) Unexpected adrenergic effect of chlorpromazine: eating elicited by injection into rat hypothalamus. Science 165:609–611
Loosen P, Prange AJ Jr (1982) Serum thyrotropin response to thyrotropin-releasing hormone in psychiatric patients: a review. Am J Psychiatry 139:405–416
MacLean PD (1957) Chemical and electrical stimulation of the hippocampus in unrestrained animals. II. Behavioral findings. Arch Neurol Psychiatry 78:128–142
MacLean PD (1958) Contrasting junctions of limbic and neocortical systems of the brain and their relevance to psychophysiological aspects of medicine. Am J Med 25:611–626
MacLean PD, Delgado JMR (1953) Electrical and chemical stimulation of frontoemporal portion of limbic system in the waking animal. EEG Clin Neurophysiol 5:91–100
Marquardt DL, Motulsky JJ, Wasserman JI (1982) Rat lung cholinergic receptor: characterization and regulation by corticosteroids. J Appl Physiol 43:731–736
Martin JB, Durand D, Gurd W, Gaille G, Audet J, Brazeau P (1978) Neuropharmacologic regulation of episodic growth hormone and prolactin secretion in the rat. Endocrinology 102:106–113
Martin JR, Overstreet DH, Driscoll P, Battig K (1981) Effects of scopolamine, pilocarpine and oxotremorine on the exploratory behavior of two psychogenetically selected lines of rats in a complex maze. Psychopharmacology 72:135–142
McEwen BS, Pfaff C (1985) Hormonal effects of hypothalamic neurons: a analyzing gene expression and neuromodulatory action. Trends Neurosci 8:105–110
Mendelson WB, Jacobs LS, Sitaram N, Wyatt RJ, Gillin JC (1978) Methscopolamine inhibition of sleep related growth hormone secretion. J Clin Invest 61:1683
Mendelson WB, Lantigua RA, Wyatt RJ, Gillin JC, Jacobs LS (1981) Piperidine enhances sleep-related and insulin-induced growth hormone secretion: further evidence for a cholinergic secretory mechanism. J Clin Endocrinol Metab 52(3):409–415
Myers RD (1964) Emotional and autonomic responses following hypothalamic chemical stimulation. Can J Psychol 18:6–14
Narabayashi H, Nugao T, Saito Y, Yosihida M, Nagahat M (1963) Sterotaxic amygdalotomy for behavioral disorders. Arch Neurol 9:1–16
Netherton RA, Overstreet DH (1981) Sex differences in effects of oxotremorine on hypothermia. Presented at international pharmacology meeting, Tokyo
Olds J (1960) Differentiation of reward systems in the brain by selfstimulation technics. In: Ramey SR, O'Doherty DS (eds) Electrical studies on the unanesthetized brain. Hoeber, New York
Overstreet DH (1986) Selective breeding for increased cholinergic function: development of a new animal model of depression. Biol Psychiatry 21:49–58
Overstreet DH, Russell RW (1982) Selective breeding for sensitivity to DFP: effects of cholinergic agonists and antagonists. Psychopharmacology 78:150–154
Overstreet DH, Russell RW (1984) Selective breeding for differences in cholinergic function: sex differences in the genetic regulation of sensitivity to the anticholinesterase, DFP. Behav Neural Biol 40:227
Overstreet DH, Russell RW, Kerni W, Netherton RA (1981) The influence of ovariectomy on the sex-dependent effects of the anticholinesterase, DFP. Psychopharmacology 74:393–394

Overstreet DH, Janowsky DS, Gillin JC, Shimoni PJ, Sutin EL (1987a) Stress-induced immobility in rats with cholinergic supersensitivity. Biol Psychiatry (to be published)
Overstreet DH, Booth RA, Dana R, Risch SC, Janowsky DS (1987b) Corticosterone supersensitivity to arecoline in rats with cholinergic supersensitivity. Psychopharmacology (to be published)
Papez JW (1937) A proposed mechanism in emotion. Arch Neurol Psychiatry 38:725–731
Pradam SN, Kamat KA (1972) Action and interaction of cholinergic agonists and antagonists on self-stimulation. Arch Int Pharmacodyn Ther 196:321–329
Prange AJ Jr (1964) The pharmacology and biochemistry of depression. Dis Nerv 25:217–221
Rainbow TC, Degroff V, Luine VN, McEwen BS (1980) Estradiol 17B increased the number of muscarinic receptors in hypothalamic nuclei. Brain Res 198:239–243
Rainbow TC, Snyder L, Beck DJ, McEwen BS (1984) Correlation of muscarinic receptor induction in the ventromedial hypothalamic nucleus with the activation of feminine sexual behavior by estradiol. Neuroendocrinology 39:476–480
Riker DK, Sastre A, Barber T, Roth RH, Riker WR (1979) Regional high affinity ^{3}H choline accumulation in cat forebrain. Selective increase in caudate-putamen after corticosteroid pretreatment. Mol Pharmacol 16:886–889
Risch SC, Janowsky DS (1984) Cholinergic-adrenergic balance in affective illness. In: Post R, Ballenger J (eds) Neurobiology of mood disorder. Williams and Wilkins, Baltimore
Risch SC, Cohen PM, Janowsky DS, Kalin NH, Insel TR, Murphy DL (1981a) Physostigmine induction of depressive symptomatology in normal volunteer subjects. J Psychiatr Res 4:89–94
Risch SC, Cohen RM, Janowsky DS, Kalin NH, Murphy DL (1981b) Plasma Beta-endorphin and cortisol elevations accompany the mood and behavioral effects of physostigmine in man. Science 209:1545–1546
Risch SC, Janowsky DS, Kalin NH, Cohen RM, Aloi JA, Murphy DL (1982) Physostigmine response as markers in affective disorder patients. In: Hanin I, Usdin E (eds) Biological markers in psychiatry and neurology. Pergamon, Oxford, pp 269–278
Risch SC, Janowsky DS, Parker D, Kalin NH, Aloi JA, Cohen RM, Judd LL, Huey LY, Murphy DL (1984) Neuroendocrine abnormalities in affective disorders: possible cholinergic mechanisms. In: Post R, Ballenger J (eds) Neurobiology of mood disorders. Williams and Wilkins, Baltimore
Risch SC, Janowsky D, Gillin JC, Kelsoe JR, Ziegler MG (1987) Muscarinic mechanisms in neuroendocrine regulation and depression (to be published)
Rowntree DW, Neven S, Wilson A (1950) The effects of diisopropylfluorophosphonate in schizophrenia and manic depressive psychosis. J Neurol Neurosurg Psychiatry 13:47–62
Rubinow DR, Post RM, Pickar D et al. (1981) Relationship between urinary free cortisol and CSF opioid-binding activity in depressed patients and normal volunteers. Psychiatr Res 5:87–93
Sachar EJ, Hellman L, Roffwarg HP et al. (1973) Disrupted 24-hour patterns of cortisol secretion in psychotic depression. Arch Gen Psychiatry 28:19–24
Sawyer CH, Everett JW (1946) Effects of various hormonal conditions in the intact rats on the synthesis of serum cholinesterase. Endocrinology 39:207–222
Schildkraut JJ (1965) The catecholamine hypothesis of affective disorders: a review of supporting evidence. Am J Psychiatry 122:509–522
Sitaram N, Dube S, Jones N et al. (1984) Cholinergic REM-induction response as a state and possible genetic vulnerability marker of depression. Clin Neuropharmacol 7 [Suppl]:524
Stamenovic BA, Varagic VM (1970) The effect of eserine on the efferent neuronal activity in the cervical sympathetic of the rat. Neuropharmacology 9:561–566
Steiner FA (1968) Influence of microelectrophoretically applied acetylcholine on the responsiveness of hippocampal and lateral geniculate neurones. Pflügers Arch 303:173–180
Suh TH, Wang CH, Lim RKS (1975) Effect of intracisternal injection of acetylcholine. Proc Soc Exp Biol Med 32:1410

Suzuki T, Abe K, Horose T (1975) Adrenal cortical secretion in response to pilocarpine in dogs with hypothalamic lesions. Neuroendocrinology 17:75–82

Sze PY, Marchi M, Towle AC, Giacobini E (1983) Increased uptake of ^{3}H choline by rat superior cervical ganglion: an effect of dexamethasone. Neuropharmacology 22:711–716

Tamminga C, Smith RC, Chang S, Naraszti JS, Davis JM (1976) Depression associated with oral choline. Lancet II:905

Targum SD, Greenberg RD, Harmon RL et al. (1984) The TRH test and thyroid hormone in refractory depression. Am J Psychiatry 141:463

Varagic V (1955) The action of eserine on the blood pressure of the rat. Br J Pharmacol 10:349–353

Varagic V, Vojvadic N (1957) Effect of guanethidine, hemicholinium and mebutamate in the hypertensive response to eserine and catecholamines. Br J Pharmacol Chemother 19:451–457

The Psychobiology of Neurotensin *

B. Levant [1] and C. B. Nemeroff [2]

Contents

* The authors are supported by NIMH MH-39415, MH-40159, MH-42088, MH-40524, and a Nanaline Duke Scholars Award to Charles B. Nemeroff

1 Department of Pharmacology, Duke University Medical Center, Durham, NC 27710, USA

2 Departments of Psychiatry and Pharmacology, Duke University Medical Center, Durham, NC 27710, USA

Current Topics in Neuroendocrinology, Vol. 8

1 Introduction

The tridecapeptide neurotensin (NT) was first identified in bovine hypothalamic extracts and characterized by Carraway and Leeman (1973, 1975, 1976) and has subsequently been found in all classes of vertebrates (Carraway and Leeman 1976; Kitabgi et al. 1976; Kataoka et al. 1979; Langer et al. 1979; Reinecke et al. 1980a; Cooper et al. 1981; Grant et al. 1982; Carraway et al. 1982; Eldred and Karten 1983), many invertebrates (Reinecke et al. 1980b; Grimmelikhuijzen et al. 1981; Price et al. 1982), and certain bacteria (Bhatnagar and Carraway 1981). It is distributed throughout the mammalian central nervous system (CNS) (Uhl and Snyder 1977a, b), gastrointestinal tract (Sundler et al. 1977; Schultzberg et al. 1980), cerebrospinal fluid (CSF), adrenals, pancreas, and plasma (Fernstrom et al. 1980). When administered systemically, the peptide has a variety of effects such as hypotension, hyperglycemia, decreased gastric acid secretion, decreased gut motility, and altered secretion of anterior pituitary hormones (Leeman and Carraway 1982). NT apparently does not cross the blood-brain barrier in appreciable quantities; however, when administered directly into the CNS, it produces a number of physiological and behavioral effects. A burgeoning body of evidence supports the role of NT as a neurotransmitter or neuromodulator. Thus far, heterogeneous CNS distribution, release of NT upon neuronal depolarization, saturable and specific binding of NT to receptors, and degradation by peptidases have all been demonstrated.

This chapter focuses on the actions of NT in the CNS, particularly the pharmacological effects of centrally administered NT and the role of NT in neuropsychiatric and other disorders.

2 Neuroanatomy

2.1 Localization and Pathways

Neurotensin has been shown to be heterogeneously distributed throughout the CNS of a variety of mammals (Uhl and Snyder 1977a; Kobayashi et al. 1977; Kahn et al. 1980; Jennes et al. 1982; Roberts et al. 1982; Emson et al. 1982; Goedert et al. 1983; Reinecke et al. 1983). Studies utilizing radioimmunoassay and immunohistochemical techniques on postmortem human brain tissue have revealed a similar distribution. High concentrations of NT are found in the zona compacta and zona reticulata of the substantia nigra, the periaqueductal gray matter, and certain hypothalamic nuclei. Relatively low concentrations of NT are found in the cerebral cortex and cerebellum (Cooper et al. 1981; Manberg et al. 1982; Ghatei et al. 1984) (Table 1).

In addition to determining the distribution of NT in the rat CNS by radioimmunoassay, the location of NT-containing cell bodies has been identified, as well as several putative NT pathways. The first NT pathway identified originates in the central amygdaloid nucleus, projecting through and terminating in the bed

Table 1. Regional distribution of NT in human brain

Area	n	pg/μg protein
Hypothalamus	17	1.835 ± 0.196
Substantia nigra pars compacta	3	0.955 ± 0.312
Periaqueductal gray	3	0.653 ± 0.328
Substantia nigra pars reticulata	3	0.556 ± 0.127
Anterior perforated substance	3	0.384 ± 0.158
Globus pallidus (medial)	3	0.179 ± 0.037
Nucleus accumbens	13	0.366 ± 0.080
Globus pallidus (lateral)	3	0.267 ± 0.009
Pons	3	0.197 ± 0.071
Hippocampus	3	0.111 ± 0.044
Septal nuclei	3	0.237 ± 0.066
Amygdala	13	0.281 ± 0.057
Thalamus, anteromedial nucleus	3	0.175 ± 0.049
Red nucleus	3	0.127 ± 0.020
Putamen	3	0.089 ± 0.017
Thalamus, anterolateral nucleus	3	0.097 ± 0.047
Cerebellar cortex	2	0.081 ± 0.020
Inferior olive	3	0.073 ± 0.010
BA 8 (frontal cortex)	3	0.069 ± 0.008
Caudate nucleus	11	0.120 ± 0.029
BA 24 (cingulate gyrus)	14	0.094 ± 0.015
BA 32 (frontal cortex)	14	0.088 ± 0.018
BA 4 (motor cortex)	3	0.012 ± 0.041
Dentate nucleus	3	0.045 ± 0.008
BA 20 (temporal cortex)	3	0.063 ± 0.021
BA 11 (frontal cortex)	3	0.037 ± 0.017
BA 38 (temporal cortex)	3	0.091 ± 0.036
Subthalamic nucleus	3	0.051 ± 0.015
BA 10 (frontal cortex)	3	0.041 ± 0.028

Reproduced with permission from Manberg et al. (1982)
Values represent means ±SEM and have not been corrected for recoveries
n, number of subjects; BA, Brodmann's areas of cerebral cortex

nucleus of the stria terminalis (Uhl and Snyder 1979; Leranth et al. 1981). Another purported NT pathway projects from the hippocampus through the cingulate cortex to the frontal cortex (Roberts et al. 1981; Bloom and Pollack 1982) though considerable controversy about this pathway exists. A third pathway has been identified originating from cells in the periaqueductal gray/nucleus of the solitary tract and projecting to the nucleus raphe magnus (Beitz 1982). In addition, a pathway from the endopiriform nucleus and piriform cortex to the anterior olfactory nucleus and the nucleus of the diagonal band of Broca has been identified (Inagaki et al. 1983). More recently, investigators utilizing combined NT immunohistochemistry and retrograde labeling techniques using horseradish peroxidase or fluorescent dye, have proposed pathways from the ventral tegmental area (VTA) to the nucleus accumbens (Kalivas and Miller 1984) and from the central nucleus of the amygdala to the parabrachial nucleus (Moga and Gray

1985). Finally, NT-immunoreactive perikarya have been identified in the subiculum with projections to the alveus, fimbria, fornix, and neuropil of the mamillary bodies (Sakamoto et al. 1986).

2.2 Colocalization with Other Neurotransmitters

Immunohistochemical techniques have demonstrated that NT is colocalized with certain subgroups of catecholaminergic neurons. Neurotensin-like immunoreactivity (NTLI) was found in the dopaminergic perikarya of the arcuate nucleus (A_{12}) and periventricular nucleus (A_{14}) of the hypothalamus (Ibata et al. 1983; Hokfelt et al. 1984). Dopamine (DA) and NT were also found to be colocalized in the A_8, A_9, and A_{10} DA neurons of the mesencephalon, particularly the A_{10} cells in the VTA (Hokfelt et al. 1984). However, destruction of DA neurons with 6-hydroxydopamine (6-OHDA) does not result in a significant change in radioimmunoassayable NT concentrations (Bissette et al. 1983) nor are NT concentrations altered in Parkinson's disease where nigrostriatal and mesolimbic DA pathways degenerate, except in the hippocampus (Bissette et al. 1985). Likewise, 1-methyl-4-phenyl-1,2,3,6-tetrahydropyridine (MPTP), which selectively destroys nigrostriatal DA neurons, significantly decreases NTLI in the frontal cortex, preoptic area, and caudate nucleus, but only by about 50% (B. Levant, G. Bissette, Y.-M. Parker, and C. B. Nemeroff, unpublished observations). While these findings appear to be discrepant, it is likely that the subpopulation of neurons in which NT and DA are colocalized are either not susceptible to the neurotoxic effects of 6-OHDA, MPTP, or Parkinson's disease, or represent only a small percentage of these DA cells.

Neurotensin-like immunoreactivity is also found in some noradrenergic cells of the medial nucleus of the solitary tract and in the majority of adrenergic cell bodies of the dorsal part of the nucleus of the solitary tract (Hokfelt et al. 1984).

Neurotensin-like immunoreactivity has also been found to be colocalized with other neuromodulatory peptides as well as with the amino acid neurotransmitter glycine. Utilizing double-staining immunohistochemistry on cat thalamus, Sugimoto et al. (1985) found cholecystokinin (CCK)-like immunoreactivity and NTLI in the anterodorsal nucleus and vasoactive intestinal peptide-like immunoreactivity and NTLI in the lateral dorsal nucleus. NT has also been shown to be colocalized with enkephalin (Li et al. 1985) and glycine (Watt et al. 1985) in subpopulations of amacrine cells of the chicken retina and with substance P in some amacrine cells of the goldfish retina (Li et al. 1986).

2.3 Synthesis, Storage, Release, and Degradation

Although no studies have been published on the biosynthesis of NT in the CNS, it, like other peptide hormones, is in all likelihood synthesized as a larger prohormone which is then processed intraneuronally into the active form. After density gradient centrifugation, NT was found preferentially in the synaptosomal frac-

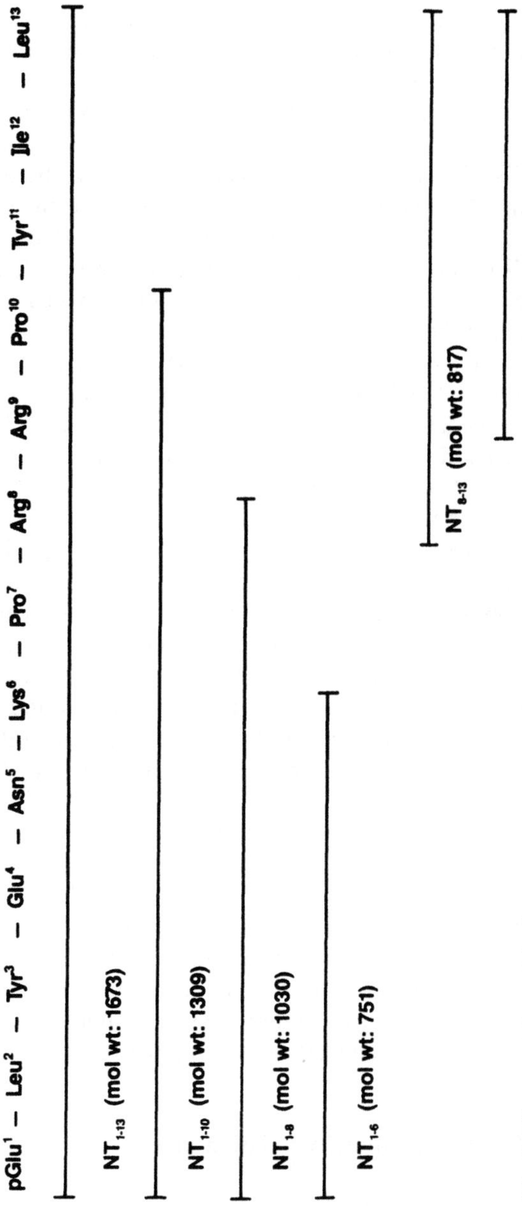

Fig. 1. Neurotensin and its fragments

tion (Kataoka et al. 1979). NT-containing synaptic vesicles have also been observed in the nerve terminals using electron microscopy (Uhl and Snyder 1977a; Uhl et al. 1977a).

Using in vitro preparations of rat hypothalamus, calcium-dependent NT release can be stimulated by depolarizing potassium concentrations (Iversen et al. 1978). Release of NT can also be stimulated by dibutyryl cyclic-AMP and dopamine (Maeda and Frohman 1981) and while morphine inhibits the release of CCK and substance P in this preparation, it does not inhibit the release of NT (Micevych et al. 1982). Potassium-stimulated, calcium-dependent NT release from in vitro preparations of rat spinal cord has also been demonstrated (Gamse and Heuberger 1983).

After synaptic release, NT is degraded by a variety of peptidases (Dupont and Merand 1978) into peptide fragments, some of which have biological activity (Fig. 1). The major site of cleavage is the Arg^{8}-Arg^{9} peptide bond (McDermott and Kidd 1984), but cleavage can also occur at the Pro^{7}-Arg^{8} (Camargo et al. 1983), Pro^{10}-Tyr^{11} (Camargo et al. 1983; Skidgel et al. 1984), and Try^{11}-Ile^{12} peptide bonds (Checler et al. 1983; Skidgel et al. 1984). Studies with isolated synaptic membranes revealed that degradative activity exists in both the membrane-bound and the soluble fractions. In the membrane-bound fraction, a dipeptidyl carboxypeptidase, angiotensin-converting enzyme, cleaves $NT_{1\text{-}10}$ to $NT_{1\text{-}8}$ at the Arg^{8}-Arg^{9} peptide bond. This enzyme is inhibited by captopril. Also located in the membrane-bound fraction, thermolysin-like metallopeptidase or "enkephalinase" cleaves NT at the Try^{11}-Ile^{12} bond. This enzyme is selectively inhibited by thiorphan and both enzymes are inhibited by 1,10-phenanthroline (Checler et al. 1983). Enzymatic activity cleaving NT at the Arg^{8}-Arg^{9} and Pro^{10}-Tyr^{11} bonds is found in the soluble fraction (Checler et al. 1983). The cleavage at the Pro^{10}-Try^{11} bond, some of which is catalyzed by endopeptidase 24.11, converts $NT_{1\text{-}13}$ to $NT_{1\text{-}10}$ (Checler et al. 1984). The cleavage of $NT_{1\text{-}13}$ into $NT_{1\text{-}8}$ and $NT_{9\text{-}13}$ at the Arg^{8}-Arg^{9} bond is conducted by a metallopeptidase (Checler et al. 1985). Inhibition of this cleavage by $dynorphin_{1\text{-}13}$ decreases NT degradation by 80% in hypothalamic slice preparations (McDermott et al. 1986). This enzyme may therefore be the major enzyme involved in NT degradation. Different cleavage patterns from those observed in mammals have been found in other vertebrates such as reptiles and fish (Griffiths et al. 1984).

2.4 Structure-Activity Relationships

While enzymatic degradation can result in inactivation of NT, many of the peptides produced possess some of the biological activities of NT. For example, the N-terminal fragments $NT_{1\text{-}10}$ and $NT_{1\text{-}8}$ both possess gastric cytoprotective activity against cold-restraint, stress-induced ulcers when administered intracisternally (i.c.) (Hernandez et al. 1984b). Intracisternally administered $NT_{1\text{-}8}$ also inhibits TRH-induced wet-dog shakes (Griffiths et al. 1982; Widdowson et al. 1983). The C-terminal fragments, $NT_{8\text{-}13}$ and $NT_{9\text{-}13}$, decrease locomotor activity and body temperature (Jolicoeur et al. 1981a, 1982a, b; Kitabgi et al. 1980; Quirion et al. 1982b) and have antinociceptive activity (Furuta et al. 1984). The

Table 2. Quantitative topographical distribution of monoiodo ^{125}I-Tyr NT binding sites in the human brain

Brain area	n	Specific [^{125}I]-NT binding (fmol/mg protein)
Cerebral cortex		
Cingulate cortex (BA 32)	5	6.6±0.6
Cingulate gyrus (BA 32)	6	9.6±0.9
Paraolfactory gyrus cortex (BA 25)	4	10.5±0.9
Frontal cortex internal layer (BA 9)	6	5.5±0.1
Frontal cortex external layer (BA 9)	6	1.9±0.2
Entorhinal cortex internal layer (BA 28)	12	7.1±0.2
Entorhinal cortex external layer (BA 28)	7	2.3±0.1
Temporal cortex internal layer (BA 21)	8	4.2±0.4
Temporal cortex external layer (BA 21)	4	1.1±0.1
White matter cortex (BA 32)	0	
Basal ganglia		
Caudate nucleus	6	7.0±0.4
Putamen	6	7.4±1.5
Posterior caudate nucleus	4	4.0±0.3
Posterior putamen	4	4.7±1.0
Nucleus accumbens	4	11.1±0.6
Internal capsule	8	1.5±0.1
Limbic system		
Ammon's horn (hippocampus)	7	1.3±0.2
Fascia dentata	4	2.2±0.1
Subiculum	7	5.6±1.0
Lateral amygdaloid nucleus	5	7.5±1.0
Basal amygdaloid nucleus	8	10.5±0.8
Central amygdaloid nucleus	3	8.6±1.6
Hypothalamus		
Ventromedial hypothalamic nucleus	10	4.2±0.2
Posterior hypothalamic nucleus	10	4.3±0.6
Mamillary bodies	10	2.5±0.2
Optic nerve	3	0.8±0.1
Brain system		
Central grey	11	2.7±0.2
Red nucleus	4	3.4±0.7
Substantia nigra pars compacta α	9	8.5±0.6
Substantia nigra pars compacta β	12	5.2±0.3
Substantia nigra pars compacta lateral	8	14.1±1.2
Paranigralis nucleus	13	13.4±1.1
Substantia nigra pars reticulata	10	1.3±0.2
Ventral tegmental area	6	13.0±0.5
Superior colliculus	4	2.7±0.5
Inferior colliculus	8	1.5±0.2
Supratrochlear nucleus	8	2.1±0.4
Cerebellar peduncle	6	1.0±0.2
Cerebral peduncle	6	0.9±0.1
Locus ceruleus	7	2.3±0.3
Parabrachial nucleus	7	3.0±0.4
Trigeminal nerve	7	5.6±0.6
Medial lemniscus	7	3.1±0.5
Inferior olive	4	2.4±0.7
Ambiguous nucleus	4	2.1±0.4
Dorsal cochlear nucleus	4	1.8±0.6

Reproduced with permission from Sarrieau et al. (1985)
Autoradiograms at various levels were obtained with 0.1 nM [^{125}I]-NT. The values represent the mean ±SEM of several readings of five different human brains
n, number of readings; BA, Brodmann's area of cerebral cortex

synthetic fragments $NT_{8\text{-}10}$, $NT_{8\text{-}11}$, $NT_{9\text{-}11}$, and $NT_{9\text{-}13}$ also produce antinociceptive effects upon i.c. administration (Furuta et al. 1984). In addition, $NT_{8\text{-}13}$, as well as $NT_{1\text{-}8}$, inhibits TRH-induced wet-dog shakes (Griffiths et al. 1982; Widdowson et al. 1983). The fragments $NT_{1\text{-}6}$ and $NT_{10\text{-}13}$ have not exhibited biological activity in any of the paradigms tested (Hernandez et al. 1984b; Jolicoeur et al. 1981a, 1982a, b; Kitabgi et al. 1980; Quirion et al. 1982b). These experiments, taken together with binding studies using native and substituted NT fragments, show that the biological activity of the peptide appears to be dependent on the presence of at least one of the two Arg residues in positions 8 and 9 (Kitabgi et al. 1980; Goedert et al. 1984b).

2.5 Receptors

Neurotensin receptors, which appear to be composed of two different 50000-MW subunits linked by a noncovalent bond (Mazella et al. 1985), are differentially distributed throughout the CNS. Saturable, reversible, and high-affinity binding of $[^{125}I]$-NT and $[^{3}H]$-NT were first observed in membrane preparations from rat brain (Uhl et al. 1977b; Kitabgi et al. 1977; Lazarus et al. 1977). The distribution of NT receptors has been visualized by autoradiography with $[^{3}H]$-NT in sectioned rat brain. These studies revealed a heterogeneous distribution of NT receptors which, in many regions, parallels the distribution of NTLI. The areas of highest receptor density include the zona compacta of the substantia nigra, various amygdaloid nuclei, and the internal plexiform layer of the olfactory bulb (Young and Kuhar 1979, 1981; Quirion et al. 1982a). A subsequent investigation using $[^{3}H]$-NT binding on membranes from different regions of rat brain revealed that NT receptors are more uniformly distributed throughout the brain than NTLI (Goedert et al. 1984b). Disparities in the density of NT receptors and NTLI are most notable in the cortex, where NTLI is low and NT receptor density is high (Lazarus et al. 1977; Uhl et al. 1977) and in the striatum where the reverse is true (Goedert et al. 1983). A similar distribution of receptors was found with both $[^{125}I]$-NT and $[^{3}H]$-NT binding in human brain (Sadoul et al. 1984a; Sarrieau et al. 1985; Kanba et al. 1986) as shown in Table 2.

3 Neurochemistry

Perhaps the most interesting, complex, and, to date, clinically relevant neurotransmitter interaction of NT is with the CNS DA systems (Nemeroff et al. 1980, 1982b; Haubrich et al. 1982; Quirion 1982b; Reches et al. 1983; Kalivas 1985; Nemeroff and Cain 1985; Nemeroff 1986; Kilts et al. 1986). NT and DA have been shown to have a largely overlapping distribution in the CNS (Anderson et al. 1984) and are colocalized in a number of brain nuclei, particularly the arcuate and periventricular nuclei of the hypothalamus and the A_8, A_9, and A_{10} DA perikarya (Ibata et al. 1983, 1984a; Hokfelt et al. 1984; Kalivas and Miller 1984). NT binding sites have been found on DA cell bodies (Palacios and Kuhar 1981;

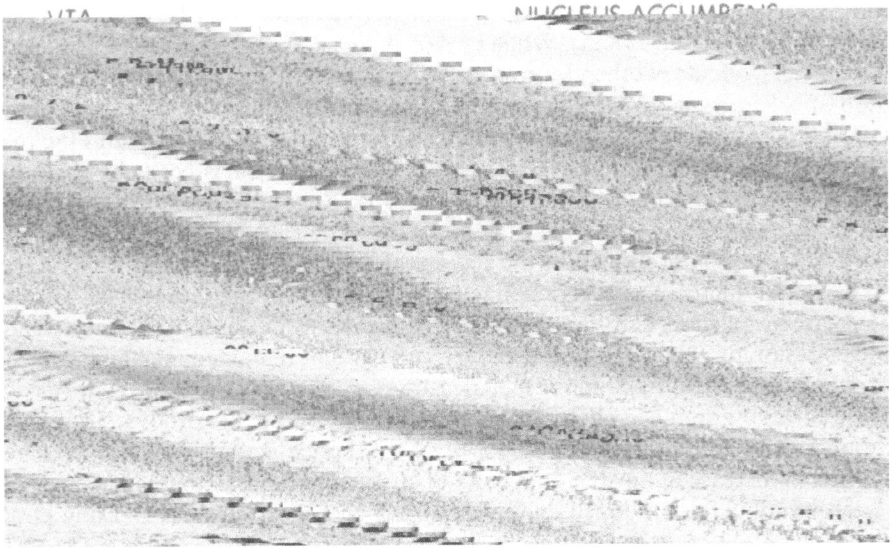

Fig. 2. Schematic drawing of selected anatomical relationships between neurotensin (*NT*) and dopamine (*DA*). Neurotensin-containing cell bodies and their projections parallel those of the mesolimbic DA system. Neurotensin receptors are present on DA cell bodies in both the A_9 and A_{10} region and on DA terminals in the striatum. Although it has not been definitely shown, for convenience, we have indicated here that both NT receptors and DA receptors are colocalized on the same cells in the striatum and nucleus accumbens. *VTA*, ventral tegmental area. (From Nemeroff and Cain 1985, with permission)

Quirion et al. 1982c) and on some DA nerve terminals (Quirion et al. 1982; Goedert et al. 1984a) (Fig. 2). Furthermore, the density of NT binding sites in these DA terminal areas appears to be regulated by DA neurotransmission (Herve et al. 1986). Lesions of the VTA with 6-OHDA resulted in decreased density of [^{125}I]-NT binding sites in the mesencephalon and striatum while the number of binding sites increased in the lateral prefrontal cortex. Similar preliminary studies have shown that dopaminergic input to the rat subcortical limbic forebrain appears to increase the number of NT receptors while decreasing their affinity for [^{3}H]-NT (Agnati et al. 1983). Likewise, in the rat striatum, DA receptors rendered supersensitive by intranigral injections of 6-OHDA also appear to decrease the affinity of striatal NT binding sites (Fuxe et al. 1986).

Neurotensin appears to have a stimulatory effect on DA neurons. In electrophysiological studies, iontophoretically applied NT produced excitation in midbrain DA neurons (Andrade and Aghajanian 1981). NT has also been shown to increase tyrosine hydroxylase activity (Reches et al. 1982) and increase DA release in the striatum (Starr 1982; DeQuitt and Emson 1983; Okuma et al. 1983), the hypothalamus (Myers and Lee 1984a), and the nucleus accumbens (Myers and Lee 1984b) by a mechanism which appears to involve an increase in intracellular calcium (Battaini et al. 1986). Accordingly, i.c. and intracerebroventricular (i.c.v.) NT increase DA turnover as determined by increased concentrations of the DA metabolites dihydroxyphenyl acetic acid and homovanillic acid, especially in the mesolimbic DA terminal fields (Garcia-Sevilla et al. 1978; Reches et al. 1982, 1983; Widerlöv et al. 1982a). While i.c. NT increases DA turnover in the nucleus accumbens, olfactory tubercles, and striatum (Nemeroff et al. 1983a), microinjection of NT into the VTA results in a selective increase in DA turnover only in the nucleus accumbens and olfactory tubercles indicating that NT activates the mesolimbic DA system (Kalivas et al. 1985). Likewise, NT has been shown to increase DA turnover in the projection from the VTA to the diagonal band of Broca (Kalivas et al. 1985).

Neurotensin also influences the activity of a variety of other neurotransmitter systems. An increase in acetylcholine (Ach) turnover in the diencephalon and a decrease in Ach concentration in the partietal cortex was observed upon i.c.v. administration of NT, while Ach turnover and content remained unchanged in the frontal cortex (Malthe-Sorenssen et al. 1978).

Centrally administered NT increases serotonin (5-HT) turnover in several brain areas (Garcia-Sevilla et al. 1978). Serotonergic activity has also been reported to increase in the diagonal band of Broca and decrease in the medial forebrain bundle and periaqueductal gray after i.c.v. injection of NT (Long et al. 1981).

As with 5-HT, i.c.v. NT increases norepinephrine (NE) turnover in some brain areas (Garcia-Sevilla et al. 1978). NT has also been shown to increase spontaneous and potassium-induced NE release from a hypothalamic slice preparation (Okuma and Osumi 1982). Further in vivo studies revealed that a 6-OHDA lesion of the ascending NE pathways results in a decrease in both NE and NT in the hypothalamus (Ibata et al. 1984b). Furthermore, i.c.v. NT has been observed to alter the concentration of NE in several brain areas. Increases in NE concentration have been reported in the hypothalamus, nucleus accumbens, and amygdala (Fuxe et al. 1984; Drumheller et al. 1986), while a depletion of NE stores was observed in preoptic suprachiasmatic and pericentricular noradrenergic nerve terminals (Fuxe et al. 1984). It is, however, important to note that NT effects on DA neurons have been scrutinized most closely and that the doses of centrally administered NT that alter DA neural activity are, in general, considerably lower that those affecting other neurotransmitter systems.

4 Pharmacological Effects

4.1 Effects on Behavior

4.1.1 Locomotor Activity

Neurotensin interacts with mesolimbic DA projections to produce changes in locomotor behavior. In the open-field paradigm, i.c.v. NT has been observed to decrease locomotor activity in rats (Rinkel et al. 1983; Elliott et al. 1986), but tolerance to this effect develops after repeated NT administration (Rinkel et al. 1983). Intracerebroventricular NT has also been shown to decrease locomotor activity in mice (Meisenberg and Simmons 1985). Conversely, microinjection of NT into the VTA produces a dose-related increase in spontaneous motor activity consisting of an increase in exploratory behaviors such as locomotion, sniffing, and rearing with a concomitant decrease in sleeping and resting behavior (Kalivas et al. 1981). This increase in motor activity is associated with an increase in DA utilization in the nucleus accumbens, a major terminal field of the mesolimbic DA projection (Kalivas et al. 1983). Likewise, Cador et al. (1985) found that while microinjection of NT into the VTA produced an increase in rearing and locomotor activity, injection of NT into the substantia nigra and hippocampus had no effect on spontaneous activity, further supporting the view that the effects of the peptide on spontaneous motor activity is mediated by the A_{10} mesolimbic pathways. Furthermore, daily injections of NT into the VTA result in sensitization to subsequent NT effects, i.e., enhanced locomotor activity responses and DA turnover (Kalivas and Taylor 1985). These enhanced responses persist for at least 7 days after the cessation of NT administration and were observed to only occur with microinjections into the VTA and not the adjacent nuclei. This potentiation of locomotor response seen with daily administration of NT did not occur when NT was administered every 48 h nor did daily injections of NT into the VTA alter the locomotor response to low doses of *d*-amphetamine (Elliott and Nemeroff 1986).

While NT produces an increase in locomotor activity when injected into the VTA, the site of origin of the mesolimbic DA neurons, microinjection of NT into the nucleus accumbens, a major terminal of the mesolimbic DA projection, has the opposite effect (Griffiths et al. 1981). When injected into the nucleus accumbens, NT produced a dose-dependent antagonism of DA-induced locomotion and rearing (Kalivas et al. 1984). Moreover, bilateral injection of NT into the nucleus accumbens blocked the effect of NT applied to the VTA (Kalivas et al. 1982a). Thyrotropin-releasing hormone (TRH) produces an increase in locomotor activity when injected into the nucleus accumbens (Griffiths et al. 1981); this effect is antagonized by NT (Griffiths et al. 1983). NT thus only produces increases in locomotor activity when applied to the DA perikarya in the VTA and decreases motor activity when applied to the terminal field of that projection or when injected i.c.v. (Fig. 3). DA autoreceptors may also be involved in this process although, in a preliminary study, i.c. NT did not attentuate the hypolocomotion produced by the putative DA receptor agonist *N*-*n*-propyl-3-(3-hydroxyphenyl)-piperidine (3-PPP) (Cain and Nemeroff 1985). Thus, many of the effects of NT on motor behavior appear to involve interactions with the mesolimbic DA circuit. In accordance with its modulatory role in the mesolimbic DA projection, NT antagonizes many of the behavioral effects induced by *d*-amphetamine. Bilateral injections of NT into the nucleus accumbens antagonized the increases in locomotor activity and rearing observed after parenteral *d*-amphetamine administration; in contrast, intracaudate injection of NT did not block *d*-amphetamine-induced stereotypy (Ervin et al. 1981). NT also antagonized the increase in locomotor activity produced by methylphenidate and cocaine in both mice and rats (Nemeroff et al. 1983a), but has no effect on scopolamine-induced (Skoog et al. 1986), or caffeine-induced (Skoog et al. 1986) increases in locomotion. NT antagonization of drug-induced locomotor hyperactivity thus seems limited to those pharmacological agents that activate the mesolimbic DA system.

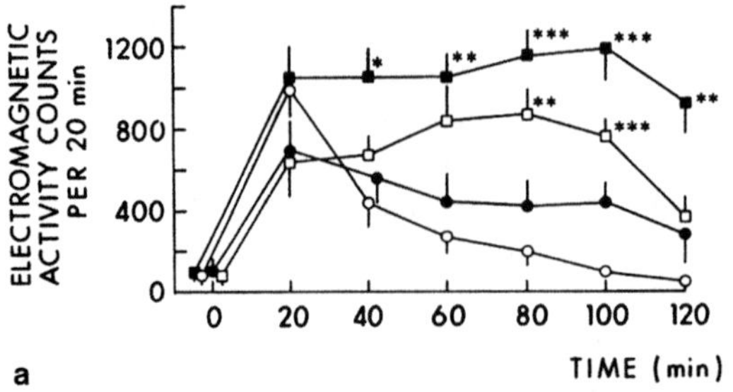

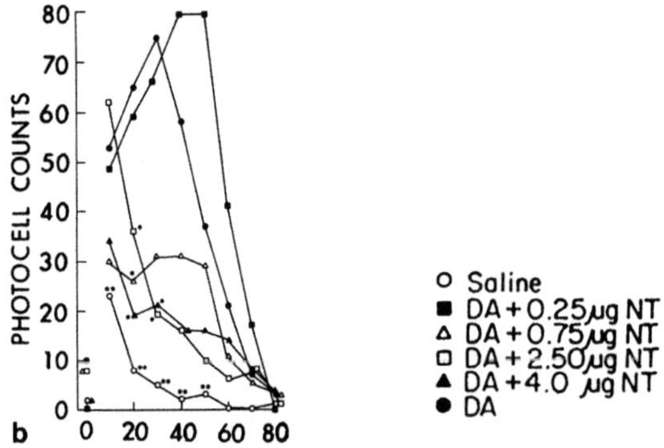

Fig. 3a, b. Effect of neurotensin (NT) on locomotion following microinjection into the ventral tegmental area and the nucleus accumbens. **a** Intra-ventral tegmental infusion of NT. Motor activity was integrated over 20 min intervals. Data is shown as mean ±SEM. Time zero is the mean counts in the 20 min just prior to injection. ○, saline, ($n=22$); ●, 1 μg NT ($n=9$); □, 2 μg NT ($n=8$); ■, 5 μg NT ($n=14$). *$p<0.05$; **$p<0.01$; ***$p<0.001$. Adapted from Kalivas et al. (1981). **b** NT injection into the nucleus accumbens: antagonism of dopamine (DA)-induced locomotion. All rats were given one trial with 30 μg DA/side plus one or two trials with NT+DA or saline. All statistical comparisons are made with DA treatment and probability values increase according to the Bonferroni method. Data are expressed as mean photocell counts (SEM bars deleted for illustrative clarity). *$p<0.05$; **$p<0.005$. (From Kalivas et al. 1984, with permission)

4.1.2 Feeding

Centrally administered NT has been shown to decrease food consumption in rats. Although stress-induced eating in the tail pinch paradigm was unaffected (Morley and Levine 1981), i.c.v. NT caused a dose-related decrease in food intake in fasted rats (Luttinger et al. 1982a; Hawkins 1986a) (Fig. 4). This effect persisted for 90 min following the injection of a large dose (100 μl) and no tolerance was observed after NT injections on 6 consecutive days (Hawkins 1986a). Furthermore, NT did not exhibit activity in the conditioned taste-aversion paradigm to novel flavors (Luttinger et al. 1982a) nor did it affect water intake (Stanley et al. 1982). Injection of NT into discrete brain regions revealed that NT produces aphagia when injected into the paraventricular (Stanley et al. 1984) or ventromedial (Hawkins et al. 1986; Hawkins 1986a) hypothalamus and VTA (Hawkins 1986b) but

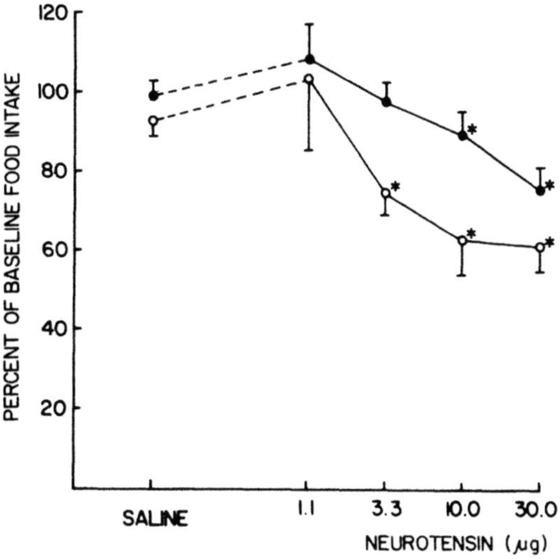

Fig. 4. Effect of i.c.v. administration of neurotensin on food intake in the 24-h food-deprived rat. Food intake was calculated as a percentage of the mean of the previous two baseline intakes. Mean ($\pm$SEM) 1 h food intake in vehicle-treated rats was 9.8 ± 0.8 g, mean 4 h intake was 19.8 ± 0.9 g, $n = 48$. ○, 1 h food intake; ●, 4 h food intake. *$p < 0.05$ compared to saline-injected controls (Dunnett's test for multiple comparisons). (From Luttinger et al. 1982a, with permission)

has no effect on food intake when injected into the lateral hypothalamus (Hawkins et al. 1986) or the nucleus accumbens (Hawkins 1986a). NT may exert its aphagic effects through an action on DA systems. It is of interest that the NT-induced reduction in food intake is antagonized by TRH which, when administered alone, also produces a decrease in food intake (Luttinger et al. 1982a). This is not surprising because TRH blocks many of the effects of NT including antinociception (vide infra).

4.1.3 Learning, Memory, and Reward

At present, investigations of the effects of NT on learning and memory have been performed in conditioned avoidance and reward paradigms. In conditioned avoidance procedures, NT has been shown to produce dose-related inhibition of extinction of active conditioned avoidance responses in a pole-jump active avoidance paradigm and to decrease passive avoidance latencies in a two-compartment, step-through situation (van Wimersma Greidanus et al. 1982). Conversely, Luttinger et al. (1980, 1982b) observed a decrease in avoidance behavior with no alteration in escape behavior in a discrete-trial conditioned avoidance paradigm utilizing a shuttle box apparatus. The decreases in avoidance behavior were antagonized by TRH (Luttinger et al. 1982b). The effects of β-endorphin, which also decreased avoidance behavior in this paradigm, were not antagonized by TRH, suggesting that these two peptides exert their effects on conditioned avoidance through distinct mechanisms.

Rats implanted with a stimulating electrode in the A_{10} area exhibited a dose-dependent decrease in the rate of electrical self-stimulation after injection of NT into the nucleus accumbens (Luttinger et al. 1980). This is further evidence that intra-accumbens NT decreases activity of the mesolimbic DA system. In the conditioned place preference paradigm, rats were observed to prefer to spend time in places associated with the effects of i.c.v. NT. Injection of NT into the VTA was also shown to be rewarding in the conditioned place preference paradigm, suggesting that NT exerts rewarding effects in this brain area (Glimcher et al. 1982, 1984). Furthermore, in an operant paradigm, rats implanted with cannulae into the VTA will self-administer NT by pressing a lever while ignoring an identical, but inactive lever (Glimcher et al. 1987).

4.2 Antinociception

Intracerebroventricularly and intracisternally administered NT produces antinociception in mice and rats in a variety of test paradigms, including the hot plate test, the acetic acid writhing test (Clineschmidt and McGuffin 1977), the tail immersion test (Nemeroff et al. 1979), and the foot shock test (van Wimersma Greidanus et al. 1982) (Fig. 5). This antinociceptive effect is dose-dependent, the effective dose varying with the test method employed (Clineschmidt and McGuffin 1977; Nemeroff et al. 1979; Clineschmidt et al. 1979) and with the route of administration (Martin et al. 1981). This effect of NT is antagonized by TRH (Osbahr et al. 1981).

In studies with rats implanted with bilateral injection cannulae in a variety of CNS sites, Kalivas et al. (1982b) found that NT produces its antinociceptive effects in several brain loci which are distinct from those mediating NT-induced hypothermia (see Sect. 4.4.1). These areas, which all contain high concentrations of NT, include the central nucleus of the amygdala, the rostral mesencephalic periaqueductal gray, and the ventral thalamus. Further microinjection studies of NT injection into the amygdala revealed that the antinociceptive effects of the peptide are abolished by stria terminalis lesions (Kalivas et al. 1982a). Moreover, destruction of the nucleus raphe magnus, comprised of serotonergic perikarya, also abolishes the antinociceptive effects observed after NT administration (Behbehani and Pert 1984). This suggests that NT-induced antinociception is associated with decreased excitation of the periaqueductal gray which activates an inhibitory system in the nucleus raphe magnus.

The receptor and neurotransmitter systems involved in the antinociceptive effects of NT remain obscure. Clineschmidt et al. (1979) found that muscarinic, α- and β-adrenergic, histaminergic, serotonergic, dopaminergic, and opiate receptor antagonists did not alter NT-induced antinociception. However, the question of involvement of opioid systems in mediating NT-induced antinociception remains unanswered, due to contradictory results obtained in different laboratories. Many investigators have documented the failure of naloxone to antagonize this effect in mice (Clineschmidt and McGuffin 1977; Nemeroff et al. 1979; Osbahr et al. 1981). In contrast, naloxone has been shown to antagonize NT-induced antinociception in rats (van Wimersma Greidanus et al. 1982; Yaksh et al. 1982). The complex nature of NT-opioid interactions is further exemplified by the following data obtained in our laboratory. Mice rendered tolerant to morphine exhibited a decrease in antinociceptive response to i.c. NT as measured in the hot plate test (Luttinger et al. 1983a). Surprisingly, NT has been observed to antagonize the antinociceptive effects of enkephalin and tuftsin (Herman et al. 1984).

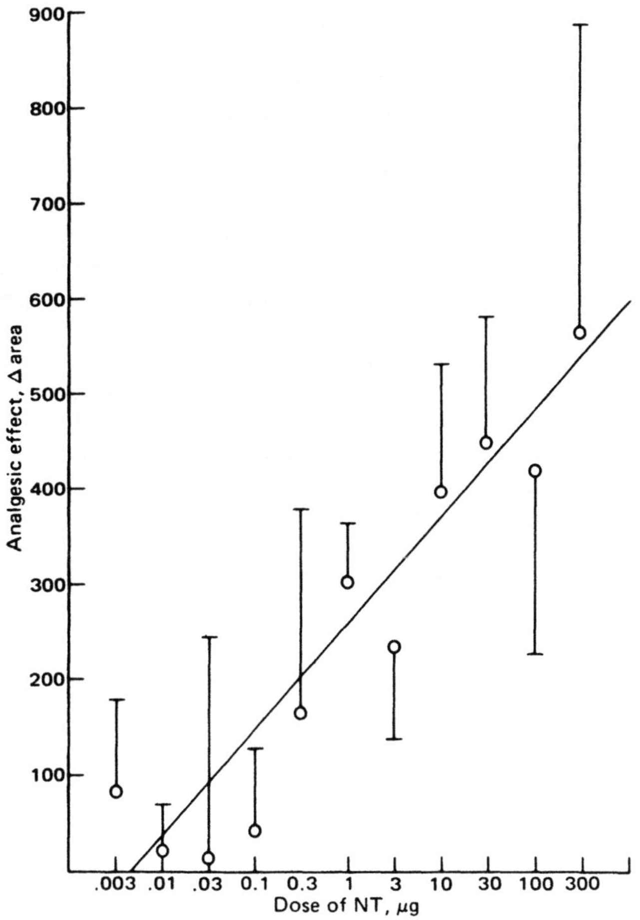

Fig. 5. Effect of various doses of neurotensin (*NT*) on the tail withdrawal response of mice (48 °C water bath). Mice were tested every 15 min after injection over a 2-h period. The *ordinate* represents the area under the curve (±SEM) described by the duration and magnitude of the antinociceptive effect after the scores from control mice were subtracted from those of the NT-treated animals. All doses >300 ng produced a significant effect ($r=0.3397$). (From Nemeroff et al. 1979, with permission)

Similarly discrepant effects have been observed with CNS DA systems. While haloperidol failed to alter NT-induced antinociception (Clineschmidt et al. 1979; Hernandez et al. 1986), chlorpromazine purportedly potentiated the antinociceptive effects of NT. The direct DA receptor agonist apomorphine and the indirect DA agonists methylphenidate and *d*-amphetamine antagonize NT-induced antinociception (Hernandez et al. 1986). The involvement of DA systems in mediating these effects is unclear at this time.

4.3 Interactions with Ethanol and Barbiturates

Neurotensin potentiates the sedative-hypnotic and hypothermic effects of barbiturates and ethanol in mice. Although NT has no sedative effects when administered alone, i.c. NT was shown to be the only peptide of 48 tested to potentiate pentobarbital-induced sedation in mice (Bissette et al. 1978). The peptide also increases ethanol-induced sleeping time by 75% (Luttinger et al. 1981). NT will also induce narcosis when administered in conjunction with subthreshold doses of ethanol (Luttinger et al. 1983 b). While pentobarbital metabolism is significantly reduced after i.c. NT administration in the mouse (Nemeroff et al. 1977), NT does not affect ethanol metabolism as blood ethanol concentrations remained unchanged (Luttinger et al. 1981; Frye et al. 1981). Similarly, i.c. NT has been shown to increase ethanol-induced impairment of the aerial righting reflex in mice but has no effect when administered alone (Frye et al. 1981; Luttinger et al. 1983 b).

While NT itself induces hypothermia (Mason et al. 1980), it also potentiates the hypothermic effects produced by ethanol (Luttinger et al. 1981, 1983). Although this synergistic effect has been observed in acute studies (Fig. 6), no clear interactions are seen with chronic administration. When rats were infused with NT i.c.v. for 14 h, tolerance to its hypothermic effects occurred. These NT-toler-

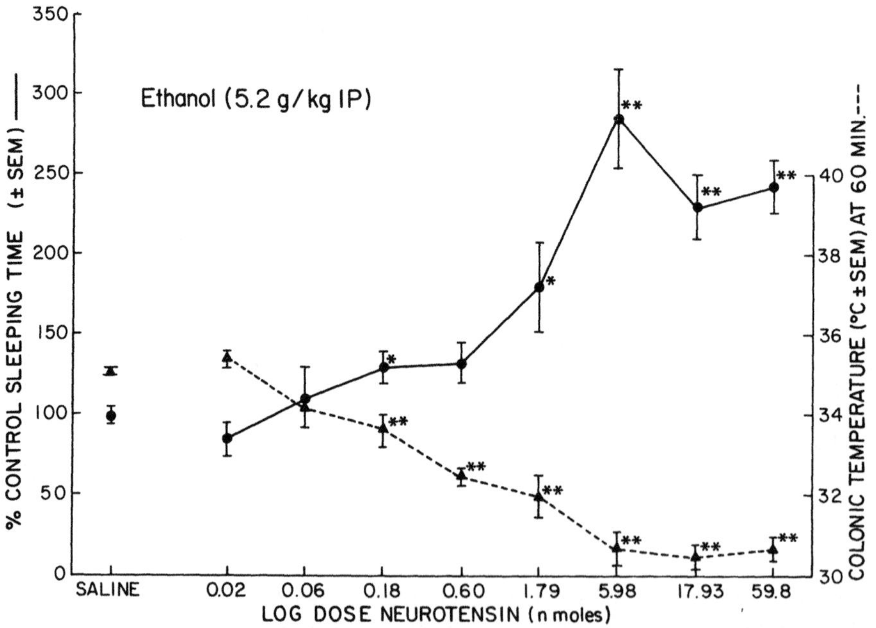

Fig. 6. Effect of i.c. neurotensin (NT) on the duration of sleep and colonic temperature after i.p. injection of ethanol (5.2 g/kg) in mice. The ethanol and vehicle-treated mice slept for 75 ± 5 min ($n = 96$). Experimental values were compared with those of controls by Dunnett's test for multiple comparisons (two-tailed). Each point represents the mean of eight or more mice. ———, effect of NT on duration of ethanol-induced sleep; ----, effect of NT on ethanol-induced hypothermia. $^*p < 0.05$; $^{**}p < 0.01$ compared with ethanol plus saline treatment. (From Luttinger et al. 1981, with permission)

ant rats were then treated with a hypothermia-inducing dose of ethanol which produced hypothermia similar to that observed in control animals; however, no potentiation of the ethanol effect occurred (Wenger et al. 1981). Thus there is no cross-tolerance between NT and ethanol. These results also suggest that despite the synergistic effects of NT and ethanol, ethanol-induced hypothermia is not NT-mediated.

The synergistic effects of NT with ethanol appear to be distinct from those of β-endorphin and bombesin, which also potentiate ethanol-induced sleeping time and impairment of the aerial righting reflex. β-Endorphin and bombesin decrease the tremor and susceptibility to audiogenic seizures associated with ethanol withdrawal; NT does not. Moreover, the ethanol-potentiating effects of NT are blocked by TRH, but not by naloxone, which antagonizes the effects of β-endorphin (Frye et al. 1981).

Finally, sensitivity to the ethanol-potentiating effects of NT appears to be genetically determined. When strains of mice bred for high and low ethanol sensitivity were treated with uniform doses of ethanol, low doses of NT increased the ethanol effect only in the ethanol-sensitive mice, as determined by loss of righting response. High doses of NT potentiated the ethanol effect in both strains. NT potentiation of pentobarbital-induced sleeping time was identical in both strains. These differences in response to the synergistic effect of NT and ethanol suggests that there may be genetic differences in the NT systems of the two strains or that NT may be producing its ethanol-potentiating effects by interacting with neurotransmitter systems that are genetically different in the two mouse strains (Erwin et al. 1987).

4.4 Miscellaneous Pharmacological Effects

4.4.1 Thermoregulation

Intracerebroventricular and intracisternal NT injections produce hypothermia in a variety of mammals (Lipton et al. 1977; Nemeroff et al. 1977, 1979; Prange et al. 1979). This dose-dependent effect (Bissette et al. 1976; Jolicoeur et al. 1981 a; Chandra et al. 1981; Martin et al. 1981), is more pronounced at 4 °C than at 25 °C ambient temperature (Nemeroff et al. 1979). When rats were placed in different temperatures following i.c.v. injection of NT, their body temperature tended to follow the ambient temperature, suggesting that NT has a poikilothermic action (Lee and Myers 1983). However, our group (Mason et al. 1980) was unable to confirm these findings in the mouse.

Direct injections of NT into certain brain regions also produce hypothermia (Martin et al. 1980; Kalivas et al. 1982c). These brain regions, however, are distinct from those where NT produces antinociception (Kalivas et al. 1982c). Intracerebroventricular administration of NT has also been reported to potentiate apomorphine-induced hypothermia (Jolicoeur et al. 1983). The hypothermic actions of NT are antagonized by TRH (Osbahr et al. 1981; Hernandez et al. 1984a) as well as by prior cold exposure (Merritt et al. 1983). Moreover, NT antagonizes the hyperthermic effects of prostaglandin E_2 (Mason et al. 1980). The mechanism by which NT produces hypothermia is as yet unclear; however, it appears to involve a cellular process in the hypothalamus including calcium (Lee et al. 1983), and does not appear to involve NT-induced DA release in the hypothalamus (Myers and Lee 1984a).

4.4.2 Effects on Muscle Tone

Several indirect measures have been used to determine the effect of i.c. NT on muscle tone. In mice, NT produces muscle relaxation as assessed in the Julou-Courvoisier test, a procedure used to screen novel agents for antipsychotic drug activity (Osbahr et al. 1979), and in the traction test without affecting the grasping response (Jolicoeur et al. 1981 b). NT has also been shown to induce catalepsy in mice (Snidjers et al. 1982). Intracisternally administered NT, however, had no effect on the electromyogram recorded from the biceps femoris muscle in the rat (Yarbrough and McGuffin 1979). To date, no studies to determine the mechanism of this NT-induced decrease in muscle tone have been performed.

4.4.3 Gastric Cytoprotective Effects

Intracerebroventricularly administered NT, like peripherally administered NT, has been reported to decrease gastric acid secretion (Osumi et al. 1978; Shiraishi et al. 1980) but was not observed after NT injection into the lateral hypothalamus by Tepperman and Evered (1980) or by our group (Nemeroff et al. 1981 a), in spite of the fact that i.c. NT prevents the development of cold-restraint stress-induced gastric ulcers in rats (Orlando et al. 1981; Nemeroff et al. 1981 a, 1982 a).

4.4.4 Effects on Blood Pressure

When administered centrally, as well as peripherally, NT produces hypotension (Rioux et al. 1981). This rapidly occurring central hypotensive effect is dose-dependent and appears to have a site of action distinct from that of NT in the periphery, as no cross-tolerance develops between i.v. and i.c.v. NT. The hypotensive action of NT in the CNS may involve potentiation of barbiturate effects (see Sect. 4.3) as these experiments were conducted in barbiturate-anesthetized rats rather than in conscious rats (Nemeroff et al. 1977). In a subsequent study in unanesthetized rats, i.c.v. NT produced hypertension, rather than hypotension (Sumners et al. 1982; Shido and Nagasaka 1985). Interestingly, this pressor effectis antagonized by TRH (Sumners et al. 1982).

4.4.5 Endocrine Effects

Centrally administered NT alters the secretion of a variety of anterior pituitary hormones. In anesthetized male rats, i.c.v. NT decreases the secretion of thyroid-stimulating hormone (TSH), growth hormone (GH), and prolactin (PRL) (Maeda and Frohman 1978). In thyroidectomized rats with high basal plasma concentrations of TSH, i.c. NT produces a significant reduction in TSH levels (Nemeroff et al. 1980). TRH, 5-HT, and PGE_2-stimulated PRL release are also inhibited by NT (Maeda and Frohman 1978; Tojo et al. 1986). Conversely, i.c. NT increases somatostatin secretion (Abe et al. 1981). In unrestrained, ovariectomized female rats i.c.v. NT produced a decrease in luteinizing hormone (LH) (Vijayan and McCann 1979; Motta and Martini 1981) and PRL secretion (Vijayan and McCann 1979) and an increase in GH secretion. TSH secretion remained unchanged (Vijayan and McCann 1980). Direct application of NT to different parts of the brain appears to produce different endocrine responses. While i.c.v. NT was observed to decrease LH release in female rats, microinjection into the medial preoptic area produces an increase in LH secretion (Ferris et al. 1984).

4.5 Similarity to Antipsychotic Drugs

NT shares a variety of properties with DA antagonists which are clinically efficacious antipsychotic drugs (Nemeroff 1980; Nemeroff et al. 1981 b; Nemeroff and Cain 1985; Nemeroff 1986). The pharmacological and neurochemical effects of centrally administered NT and systemically administered neuroleptics are remarkably similar. Both NT and neuroleptics increase DA turnover in the nucleus accumbens, the olfactory tubercles, and, to a lesser extent, the caudate nucleus as assessed by a variety of methods (Widerlöv et al. 1982 a; Nemeroff et al. 1983 a; Reches et al. 1982, 1983). Likewise, NT and neuroleptics potentiate barbiturate and ethanol sedation, induce hypothermia, decrease locomotor activity, cause catalepsy and muscle relaxation, and inhibit avoidance (but not escape) in the conditioned avoidance behavioral paradigm. Furthermore, like the antipsychotic drugs, NT blocks the hyperactivity produced by *d*-amphetamine and other psychostimulants.

Although NT shares several properties with neuroleptic drugs, some differences exist. While neuroleptics inhibit the binding of radiolabeled DA agonists and antagonists to CNS membrane receptors, NT fails to alter binding of the DA antagonist [^{3}H]-spiperone in the rat CNS (Nemeroff et al. 1982 b, 1983 a; Reches et al. 1982). Furthermore, while haloperidol blocks *d*-amphetamine-induced hyperactivity and stereotypy, NT blocks only the former. In this way, NT is similar to the atypical antipsychotic drugs, such as clozapine, which also fail to antagonize *d*-amphetamine-induced stereotypy (Costall and Naylor 1976; Bentall and Herberg 1980). Finally, NT produces antinociception while neuroleptics do not.

5 Alterations in CNS Neurotensin Concentration

5.1 Effects of Psychoactive Drugs

Chronic treatment with antipsychotic drugs alters NTLI concentrations in certain brain areas. After treating rats with haloperidol (2 mg/kg/day) for 3 weeks, Govoni et al. (1980) observed a 100% increase in NTLI in the nucleus accumbens and a somewhat smaller increase in the striatum, terminal fields of the mesolimbic and nigroneostriatal DA projections. Upon acute administration of haloperidol, increases in NTLI were seen 16 h after the injection and persisted for at least 8 h. Similar increases in NTLI in the nucleus accumbens were also observed after chronic treatment with the clinically effective antipsychotics chlorpromazine, trifluoroperazine, and pimozide, but two clinically ineffective phenothiazines, promazine and promethazine, failed to produce any increase in NT in this brain area.

Our group (Kilts et al. 1984) assayed NT in 30 brain nuclei of rats treated chronically with either haloperidol or clozapine; the latter antipsychotic purportedly produces no acute extrapyramidal effects or tardive dyskinesia (Rupniak et al. 1984). Both drugs increased NTLI in the nucleus accumbens and decreased NTLI in the medial prefrontal cortex, the anterior cingulate cortex, and the bed nucleus of the stria terminalis. These common alterations in NTLI, particularly in the nucleus accumbens, medial prefrontal cortex, and anterior cingulate cortex reflect the shared properties of haloperidol and clozapine as effective antipsychotic agents and their actions at limbic and cortical DA terminal fields. Haloperidol, however, also increased NT int he caudate and medial preoptic nucleus, whereas clozapine increased NT in the cortical amygdaloid nucleus and the pars compacta of the substantia nigra, indicating different functional interactions between NT and DA in distinct brain regions. The haloperidol-induced NTLI increase in the caudate, a terminal area of the nigrostriatal DA projection, may reflect the ability of this drug to cause extrapyramidal effects and may be a function of its ability to cause striatal DA receptor supersensitivity.

The effects of antagonists of DA receptor subtypes has also been investigated. Frey et al. (1986) treated rats chronically with several nonselective DA antagonists, the selective D2 antagonists haloperidol and sulpiride, and the selective D1 antagonist SCH 23390. While all of the drugs tested produced an increase in NTLI in the nucleus accumbens and striatum, tolerance to this effect occurred in animals treated with D1 and nonselective DA antagonists. This suggests that the extrapyramidal effects produced by selective D2 antagonists may be related to their ability to maintain increased NT concentrations in the basal ganglia.

In addition to altering NTLI in certain brain areas, chronic treatment with neuroleptics has been shown to produce an increase in the number of NT receptors in the rat substantia nigra (Uhl and Kuhar 1984).

Brain NT concentrations have been observed to be altered by opiates as well as neuroleptic drugs. Morley et al. (1980) observed an increase in thalamic NTLI in morphine-addicted rats. Thalamic NT concentrations decreased to normal levels during withdrawal. Treatment with naloxone alone produced a decrease in thalamic NT when compared to controls. The results suggest that changes in the concentration of thalamic NT, along with observed changes in the concentrations of somatostatin, CCK, and substance P, may modulate the release of hypothalamic trophic factors in the generation of morphine addiction and withdrawal syndromes. The mechanism for this modulation, however, remains unclear. While morphine inhibits potassium-stimulated calcium-dependent release of CCK and substance P from cat hypothalamic slices, neither morphine nor naloxone has an effect on potassium-stimulated NT release (Micevych et al. 1982).

Like neuroleptics and opiates, phencyclidine, a psychotomimetic, has been observed to alter NTLI in certain brain areas. In rats treated with acute and chronic phencyclidine, NTLI decreased in the frontal cortex but remained unchanged in the caudate-putamen, limbic forebrain, hypothalamus, and midbrain (Wachi et al. 1985).

5.2 Neurotensin and Psychiatric Disorders

Neurotensin concentrations have been measured in plasma, in CSF, and in various postmortem brain regions of patients with a variety of psychiatric disorders. Our group (Widerlöv et al. 1982b) measured NTLI in the CSF of drug-free schizophrenic patients and normal controls and found a subgroup of patients with low CSF NT concentrations. After treatment with neuroleptic drugs for more than 1 month, the CSF NT concentrations of this subgroup of patients rose to normal concentrations. In a second study in collaboration with Widerlöv, we again found schizophrenic patients to have lower CSF NT concentrations than age- and sex-matched controls. NTLI has been found to be increased in postmortem brain samples from one area of the frontal cortex (BA 32) of schizophrenic patients (Nemeroff et al. 1983b). No change in NTLI was found in other brain regions. NTLI has also been measured in postmortem brain samples from patients with Huntington's chorea and Parkinson's disease. In the Huntington's chorea samples, NTLI was observed to be increased in the caudate nucleus (Nemeroff et al. 1983a) and in the globus pallidus (Emson et al. 1985). NT has been reported to be decreased in the hippocampus of patients with Parkinson's disease (Bissette et al. 1985), though Emson et al. (1985) found no such changes. In addition, autoradiography of the substantia nigra from patients with Parkinson's disease reveals that [^{3}H]-NT binding is significantly reduced compared to controls, indicating a depletion of nigral NT receptors (Uhl et al. 1984; Sadoul et al. 1984b) which is expected considering the loss of nigral DA neurons.

While diseases affecting the DA systems have been of principle interest, NT concentrations have been observed to be altered in a few other conditions. Yates et al. (1985) found an increase in NT concentration in the caudate, temporal cortex, and frontal cortex of postmortem brain samples from patients with Down's syndrome. An increase in plasma NT has also been observed during hot flashes in menopausal women (Kronenberg and Carraway 1985). NT concentrations have been found to be unaltered in the amygdala and CSF in depression (Biggins et al. 1983; Manberg et al. 1983), in CSF in anorexia/bulemia (Manberg et al. 1983), and in plasma in premenstrual syndrome (Chuong et al. 1985). Several studies have also found no alteration in the NT concentration of the amygdala and other brain regions in Alzheimer's disease (Biggins et al. 1983; Yates et al. 1985; Emson et al. 1985), although our group (Nemeroff et al. 1983c) observed a 30% decrease in NT concentration in the amygdala of postmortem samples from Alzheimer's disease patients.

6 Conclusion

It is clear from the data presented in this chapter that NT fulfills some of the classical neurotransmitter criteria. In addition, centrally administered NT produces a variety of behavioral and physiological effects such as antinociception and potentiation of ethanol and barbiturate effects. Although it is not possible to determine whether these pharmacological effects of NT underlie its normal physiological actions, NT definitely seems to be involved in modulating dopaminergic activity, as demonstrated by changes in DA activity in response to exogenously applied NT, in endogenous NT concentration after treatment with DA antagonists, and in diseases that probably involve the DA system. Likewise, NT may play an important role in psychiatric disorders such as schizophrenia. Further progress in the study of this peptide is thus promising in increasing our understanding not only of the role of NT in the CNS, but also of other neurotransmitter systems and the pathogenesis of schizophrenia.

Acknowledgement. We acknowledge the excellent secretarial assistance of Shelia Walker.

References

Abe I, Chichara K, Chiba T, Shigeru M, Fujita A (1981) Effect of intraventricular injection of neurotensin and other various bioactive peptides on plasma immunoreactive somatostatin levels in rat hypophysial portal blood. Endocrinology 108:1939–1943

Agnati LF, Fuxe K, Benfenati F, Battistini N (1983) Neurotensin in vitro markedly reduces the affinity in subcortical limbic [^{3}H]-N-propylnorapomorphine binding sites. Acta Physiol Scand 119:459–461

Anderson CM, Bissette G, Nemeroff CB, Kilts CD (1984) Micotopographic distribution of neurotensin and catecholamine concentrations at the level of individual rat brain nuclei. Soc Neurosci Abstr 10:437

Andrade R, Aghajanian GK (1981) Neurotensin selectively activates dopamine neurons in the substantia nigra. Soc Neurosci Abstr 7:573

Battaini F, Govoni S, Di Giovine S, Trabucchi M (1986) Neurotensin effect on dopamine release and calcium transport in rat striatum: interactions with diphenylalkylamine calcium antagonists. Naunyn Schmiedebergs Arch Pharmacol 332:267–270

Behbehani MM, Pert A (1984) A mechanism for the analgesic effect of neurotensin as revealed by behavioral and electrophysiological techniques. Brain Res 324:35–42

Beitz AJ (1982) The sites of origin of brain stem neurotensin and serotonin projections to the rodent nucleus raphe magnus. J Neurosci 2:829–842

Bentail ACC, Herberg LJ (1980) Blockade of amphetamine-induced locomotor activity and stereotypy in rats by spiroperidol but not by an atypical neuroleptic, thioridazine. Neuropharmacology 19:699–703

Bhatnagar YM, Carraway R (1981) Bacterial peptides with C-terminal similarities to bovine neurotensin. Peptides 2:51–59

Biggins JA, Perry EK, McDermott JR, Smith AI, Perry RH, Edwardson JA (1983) Post mortem levels of thyrotropin-releasing hormone and neurotensin in the amygdala in Alzheimer disease, schizophrenia and depression. J Neurol Sci 58:117–122

Bissette G, Nemeroff CB, Loosen PT, Prange AJ Jr, Lipton MA (1976) Hypothermia and intolerance to cold induced by intracisternal administration of the hypothalamic peptide neurotensin. Nature 262:607–609

Bissette G, Nemeroff CB, Loosen PT, Breese GR, Burnett GB, Lipton MA, Prange AJ Jr (1978) Modification of pentobarbital-induced sedation by natural and synthetic peptides. Neuropharmacology 17:229–237

Bissette G, Jennes L, Prange AJ Jr, Breese GR, Nemeroff CB (1983) Neurotensin and dopamine are not co-localized in rat brain. Soc Neurosci Abstr 9:290
Bissette G, Nemeroff CB, Decker MW, Kizer JS, Agid Y, Javoy-Agid F (1985) Alterations in regional brain concentrations of neurotensin and bombesin in Parkinson's disease. Ann Neurol 17:324–329
Bloom SR, Polak JM (1982) Aspects of neurotensin physiology and pathology. Ann NY Acad Sci 400:105–116
Cador M, Kelley AE, Le Moal M, Stinus L (1985) Behavioral analysis of the effect of neurotensin injected into the ventral mesencephalon on investigatory and spontaneous motor behavior in the rat. Psychopharmacology 85:187–196
Cain ST, Nemeroff CB (1985) Hypolocomotion induced by activation of dopamine autoreceptors effects of neurotensin. Soc Neurosci Abstr 11:340
Camargo ACM, Caldo H, Emson PC (1983) Degradation of neurotensin by rabbit brain endo-oligopeptidase A and endo-oligopeptidase B. Biochem Biophys Res Commun 116:1151–1159
Carraway R, Leeman SE (1973) The isolation of a new hypotensive peptide, neurotensin, from bovine hypothalami. J Biol Chem 248:6854–6861
Carraway RE, Leeman SE (1975) The amino acid sequence of a hypothalamic peptide, neurotensin. J Biol Chem 250:1907–1911
Carraway R, Leeman SE (1976) Characterization of radioimmunoassayable neurotensin in the rat: its differential distribution in the central nervous system, small intestine, and stomach. J Biol Chem 251:7045–7052
Carraway R, Ruane SE, Kim HR (1982) Distribution and immunochemical character of neurotensin-like material in representative vertebrates and invertebrates: apparent conservation of the COOH-terminal region during evolution. Peptides 3:115–123
Chandra A, Chov HC, Chang C, Lin MT (1981) Effects of intraventricular administration of neurotensin and somatostatin on thermoregulation in the rat. Neuropharmacology 20:715–718
Checler F, Vincent JP, Kitabgi P (1983) Neurotensin analogs [D-Tyr^{11}] and [D-Phe^{11}] neurotensin resist degradation by brain peptides in vitro and in vivo. J Pharmacol Exp Ther 227:743–748
Checler F, Emson PC, Vincent JP, Kitabgi P (1984) Inactivation of neurotensin by rat brain synaptic membranes. Cleavage at the Pro^{10}-Tyr^{11} bond by endopeptidase 24.11 (enkephalinase) and a peptidase different from proline-endopeptidase. J Neurochem 43:1295–1301
Checler F, Vincent JP, Kitabgi P (1985) Inactivation of neurotensin by rat brain synaptic membranes partly occurs through cleavage at the Arg^8-Arg^9 peptide bond by a metalloendopeptidase. J Neurochem 45:1509–1513
Chuong CJ, Coulam CB, Kao PC, Bergstralh EJ, Go VLW (1985) Neuropeptide levels in premenstrual syndrome. Fertil Steril 44:760–765
Clineschmidt BV, McGuffin JC (1977) Neurotensin administered intracisternally inhibits responsiveness of mice to noxious stimuli. Eur J Pharmacol 46:395–396
Clineschmidt BV, McGuffin JC, Bunting PB (1979) Neurotensin: antinocisponsive action in rodents. Eur J Pharmacol 54:129–139
CooperPE, Fernstrom MH, Rorstad OP, Leeman SE, Martin JB (1981) The regional distribution of somatostatin, substance P and neurotensin in human brain. Brain Res 218:219–232
Costall B, Naylor RJ (1976) A comparison of the abilities of typical neuroleptic agents and of thioridazine, clozapine, sulpiride, and metaclopramide to antagonize the hyperactivity induced by dopamine applied intracerebrally to areas of the extrapyramidal and mesolimbic systems. Eur J Pharmacol 40:9–19
DeQuidt M, Emson PC (1983) Neurotensin facilitates dopamine release in vitro from rat striatal slices. Brain Res 274:376–380
Drumheller AL, St Pierre S, Jolicoeur FB (1986) Effects of neurotensin on regional concentrations of norepinephrine in rat brain. Brain Res Bull 16:755–758
Dupont A, Merand Y (1978) Enzymic inactivation of neurotensin by hypothalamic and brain extracts of the rat. Life Sci 22:1623–1630

Eldred WD, Karten HJ (1983) Characterization and quantification of peptidergic amacrine cells in the turtle retina: enkephalin, neurotensin and glucagon. J Comp Neurol 221:371–381
Elliott PJ, Nemeroff CB (1986) Repeated neurotensin administration in the ventral tegmental area: effects on baseline and *d*-amphetamine-induced locomotor activity. Neurosci Lett 68:239–244
Elliott PJ, Chan J, Parker Y-M, Nemeroff CB (1986) Behavioral effects of neurotensin in the open field: structure-activity studies. Brain Res 381:259–265
Emson PC, Goedert M, Horsfield P, Rioux F, St. Pierre S (1982) The regional distribution and chromatographic characterization of neurotensin-like immunoreactivity in the rat central nervous system. J Neurochem 28:992–999
Emson PC, Horsfield PM, Goedert M, Rosser MN, Hawkes CH (1985) Neurotensin in human brain: regional distribution and effects of neurological illness. Brain Res 347:239–244
Ervin GN, Birkemo LS, Nemeroff CB, Prange AJ Jr (1981) Neurotensin blocks certain amphetamine behaviors. Nature 291:73–76
Erwin VG, Korte A, Marty M (1987) Neurotensin selectivity alters ethanol-induced anesthesia in LS/Ibg and SS/Ibg lines of mice. Brain Res 400:80–90
Ferris CF, Pan JX, Singer EA, Boyd ND, Carraway RE, Leeman SE (1984) Stimulation of luteinizing hormone release after steriotaxic microinjections of neurotensin into the medial preoptic area of rats. Neuroendocrinology 38:145–151
Fernstrom MH, Carraway RE, Leeman SE (1980) Neurotensin. In: Martini L, Ganong WF (eds) Frontiers in neuroendocrinology, vol 6. Raven, New York
Frey P, Fuxe K, Eneroth P, Agnati LF (1986) Effects of acute and long-term treatment with neuroleptics on regional telencephalic neurotensin levels in the male rat. Neurochem Int 8(3):429–434
Frye GD, Luttinger D, Nemeroff CB, Vogel RA, Prange AJ Jr, Breese GR (1981) Modification of the actions of ethanol by centrally active peptides. Peptides 2[Suppl 1]:99–106
Furuta S, Kisara K, Sakurada S, Sakurada T, Sasaki V, Suzuki K (1984) Structure-antinociceptive activity studies with neurotensin. Br J Pharmacol 83:43–48
Fuxe K, Agnati LF, Andersson K, Eneroth P, Harfstrand A, Goldstein M, Zoli M (1984) Studies on neurotensin-catecholamine interactions in the hypothalamus and in the forebrain of the male rat. Neurochem Int 6:737–750
Fuxe K, Agnati LF, Battistini N, Giardino L, Benfenati F, Martire M, Ruggeri M (1985) Further evidence for the existence of interactions between receptors for dopamine and neurotensin. Dopamine reduces the affinity and increases the number of [^{3}H]-neurotensin binding sites in the subcortical limbic forebrain of the rat. Acta Physiol Scand 124:125–128
Fuxe K, Agnati LF, Martire M, Neumeyer A, Benfenati F, Frey P (1986) Studies of neurotensin-dopamine receptor interactions in striatal membranes of the male rat. The influence of 6-hydroxydopamine-induced dopamine receptor supersensitivity. Acta Physiol Scand 126:147–149
Gamse R, Heuberger S (1983) Release of neurotensin from rat spinal cord in vitro. Neurosci Lett 36:87–91
Garcia-Sevilla JA, Magnusson T, Carlsson A, Leban J, Folkers K (1978) Neurotensin and its amide analogue [Gln^4]-neurotensin: effects on brain monoamine turnover. Naunyn Schmiedebergs Arch Pharmacol 305:213–218
Ghatei MA, Bloom SR, Langerin H, McGregor GP, Lee YC, Adrian TE, O'Shaughessy DJ, Blank MA, Uttenthal LO (1984) Regional distribution of bombesin and seven other regulatory peptides in the human brain. Brain Res 293:101–109
Glimcher P, Margolin D, Hoebel BG (1982) Rewarding effects of neurotensin in the brain. Ann NY Acad Sci 400:422–424
Glimcher PW, Margolin DH, Giovino AA, Hoebel BG (1984) Neurotensin: a new "reward peptide". Brain Res 291:119–124
Glimcher PW, Giovino AA, Hoebel BG (1987) Neurotensin self-injection in the ventral tegmental area. Brain Res (to be published)

Goedert M, Mantyh PW, Hunt SP, Emson PC (1983) Mosaic distribution of neurotensin-like immunoreactivity in the cat striatum. Brain Res 274:176–179
Goedert M, Pittaway K, Emson PC (1984a) Neurotensin receptors in the rat striatum: lesion studies. Brain Res 299:164–168
Goedert M, Pittaway K, Williams BJ, Emson PC (1984b) Specific binding of tritiated neurotensin to rat brain membranes: characterization and regional distribution. Brain Res 304:71–81
Govoni S, Hong JS, Yang H-YT, Costa E (1980) Increase of neurotensin content elicited by neuroleptics in nucleus accumbens. J Pharmacol Exp Ther 215:413–417
Grant LD, Nemeroff CB, Bissette G (1982) Neuroanatomical distribution of neuroactive peptides in mammalian species. In: Nemeroff CB, Dunn AJ (eds) Peptides, hormones and behavior. Spectrum, New York
Griffiths EC, Slater P, Webster VAD (1981) Behavioural interactions between neurotensin and thyrotropin releasing hormone in rat central nervous system. J Physiol 320:90
Griffiths EC, Widdowson P, Slater P (1982) Antagonism of TRH-induced wet-dog shaking in rats by neurotensin and a neurotensin fragment. Neurosci Lett 31:17
Griffiths EC, Robertson R, Slater P (1983) Neurotensin antagonises locomotor stimulant actions of TRH and RX77368. Proc Br Pharmacol Soc Abstr, Brit J Pharmacol 80[suppl]:533
Griffiths EC, McDermott JR, Smith AI (1984) A comparative study of neurotensin inactivation by brain peptidases from different vertebrate species. Comp Biochem Physiol 77C:363–366
Grimmelikhuijzen CJP, Carraway RE, Rokaeus A, Sundler F (1981) Neurotensin-like immunoreactivity in the nervous system of hydra. Histochemistry 72:199–209
Hawkins MF (1986a) Central nervous system neurotensin and feeding. Physiol Behav 36:1–8
Hawkins MF (1986b) Aphagia in the rat following microinjection of neurotensin into the ventral tegmental area. Life Sci 38:2383–2388
Hawkins MF, Barkenmeyer CA, Tulley RT (1986) Synergistic effects of dopamine agonists and centrally administered neurotensin on feeding. Pharmacol Biochem Behav 24:1195–1201
Haubrich DR, Martin GE, Pflueger B, Williams M (1982) Neurotensin effects on brain dopaminergic systems. Brain Res 231:216–221
Herman ZS, Jez W, Siemion IZ, Nawrocka E (1984) Interaction on the antinociceptive effect between neurotensin and enkephalins or tuftsin. Acta Physiol Pol 35:5–6
Hernandez D, Nemeroff CB, Valderrama MM, Prange AJ Jr (1984a) Neurotensin-induced antinociception and hypothermia in mice: antagonism by TRH and structural analogs of TRH. Regul Pept 8:41–49
Hernandez DE, Richardson CM, Nemeroff CB, Orlando RC, St Pierre S, Rioux F, Prange AJ Jr (1984b) Evidence for biological activity of two N-terminal fragments of neurotensin, $\text{neurotensin}_{1\text{-}8}$ and $\text{neurotensin}_{1\text{-}10}$. Brain Res 301:153–156
Hernandez DE, Drago F, Mason GA, Stanley DA, Prange AJ Jr (1986) Effect of dopamine agonists and antagonists on neurotensin-induced antinociception. Pharmacol Biochem Behav 24:425–428
Herve D, Tassin JP, Studler JM, Dana C, Kitabgi P, Vincent JP, Glowinski J, Rostene W (1986) Dopaminergic control of ^{125}I-labeled neurotensin binding site density in corticolimbic structures of the rat brain. Proc Natl Acad Sci USA 83:6203–6207
Hokfelt T, Everitt BJ, Theodorsson-Norheim E, Goldstein M (1984) Occurrence of neurotensin-like immunoreactivity in subpopulations of hypothalamic, mesencephalic, and medullary catecholamine neurons. J Comp Neurol 222:543–559
Ibata Y, Okamura FH, Kawakami T, Tanaka M, Obata HL, Tsuto OT, Terubayashi H, Yanaihara C, Yanaihara N (1983) Coexistence of dopamine and neurotensin in hypothalamic arcuate and periventricular neurons. Brain Res 269:177–179
Ibata Y, Kawakami F, Fukui K, Obata-Tsuto HL, Tanaka M, Kubo T, Okamura H, Morimoto N, Yanaihara C, Yanaihara N (1984a) Light and electron microscopic immunocytochemistry of neurotensin-like immunoreactive neurons in the rat hypothalamus. Brain Res 302:221–230

Ibata Y, Kawakami G, Fukui K, Okamura H, Obata-Tsuto HL, Tsuto T, Terubayashi H (1984 b) Morphological survey of neurotensin-like immunoreactive neurons in the hypothalamus. Peptides 5:109–120

Inagaki S, Shinoda K, Kubota Y, Shiosaka S, Marsuzaki T, Tohyama M (1983) Evidence for the existence of a neurotensin-containing pathway from the endopiriform nucleus and the adjacent prepiriform cortex to the anterior olfactory nucleus and nucleus of diagonal band (Broca) of the rat. Neuroscience 8:487–493

Iversen LL, Iversen SD, Bloom F, Douglas C, Brown M, Vale W (1978) Calcium-dependent release of somatostatin and neurotensin from rat brain in vitro. Nature 273:161

Jennes L, Stumpf WE, Kalivas PW (1982) Neurotensin: topographical distribution in rat brain by immunohistochemistry. J Comp Neurol 210:211–224

Jolicoeur FB, Barbeau A, Rioux F, Quirion R, St. Pierre S (1981 a) Differential neurobehavioral effects of neurotensin and structural analogues. Peptides 2:171–175

Jolicoeur FB, Rondeau D, St. Pierre S, Rioux F, Barbeau A (1981 b) Peptides and the basal ganglia. In: Gessa GL, Corsini GU (eds) Clinical pharmacology of apomorphine and other dopamimetics. Raven, New York

Jolicoeur FB, Barbeau A, Quirion R, Rioux F, St. Pierre S (1982 a) Pharmacological evidence for a heterogeneity of receptors in developing various central and peripheral effects of neurotensin. Ann NY Acad Sci 400:440–441

Jolicoeur FB, Rioux F, St. Pierre S, Barbeau A (1982 b) Structure-activity studies of neurotensin's neurobehavioral effects. In: Collu R, Barbeau A, Ducharme JR, Tollis G (eds) Brain neurotransmitters and hormones. Raven, New York

Jolicoeur FB, De Michele G, Barbeau A (1983) Neurotensin affects hyperactivity but not stereotypy induced by pre- and post-synaptic dopaminergic stimulation. Neurosci Biobehav Rev 7:385–390

Kahn D, Abrams GM, Zimmerman EA, Carraway R, Leeman SE (1980) Neurotensin neurons in the rat hypothalamus: an immunocytochemical study. Endocrinology 107:47–54

Kalivas PW (1985) Interactions between neuropeptides and dopamine neurons in the ventromedial mesencephalon. Neurosci Biobehav Rev 9:573–587

Kalivas PW, Miller JS (1984) Neurotensin neurons in the ventral tegmental area project to the medial nucleus accumbens. Brain Res 300:157–160

Kalivas PW, Taylor S (1985) Behavioral and neurochemical effect of daily injections with neurotensin into the ventral tegmental area. Brain Res 358:70–76

Kalivas PW, Nemeroff CB, Prange AJ Jr (1981) Increase in spontaneous motor activity following infusion of neurotensin into the ventral tegmental area. Brain Res 229:525–529

Kalivas PW, Nemeroff CB, Prange AJ Jr (1982 a) Neuroanatomical site specific modulation of spontaneous motor activity by neurotensin. Eur J Pharmacol 78:471–474

Kalivas PW, Gau B, Nemeroff CB, Prange AJ Jr (1982 b) Antinociception after microinjection of neurotensin into the central amygdaloid nucleus of the rat. Brain Res 243:279–286

Kalivas PW, Jennes L, Nemeroff CB, Prange AJ Jr (1982 c) Neurotensin: topographic distribution of brain sites involved in hypothermia and antinociception. J Comp Neurol 210:225–238

Kalivas PW, Burgess SK, Nemeroff CB, Prange AJ Jr (1983) Behavioral and neurochemical effects of neurotensin microinjection into the ventral tegmental area of the rat. Neuroscience 8:495–505

Kalivas PW, Nemeroff CB, Prange AJ Jr (1984) Neurotensin microinjection into the nucleus accumbens antagonizes dopamine-induced increase in locomotion and rearing. Neuroscience 11:919–930

Kalivas PW, Jennes L, Miller JS (1985) A catecholaminergic projection from the ventral tegmental area to the diagonal band of Broca: modulation by neurotensin. Brain Res 326:229–238

Kanba KS, Kanba S, Okazaki H, Richelson E (1986) Binding of [^{3}H]neurotensin in human brain: properties and distribution. J Neurochem 46:946–952

Kataoka K, Mizumo N, Frohman LA (1979) Regional distribution of immunoreactive neurotensin in monkey brain. Brain Res Bull 4:57–60

Kilts CD, Anderson C, Bissette G, Ely T, Davidson J, Nemeroff C (1984) Differential effects of antipsychotic drugs on the neurotensin content of individual rat brain nuclei. Proc Am Coll Neuropsychopharmacol Abst 21:102

Kilts CD, Bissette G, Cain ST, Skoog KM, Nemeroff CB (1986) Neuropeptide-dopamine interactions: focus on neurotensin. In: Woodruff GN, Roberts PJ (eds) Dopamine '84. Macmillan, London

Kitabgi P, Carraway R, Leeman SE (1976) Isolation of a tridecapeptide from bovine intestinal tissue and its partial characterization as neurotensin. J Biol Chem 251:7053–7058

Kitabgi P, Carraway R, Van Rietschoten J, Granier C, Morgat JL, Mendez A, Leeman S, Freychet P (1977) Neurotensin: specific binding to synaptic membranes from rat brain. Proc Natl Acad Sci USA 74:1846–1850

Kitabgi P, Poustis C, Granier C, Van Rietschoten J, Rivier J, Morgat JL, Freychet P (1980) Neurotensin binding to extraneural and neural receptors: comparison with biological activity and structure-activity relationships. Mol Pharmacol 18:11–19

Kobayashi RM, Brown MR, Vale W (1977) Regional distribution of neurotensin and somatostatin in rat brain. Brain Res 126:584–588

Kronenberg F, Carraway RE (1985) Changes in neurotensin-like immunoreactivity during menopausal hot flashes. J Clin Endocrinol Metab 60:1081–1086

Langer M, Van Noorden S, Polak JM, Pearse AGE (1979) Peptide hormone-like immunoreactivity in the gastrointestinal tract and endocrine pancreas of eleven teleost species. Cell Tissue Res 199:493–508

Lazarus LH, Brown MR, Perrin HH (1977) Distribution, localization, and characterization of neurotensin binding sites in the rat brain. Neuropharmacology 16:625–629

Lee TF, Myers RD (1983) Analysis of the thermolytic action of ICV neurotensin in the rat at different ambient temperatures. Brain Res Bull 10:661–665

Lee TF, Hepler JR, Myers RD (1983) Evaluation of neurotensin's thermolytic action by ICV infusion with receptor antagonists and a Ca^{++} chelator. Pharmacol Biochem Behav 19:477–481

Leeman SE, Carraway RE (1982) Neurotensin: discovery, isolation, characterization, synthesis and possible physiological roles. Ann NY Acad Sci 400:1–16

Leranth CS, Jew JY, Williams TH, Palkovits M (1981) Stria terminalis axons ending on substance P and neurotensin-containing cells of rat central amygdaloid nucleus: an electron microscopic immunocytochemical study. Neuropeptides 1:261–272

Li H-B, Watt CB, Lam DM-K (1985) The coexistence of two neuroactive peptides in a subpopulation of retinal amacrine cells. Brain Res 345:176–180

Li H-B, Marshak DW, Dowling JE, Lam DM-K (1986) Colocalization of immunoreactive substance P and neurotensin in amacrine cells of the goldfish retina. Brain Res 366:307–313

Lipton MA, Bissette G, Nemeroff CB, Loosen PT, Prange AJ Jr (1977) Neurotensin: a possible mediator of thermoregulation in the mouse. In: Lomax P, Schombaum E, Cox B (eds) Drugs, biogenic amines, and body temperature – third symposium on the pharmacology of thermoregulation. Karger, Basel

Long JB, Kalivas PW, Youngblood WW, Prange AJ Jr, Kizer JS (1981) Selective effects of centrally administered neurotensin on in vivo tryptophan hydroxylase activity (THA) and serotonin (5HT) levels in discrete rat brain areas. Pharmacologist 23:133

Luttinger D, Wiggins R, King RA, Nemeroff CB, Prange AJ Jr (1980) Inhibition of conditioned avoidance responding and rewarding electrical self-stimulation of the brain by centrally administered neurotensin. Soc Neurosci Abstr 6:843

Luttinger D, Nemeroff CB, Mason GA, Frye GD, Breese GR, Prange AJ Jr (1981) Enhancement of ethanol-induced sedation and hypothermia by centrally administered neurotensin, β-endorphin, and bombesin. Neuropharmacology 20:305–309

Luttinger D, King RA, Sheppard D, Strupp J, Nemeroff CB, Prange AJ Jr (1982a) The effect of neurotensin on food consumption in the rat. Eur J Pharmacol 81:499–504

Luttinger D, Nemeroff CB, Prange AJ Jr (1982b) The effects of neuropeptides on discrete-trial conditioned avoidance responding. Brain Res 237:183–192

Luttinger DE, Burgess SK, Nemeroff CB, Prange AJ Jr (1983a) The effect of chronic morphine treatment on neurotensin-induced antinociception. Psychopharmacology 81:10–13

Luttinger D, Frye GD, Nemeroff CB, Prange AJ Jr (1983b) The effects of neurotensin, β-endorphin and bombesin on ethanol-induced behaviors in mice. Psychopharmacology 79:357–363
Maeda K, Frohman LA (1978) Dissociation of systemic and central effects of neurotensin on the secretion of growth hormone, prolactin, and thyrotropin. Endocrinology 103:1903–1909
Maeda K, Frohman LA (1981) Neurotensin release by rat hypothalamic fragments in vitro. Brain Res 210:261–269
Malthe-Sorenssen D, Wood PL, Cheney DL, Costa E (1978) Modulation of the turnover rate of acetylcholine in rat brain by intraventricular injections of thyrotropin-releasing hormone, somatostatin, neurotensin and angiotensin II. J Neurochem 31:685–691
Manberg PJ, Youngblood WW, Nemeroff CB, Rossor M, Iversen LL, Prange AJ Jr, Kizer JS (1982) Regional distribution of neurotensin in human brain. J Neurochem 38:1777–1780
Manberg PJ, Nemeroff CB, Bissette G, Prange AJ Jr, Gerner RH (1983) Cerebrospinal fluid levels of neurotensin-like immunoreactivity in normal controls and in patients with affective disorder, anorexia nervosa and premenstrual syndrome. Soc Neurosci Abstr 9:1054
Martin GE, Bacino CB, Papp NL (1980) Hypothermia elicited by the intracerebral microinjection of neurotensin. Peptides 1:333–339
Martin GE, Naruse T, Papp NL (1981) Antinociceptive and hypothermic actions of neurotensin administered centrally in the rat. Neuropeptides 1:447–454
Mason GA, Nemeroff CB, Luttinger D, Hatley OL, Prange AJ Jr (1980) Neurotensin and bombesin: differential effects on body temperature of mice after intracisternal administration. Regul Peptides 1:53–60
Mazella J, Kitabgi P, Vincent J-P (1985) Molecular properties of neurotensin receptors in rat brain. J Biol Chem 260:508–514
McDermott JR, Kidd AN (1984) Ion-exchange gel-filtration and reversed-phase high-performance liquid chromatography in the isolation of neurotensin-degrading enzymes from rat brain. J Chromatogr 290:231–239
McDermott JR, Virmani MA, Turner JD, Kidd AM (1986) Peptidases involved in the catabolism of neurotensin: inhibitor studies using superfused rat hypothalamic slices. Peptides 7:225–230
Meisenberg G, Simmons WH (1985) Motor hypoactivity induced by neurotensin and related peptides in mice. Pharmacol Biochem Behav 22:189–193
Merritt WD, Bissette G, Luttinger D, Prange AJ Jr, Nemeroff CB (1983) Prior cold exposure antagonizes neurotensin-induced hypothermia in mice. Soc Neurosci Abstr 9:141
Micevych PE, Yaksh TL, Go VLW (1982) Opiate-mediated inhibition of the release of cholecystokinin and substance P, but not neurotensin from cat hypothalamic slices. Brain Res 250:283–289
Moga MM, Gray TS (1985) Peptidergic efferents from the intercalated nuclei of the amygdala to the parabrachial nucleus in the rat. Neurosci Lett 61:13–18
Morley JE, Levine AS (1981) Bombesin inhibits stress-induced eating. Pharmacol Biochem Behav 14:149–151
Morley JE, Yamada T, Walsh JH, Lamers CB, Wong H, Shulkes A, Damassa DA, Gordon J, Carlson HE, Hershman JM (1980) Morphine addiction and withdrawal alters brain peptide concentrations. Life Sci 26:2239–2244
Motta M, Martini L (1981) Neurotensin inhibits LH release (41235). Proc Soc Exp Biol Med 168:62–64
Myers RD, Lee TF (1984a) Neurotensin perfusion of rat hypothalamus: dissociation of dopamine release from body temperature change. Neuroscience 12:241–253
Myers RD, Lee TF (1984b) In vivo release of dopamine during perfusion of neurotensin in the substantia nigra of the unrestrained rat. Peptides 4:955–961
Nemeroff CB (1980) Neurotensin: perchance an endogenous neuroleptic? Biol Psychiatry 15:283–302
Nemeroff CB (1986) The interaction of neurotensin with dopaminergic pathways in the central nervous system: basic neurobiology and implications for the pathogenesis and treatment of schizophrenia. Psychoneuroendocrinology 11:15–37

Nemeroff CB, Cain ST (1985) Neurotensin-dopamine interactions in the central nervous stem. Trends Pharmacol Sci 6:201–205

Nemeroff CB, Bissette G, Prange AJ Jr, Loosen PT, Barlow TS, Lipton MA (1977) Neurotensin: central nervous system effects of a hypothalamic peptide. Brain Res 128:485–496

Nemeroff CB, Osbahr AJ III, Manberg PJ, Ervin GN, Prange AJ Jr (1979) Alterations in nociception and body temperature after intracisternally administered neurotensin, β-endorphin, other endogenous peptides, and morphine. Proc Natl Acad Sci USA 76:5368–5371

Nemeroff CB, Bissette G, Manberg PJ, Osbahr AJ III, Breese GR, Prange AJ Jr (1980) Neurotensin-induced hypothermia: evidence for an interaction with dopaminergic systems and the hypothalamic-pituitary-thyroid axis. Brain Res 195:69–84

Nemeroff CB, Hernandez DE, Orlando RC, Mason GA, Prange AJ Jr (1981 a) Cytoprotection by intracisternally administered neurotensin against cold-restraint stress-induced gastric ulcers: correlation of time course with effects on gastric pH and anterior pituitary hormone levels. Soc Neurosci Abstr 7:31

Nemeroff CB, Prange AJ Jr, Luttinger D, Hernandez DE, King RA, Burgess SK (1981 b) Similarities and differences in the effects of centrally administered neurotensin and neuroleptics. Psychopharmacol Bull 17:145–147

Nemeroff CB, Hernandez DE, Orlando RC, Prange AJ (1982 a) Cytoprotective effect of centrally administered neurotensin in rats. Am J Physiol 242:G342–346

Nemeroff CB, Hernandez DE, Luttinger D, Kalivas P, Prange AJ Jr (1982 b) Interactions of neurotensin with brain dopamine systems. Ann NY Acad Sci 400:330–344

Nemeroff CB, Luttinger D, Hernandez DE, Mailman RB, Mason GA, Davis SD, Widerlöv E, Frye GD, Kilts CD, Beaumont K, Breese GR, Prange AJ Jr (1983 a) Interactions of neurotensin with brain dopamine systems: biochemical and behavioral studies. J Pharmacol Exp Ther 225:337–345

Nemeroff CB, Youngblood W, Manberg PJ, Prange AJ Jr, Kizer JS (1983 b) Regional brain concentrations of neuropeptides in Huntington's chorea and schizophrenia. Science 221:972–975

Nemeroff CB, Bissette G, Busby WH, Youngblood WW, Rossor M, Roth M, Kizer JS (1983 c) Regional concentrations of neurotensin, thyrotropin-releasing hormone, and somatostatin in Alzheimer's disease. Soc Neurosci Abstr 9:1052

Okuma Y, Osumi Y (1982) Neurotensin-induced release of endogenous noradrenaline from rat hypothalamic slices. Life Sci 30:77–84

Okuma Y, Fukuda Y, Osumi Y (1983) Neurotensin potentiates the potassium-induced release of endogenous dopamine from rat striatal slices. Eur J Pharmacol 93:27–33

Orlando RC, Hernandez DE, Nemeroff CB, Prange AJ Jr (1981) Cytoprotective effect of intracisternally administered neurotensin on cold and restraint induced gastric ulcers in the rat. Gastroenterology 80:1244

Osbahr AJ III, Nemeroff CB, Manberg PH, Prange AJ Jr (1979) Centrally administered neurotensin: activity in the Julou-Courvosier muscle relaxation test in mice. Eur J Pharmacol 54:299–302

Osbahr AJ III, Nemeroff CB, Luttinger D, Mason GA, Prange AJ Jr (1981) Neurotensin-induced analgesia in mice: antagonism by thyrotropin-releasing hormone. J Pharmacol Exp Ther 217:645–651

Osumi Y, Nagasaka Y, Wang LHF, Fujiwara M (1978) Inhibition of gastric acid secretion and mucosal blood flow induced by intraventricularly applied NT in rats. Life Sci 23:2275–2280

Palacios JM, Kuhar MJ (1981) Neurotensin receptors are located on dopamine-containing neurons in rat midbrain. Nature 294:587–589

Prange AJ Jr, Nemeroff CB, Bissette G, Manberg PJ, Osbahr AJ III, Burnett GB, Loosen PT, Kraemer GW (1979) Neurotensin: distribution of hypothermic response in mammalian and submammalian vertebrates. Pharmacol Biochem Behav 11:473–477

Price CH, Ruane SE, Carraway RE (1982) Biochemistry and physiology of neurotensin-like peptides in the brain and gut of the mollusc *Aplysia*. Ann NY Acad Sci 400:409–410

Quirion R (1983) Interactions between neurotensin and dopamine in the brain: an overview. Peptides 4:609–615
Quirion R, Gaudreau P, St. Pierre S, Rioux F (1982a) Localization of neurotensin binding sites in rat kidney. Peptides 3:765–769
Quirion R, Gaudreau P, St. Pierre S, Rioux F, Pert C (1982b) Autoradiographic distribution of [^{3}H]neurotensin receptors in rat brain. Peptides 3:757–763
Quirion R, Everist HD, Pert A (1982c) Nigrostriatal dopamine terminals bear neurotensin receptors but the mesolimbic do not. Soc Neurosci Abstr 8:582
Reches A, Burke RE, Jiang D, Wagner HR, Fahn S (1982) The effect of neurotensin on dopaminergic neurons in rat brain. Ann NY Acad Sci 400:420–421
Reches A, Burke RE, Jiant DH, Wagner HR, Fahn S (1983) Neurotensin interactions with dopaminergic neurons in rat brain. Peptides 4:43–48
Reinecke M, Almasan K, Carraway R, Helmstaedter V, Forssman WG (1980a) Distribution patterns of neurotensin-like immunoreactive cells in the gastrointestinal tract of vertebrates. Cell Tissue Res 205:383–395
Reinecke M, Carraway RE, Falkemer S, Feurley GE, Forssman WG (1980b) Occurrence of neurotensin-immunoreactive cells in the digestive tract of lower vertebrates and deuterostomian invertebrates. Cell Tissue Res 212:173–183
Reinecke M, Forssmann WG, Thiekötter G, Triepel J (1983) Localization of neurotensin-immunoreactivity in the spinal cord and peripheral nervous system of the guinea pig. Neurosci Lett 37:37–42
Rinkel GJE, Hoeke EC, van Wimersma-Greidanus TJB (1983) Elective tolerance to behavioral effects of neurotensin. Pharmacol Behav 31:467–470
Rioux F, Quirion R, Regoli D, Leblanc M-A, St. Pierre S (1980) Pharmacological characterization of neurotensin receptors in the rat isolated portal vein using analogues and fragments of neurotensin. Eur J Pharmacol 66:273–279
Rioux F, Quirion R, St. Pierre S, Regoli D, Jolicoeur F, Belanger F, Barbeau A (1981) The hypotensive effect of centrally administered neurotensin in rats. Eur J Pharmacol 69:241–247
Rivier JE, Lazarus LH, Perrin MH, Brown MR (1977) Neurotensin analogues: structure-activity relationships. J Med Chem 20:1409–1411
Roberts GW, Crow TJ, Polak JM (1981) Neurotensin: first report of a cortical pathway. Peptides 2[Suppl 1]:37–43
Roberts GW, Woodhams PL, Polak JM, Crow TJ (1982) Distribution of neuropeptides in the limbic system of the rat: the amygdala complex. Neuroscience 7:99–131
Roberts GW, Allen Y, Crow TJ, Polak JM (1983) Immunocytochemical localization of neuropeptides in the fornix of rat, monkey and man. Brain Res 263:151–155
Rupniak NMJ, Mann S, Hall MD, Jenner P, Marsden CD (1984) Comparison of the effects on striatal DA receptor function of chronic administration of haloperidol, clozapine or sulpiride to rats for up to 12 months. In: Usdin E, Carlsson A, Dahlstrom A, Engels J (eds) Catecholamines: neuropharmacology and CNS. Theoretical aspects. Liss, New York
Sadoul J-L, Kitabgi P, Rostene W, Javoy-Agid F, Agid Y, Vincent J-P (1984a) Characterization and visualization of neurotensin binding to receptor sites in human brain. Biochem Biophys Res Commun 120:206–213
Sadoul J-L, Checler F, Kitabgi P, Rostrene W, Javoy-Agid F, Vincent J-P (1984b) Loss of high affinity neurotensin receptors in substantia nigra from parkinsonian subjects. Biochem Biophys Res Commun 125:395–404
Sakamoto N, Michel J-P, Kiyama H, Tohyama M, Kopp N, Pearson J (1986) Neurotensin immunoreactivity in the human cingulate gyrus, hippocampal subiculum and mammillary bodies. Its potential role in memory processing. Brain Res 375:351–356
Sarrieu A, Javoy-Agid F, Kitabgi P, Dussaillant M, Vial M, Vincent JP, Agid Y, Rostene WH (1985) Characterization and autoradiographic distribution of neurotensin binding sites in the human brain. Brain Res 348:375–380
Schultzberg M, Hokfelt T, Nilsson G, Terenius L, Rehfield JF, Brown M, Elde R, Goldstein M, Said S (1980) Distribution of peptide- and catecholamine-containing neurons in the gastrointestinal tract of rat and guinea pig: immunohistochemical studies with

antisera to substance P, vasoactive intestinal polypeptide, enkephalins, somatostatin, gastrin/cholecystokinin, neurotensin, and dopamine β-hydroxylase. Neuroscience 5:689–744
Shido O, Nagasaka T (1985) Effects of intraventricular neurotensin on blood pressure and heat balance in rats. Jpn J Physiol 35:311–320
Shiraishi T, Inoue A, Yanaihara N (1980) Neurotensin and bombesin effects on LHA-gastrosecretory relations. Brain Res Bull 4[Suppl 5]:133–142
Skidgel RA, Engelbrecht S, Johnson AR, Erdos EG (1984) Hydrolysis of substance P and neurotensin by converting enzyme and neutral endopeptidase. Peptides 5:769–776
Skoog KM, Cain ST, Nemeroff CB (1986) Centrally administered neurotensin suppresses locomotor hyperactivity induced by *d*-amphetamine but not by scopolamine or caffeine. Neuropharmacology 25:777–782
Snidjers R, Kramarcy NR, Hurd RW, Nemeroff CB, Dunn AJ (1982) Neurotensin induces catalepsy in mice. Neuropharmacology 21:465–468
Stanley BG, Eppel N, Noebel BG (1982) Neurotensin injected into the paraventricular hypothalamus suppresses feeding in rats. Ann NY Acad Sci 400:425–427
Starr MS (1982) Influence of peptides on ^{3}H-dopamine release from superfused rat striatal slices. Neurochem Int 4:233–240
Sugimoto T, Itoh K, Yasui Y, Kaneko T, Mizuno N (1985) Coexistence of neuropeptides in projection neurons of the thalamus in the cat. Brain Res 347:381–384
Sumners C, Phillips MI, Richards EM (1982) Central pressor action of neurotensin in conscious rats. Hypertension 4:888–893
Sundler F, Hakanson R, Hammer RA, Alumets J, Carraway R, Leeman SE, Zimmerman EA (1977) Immunohistochemical localization of neurotensin in endocrine cells of the gut. Cell Tissue Res 178:313–321
Tepperman BL, Evered M (1980) Gastrin injected into the lateral hypothalamus stimulates secretion of gastric acid in rats. Science 94:9–15
Tojo K, Kato Y, Kabayama Y, Ohta H, Inoue T, Imura H (1986) Further evidence that central neurotensin inhibits pituitary prolactin secretion by stimulating dopamine release from the hypothalamus. Proc Soc Exp Biol Med 181:517–522
Uhl GR, Kuhar MJ (1984) Chronic neuroleptic treatment enhances neurotensin receptor binding in human and rat substantia nigra. Nature 309:350–352
Uhl GR, Snyder SH (1977a) Regional and subcellular distribution of brain neurotensin. Life Sci 19:1827–1832
Uhl GR, Snyder SH (1977b) Neurotensin receptor binding: regional and subcellular distributions favor transmitter role. Eur J Pharmacol 41:89–91
Uhl GR, Snyder SH (1979) Neurotensin: a neuronal pathway projecting from amygdala through stria terminalis. Brain Res 161:522–526
Uhl GR, Kuhar MJ, Snyder SH (1977a) Neurotensin: immunohistochemical localization in rat central nervous system. Proc Natl Acad Sci USA 74:4059–4063
Uhl GR, Bennett JP, Synder SH (1977b) Neurotensin: a central nervous system peptide; apparent receptor binding in brain membranes. Brain Res 130:299–313
Uhl GR, Whitehouse PJ, Price DL, Tourtelotte WW, Kuhar MJ (1984) Parkinson's disease: depletion of substantia nigra neurotensin receptors. Brain Res 308:186–190
van Wimersma Greidanus TB, van Praag MCG, Kalmann R, Rinkel GJE, Croiset G, Hoeke EC, van Egmond MAH, Fekete M (1982) Behavioral effects of neurotensin. Ann NY Acad Sci 400:319–329
Vijayan E, McCann SM (1979) In vivo and in vitro effects of substance P and neurotensin on gonadotropin and prolactin release. Endocrinology 105:64–68
Vijayan E, McCann SM (1980) Effects of substance P and neurotensin on growth hormone and thyrotropin release in vivo and in vitro. Life Sci 26:321–327
Wachi M, Okuda M, Togashi S, Fuwano S, Miyashita O (1985) Effects of phencyclidine administration on behavior and brain neurotensin-like immunoreactivity in rats. Brain Res 333:393–396
Watt CB, Li H-B, Lam DM-K (1985) The presence of three neuroactive peptides in putative glycinergic amacrine cells of an avian retina. Brain Res 348:187–191

Wenger J, Yusim J, Bloom F, Brown M (1981) Rapid induction of tolerance to the hypothermic effect of β-endorphin (βE), bombesin (BN), neurotensin (NT) and lack of cross-tolerance to the hypothermic effect of ethanol (E). Soc Neurosci Abstr 7:505

Widdowson PS, Griffiths EC, Slater P, Yajima H (1983) Effects of mammalian and avian neurotensins and neurotensin fragments on wet-dog shaking and body temperature in the rat. Regul Pept 7:357–365

Widerlöv E, Kilts CD, Mailman RB, Nemeroff CB, McCown TJ, Prange AJ Jr, Breese GR (1982a) Increase in dopamine metabolites in rat brain by neurotensin. J Pharmacol Exp Ther 222:1–6

Widerlöv E, Lindstrom LH, Besev G, Manberg PJ, Nemeroff CB, Breese GR, Kizer JS, Prange AJ Jr (1982b) Subnormal CSF levels of neurotensin in a subgroup of schizophrenic patients: normalization after neuroleptic treatment. Am J Psychiatry 139:1122–1126

Yaksh TL, Schmouss C, Micevych PE, Abay EO, Go VLM (1982) Pharmacological studies on the application, disposition and release of neurotensin in the spinal cord. Ann NY Acad Sci 400:228–243

Yarbrough GG, McGuffin JC (1979) MK-771-induced electromyographic (EMG) activity in the rat: comparison with thyrotropin releasing hormone (TRH) and antagonism by neurotensin. Eur J Pharmacol 60:41–46

Yates CM, Fink G, Bennie JG, Gordon A, Simpson J, Eskay RL (1985) Neurotensin immunoreactivity in post-mortem brain is increased in Down's syndrome but not in Alzheimer-type dementia. J Neurol Sci 67:327–335

Young WS III, Kuhar MJ (1979) Neurotensin receptors: autoradiographic localization in rat CNS. Eur J Pharmacol 59:161–163

Young WS III, Kuhar MJ (1981) Neurotensin receptor localization by light microscopic autoradiography in rat brain. Brain Res 220:273–285

Cholecystokinin and Mood

I. N. Ferrier [1] and T. J. Crow [2]

Contents

1 Introduction

Cholecystokinin (CCK) is a peptide hormone which has long been known to be present in the pancreas and gastro-intestinal tract and has recently been found to be present in brain (Morley 1982). High concentrations are found in animal and human cerebral cortex, basal ganglia, hypothalamus and the limbic lobe (hippocampus, amygdala, olfactory tubercle and associated cortices) (Geola et al. 1981). The major CCK entity in the brain is the COOH-terminal octapeptide CCK-8, although lesser amounts of the COOH-terminal tetrapeptide CCK-4 (Geola et al. 1981; Rehfeld 1978) are found. There is good evidence that CCK acts as a neurotransmitter or neuromodulator in the CNS; it is synthesised in neurones and concentrated in synaptic vesicles, its release is calcium-dependent (Emson et al. 1980 b) and it has a potent excitatory effect on postsynaptic membranes of hippocampal cells (Dodd and Kelly 1981). In addition there have been several reports of high-affinity binding sites for radiolabelled CCK in the brains of rodents and man (Hays et al. 1981).

CCK co-exists with dopamine (DA) inside neurones which comprise the mesolimbic and mesocortical DA systems but not inside those DA neurones which project to the extrapyramidal motor system (Hokfelt et al. 1980 a). CCK has been

1 Medical Research Council Neuroendocrinology Unit, Newcastle General Hospital, Westgate Road, Newcastle upon Tyne NE4 6BE, UK
2 Division of Psychiatry, Clinical Research Centre, Northwick Park Hospital, Watford Road, Harrow, Middlesex HA1 3UJ, UK

Current Topics in Neuroendocrinology, Vol. 8

reported to increase the affinity and decrease the number of central DA receptors (Fuxe et al. 1981). There have been reports that CCK modifies DA release and turnover (Fekete et al. 1981), but only those DA neurones which project to limbic areas are responsive to CCK (Skirboll et al. 1981). CCK co-exists in some neurones with a number of other transmitters or neuropeptides, e.g. adrenaline, substance P and vasopressin (Crawley 1985).

A number of behavioural effects of central CCK have been suggested. There is evidence that brain CCK peptides function as important transmitters of hunger and satiety signals and that specific hypothalamic sites are involved in the release or action of CCK (Baile and Della Ferra 1985). There are a number of unresolved problems with the hypothesis that central CCK has a role in feeding behaviour, notably the wide species variation, the difficulty of separating peripheral and central effects of CCK and the lack of evidence that the satiety effect of exogenous CCK reveals a physiological function for endogenous CCK (Smith and Gibbs 1985). In animals a wide variety of behaviours have been reported following CCK administration, including sedation, hypothermia, antagonism of tremors and stereotyped behaviours, and inhibition of conditioned avoidance responses and reduction in the rate of self-stimulation (Zetler 1985; De Witte et al. 1985). These effects occur with relatively small doses, and changes in glucose uptake in various regions of the brain following similar doses of CCK-8 have been shown (Lucignani et al. 1984). It has been suggested that some of these behavioural changes relate to the effect of CCK on the release of dopamine and that CCK has a neuroleptic-like spectrum of neuropharmacological activity, but there is considerable controversy in this area.

In view of the high concentrations of CCK in limbic structures which have been implicated in learning, memory, emotional responses and social behaviour, a role for this peptide in normal and abnormal human mood states has been suggested. It is noteworthy that those mesolimbic and mesocortical DA systems whose overactivity has been hypothesized in schizophrenia (Crow 1980) are actually CCK/DA systems. It is also becoming recognized that the measurement of neuropeptides may provide information on the integrity of specific neuronal systems and that neuropeptides are useful markers in degenerative CNS disease (Rossor and Emson 1985). For example, in Parkinson's disease a loss of CCK has been demonstrated in the nigro-striatal system which correlates well with the loss of dopaminergic neurones in these areas in this condition (Studler et al. 1982). In Huntington's disease, a marked reduction in CCK and other peptides are found in the substantia nigra (Emson et al. 1980 a) indicating damage to the striatal efferents. By contrast, in Alzheimer's disease cortex, CCK and other peptides are normal but somatostatin is reduced (Studler et al. 1982), indicating, as has been shown on other evidence, preservation of certain populations of intrinsic cortical cells in this disease.

This review focusses on studies which have examined the effects of CCK on psychological function in normal and abnormal mood states and also looks at those studies which have measured CCK in brain or CSF in psychiatric disease.

2 Effects of CCK in Normal Humans

CCK-8 has been used to suppress human feeding and can cause nausea at higher doses. Consequently most studies have used the decapeptide ceruletide, the amphibian analogue of CCK-8, which has many properties similar to CCK-8. Several reports have noted mild drowsiness in humans given either CCK-8 or ceruletide (Stacher et al. 1979). In a blind study on seven normal volunteers, an increase in fatigue ratings following 0.6 µg/kg of ceruletide was noted. No effect on memory, vigilance, retrieval or cognition was found. There was also no change in mood following ceruletide on the Profile of Mood States questionnaire, which includes such subscales as tension, depression, anger and vigour (Hommer et al. 1985).

3 CCK in Schizophrenia

3.1 Measurement in CSF and Brain

Most of the CCK immunoreactive material in the CSF originates in the CNS. CCK crosses the blood-brain barrier only poorly, and there is a lack of correlation between CSF and blood levels (Kruse-Larsen and Rehfeld 1979). Verbanck et al. (1984) have reported reduced CCK in lumbar CSF from patients with schizophrenia who were untreated – a smaller, non-significant decrease was reported in schizophrenics on medication. However, no difference in CSF levels of CCK was noted between unmedicated schizophrenics and controls in a subsequent study (Gerner and Yamada 1982). This underlines the uncertainty about the significance of the results of CSF peptide studies, since the relationship between activity in any central peptidergic system and the concentration of the same peptide in CSF is not established (Edwardson and McDermott 1982).

CCK is stable post mortem, and a number of studies have examined levels in postmortem brain from patients with a history of schizophrenia. Perry et al. (1981) noted no difference in CCK in entorhinal cortex, and Kleinman et al. (1983) reported normal CCK in amygdala, hippocampus and temporal cortex in brains from a group of psychotic patients, although only some of them appear to suffer from schizophrenia as defined in the Research Diagnostic Criteria.

More recently we have examined brains from patients who had been clinically assessed before death [Ferrier et al. 1985; Roberts et al. 1983). CCK immunoreactivity was significantly reduced in the temporal cortex in the schizophrenics. There was no change in CCK in the amygdala and hippocampus in the whole schizophrenic group, but in those patients with negative symptoms (poverty of speech and flattening of affect – the "type II" patients) (Crow 1980) there was a significant reduction (by 34% and 37% respectively). CCK binding was reduced significantly in hippocampus (by 40%) and frontal cortex (by 20%) and non-significantly in amygdala and temporal cortex (Farmery et al. 1984).

The observed alterations in CCK content of the limbic lobe may reflect changes in production, storage, secretion or destruction of intrinsic cells or afferent fibres. The major proportion of CCK in the hippocampus is intrinsic (Handelmann et al. 1981), although CCK-containing inputs may enter from surrounding regions, especially the amygdala (Roberts et al. 1982). A significant loss of CCK was found in the amygdala of type II schizophrenics: thus the loss of CCK in the hippocampus in the same group may reflect damage to the CCK-containing amygdala-hippocampal link. The decrease in the CCK content of the temporal cortex in schizophrenics could also relate to the observed changes in CCK content of the amygdala and hippocampus, since the temporal cortex is a possible source of afferents to these areas (Price 1981). However, the change in CCK in temporal cortex was found in the total schizophrenic group, whereas changes in amygdaloid and hippocampal CCK were restricted to the group of patients with negative symptoms and no correlation was observed between the CCK content of these regions.

This finding is of interest, since it represents the first evidence of a biochemical abnormality associated with negative symptoms, the hallmark of the impairments of chronic schizophrenia, whose presence may identify a group of patients with poor outcome, intellectual impairment, neurological features and, perhaps, structural changes in the brain (Crow et al. 1986). CCK-containing neurones are found in most hippocampal regions, particularly the CA1/2 region (Greenwood et al. 1981), an area where cell disarray has been reported in chronic schizophrenia (Kovelman and Scheibel 1984). It is conceivable that the reductions in CCK in hippocampus and amygdala in type II schizophrenics are markers for pathological changes in this region and that these neuropathological changes underlie the impairments of chronic schizophrenia. Recently a trophic role for CCK in the developing mammalian brain has been demonstrated (Chun et al. 1987), and it is therefore possible that reduced CCK levels in limbic lobe in early life may be implicated directly in the development of the pathological changes in that area in schizophrenia.

CCK is present in DA-rich areas of brain such as the substantia nigra, the striatum and the amygdala. Since overactivity of DA has been hypothesised to underlie schizophrenia (Crow 1980), it is interesting to speculate whether the changes in CCK might relate to changes in DA levels. The relationship between DA and CCK is complex. There is evidence that CCK and DA co-exist in a subpopulation of midbrain DA-containing neurones (Hokfelt et al. 1980 a) which innervate mesolimbic structures. CCK has been reported to decrease DA turnover in the striatum (Fuxe et al. 1981) and CCK fragments inhibit DA release in the nucleus accumbens (Hokfelt et al. 1980 b). However, it seems unlikely that the changes reported here relate to DA. The CCK content of the amygdala is highest in the central and medial nuclei (Vanderhaeghen et al. 1980) and the latter nucleus has a sparse DA innervation (Meibach and Katzman 1981). The hippocampus contains very little DA (Lindvall and Bjorklund 1978). Thus the reductions in CCK in type II schizophrenics are seen in areas which are not associated with a dense dopaminergic innervation and therefore it appears unlikely that they are related to postulated DA overactivity. The report by Reynolds that DA is increased in schizophrenia only in left amygdala has, however, re-emphasised the importance of laterality in studies of schizophrenia and indicated the need to examine neuropeptides from this aspect (Reynolds 1983).

A related question is whether neuroleptic medication could have accounted for the observed differences in CCK content of the limbic lobe in schizophrenia. There was no difference in peptide content between those patients who had received neuroleptics up till the time of death and those who were drug-free, and the abnormalities noted are localised rather than general. It also seems unlikely that the reduced CCK binding in frontal cortex and hippocampus was due to neuroleptic medication, since a drug effect would be expected to occur in all regions. Moreover, it has been reported that chronic haloperidol treatment results in an increased number of CCK binding sites in animal brain (Chang et al. 1983). However, the total number of schizophrenic patients studied was small ($n = 14$), as was the number of patients considered drug-free ($n = 5$). Hence these findings and the role of neuroleptics in their causation require replication in a larger sample.

The pathophysiological basis of reduced CCK binding in schizophrenic hippocampus and frontal cortex is uncertain at present. Because of the limited availability of tissue, it was not possible to carry out a saturation analysis to determine whether the decrease in CCK binding was due to reduction in receptor number or to a change in affinity. However, such changes in binding are usually the result of a variation in receptor numbers, as reported by Hays et al. (1981) and others. As discussed above, there are a large number of CCK nerve terminals in the hippocampus which are derived from intrinsic cells, and it may be that a population of CCK receptors is present in these cells (Zarbin et al. 1983). It is possible that the decrease in CCK binding in the hippocampus of schizophrenic patients observed in this study is a consequence of some pathological process, for example cell loss in this region of brain. This may be related to the finding by Bogerts et al. that hippocampal size is reduced in schizophrenics (Bogerts et al. 1985). Although the decrease (20%) in CCK binding in frontal cortex of schizophrenics was only half that observed in the hippocampus, this change may also be indicative of some underlying pathology. Hays et al. (1981) observed a reduction in CCK binding in the cerebral cortex of Huntington's disease patients which was greater than the degree of cell loss observed, suggesting a selective loss of neurones possessing CCK receptors in this particular condition. It is conceivable that such a mechanism operates in the schizophrenic brain, but as many biochemical indices and CCK itself are normal in schizophrenic frontal cortex this finding requires replication and much greater study.

Recently Carruthers et al. (1984) have also reported significantly reduced CCK in amygdala in schizophrenia, though no clinical subtyping was possible.

3.2 Effect of CCK and Analogues in Schizophrenia

In view of the associations between CCK and DA outlined above, a number of investigators have examined the clinical effect of CCK or its analogues in schizophrenia. Moroji and co-workers first performed open clinical studies with caerulein (ceruletide diethylamine) and ceruletide in patients with chronic schizophrenia and reported overall improvement, with a reduction in hallucinations and favourable changes in mood states (Moroji et al. 1982). In subsequent studies these authors report that administration of a single dose of 40 μg caerulein effected improvement in 13 of 20 chronic schizophrenic patients, with a reduction in both positive and negative symptoms (Moroji et al. 1985). The patients who responded were maintained well with 2-weekly injections of caerulein over the period of 1 year. These findings were confirmed in two other large open studies (Itoh et al. 1982; Kudo 1983). Most of these studies noted improvements in the anxiety and depression experienced by the patients. In preliminary open studies Nair and co-workers also found that the i.v. administration of both CCK-33 (Nair et al. 1982) and CCK-8 (Bloom et al. 1983) improved psychotic symptoms in neuroleptic-resistant chronic schizophrenic patients maintained on neuroleptic medication.

However, subsequent placebo-controlled blind studies have largely failed to replicate these positive findings. Lostra et al. (1985), in a double-blind crossover study, did not show that ceruletide has antipsychotic properties in neuroleptic-free patients suffering from schizophrenia, and Hommer et al. (1985) reported similar findings in a comparable study of patients on neuroleptics. Albus et al. (1984), in a double-blind controlled study of 20 patients with predominant negative symptoms, found no evidence for any therapeutic benefit of ceruletide. In a series of studies which included single dose crossover, rising dose and parallel group designs, Chase et al. found no therapeutic benefit of caerulein in schizophrenic patients (Chase et al. 1985). Nair et al., in a double-blind placebo-controlled trial, found a significant therapeutic benefit of weekly i.v. CCK-8 in a group of schizophrenic patients who were also receiving neuroleptics. However, it was noted that CCK-8 caused gastro-intestinal effects which may have "unblinded" the study (Nair et al. 1985).

It is conceivable that the lack of effect of CCK and analogues in the above studies is due to the fact that they are restricted in crossing the blood-brain barrier. In animals, peripherally administered CCK cannot be detected in CSF (Passaro et al. 1982). However, it has been found in animals that peripherally administered CCK and ceruletide modify DA neurone activity[10] and DA autoreceptor sensitivity (Hommer and Skirboll 1983). The finding of a reduction in CCK in schizophrenic brain, that is local rather than general (Ferrier et al. 1985) makes it unlikely that the peripheral administration of CCK would have marked clinical effects. There is some evidence that the CNS effects of ceruletide are slighter than those of CCK-8, and there is a need for further studies with this latter peptide (Lostra et al. 1985).

4 CCK in Affective Illness

4.1 Studies in CSF and Brain

Lostra et al. (1985) found no difference in CSF levels of CCK between controls and drug-free patients with primary depression according to the Research Diagnostic Criteria, and there was no difference between unipolar and bipolar patients. Gerner and Yamada (1982) similarly found no difference in CSF levels of CCK between depression and mania and normal controls. In a postmortem study, Perry et al. (1981) found that cortical levels of CCK were normal in a group of brains from patients who had died in psychiatric hospital with a diagnosis of depression.

4.2 Effect of CCK in Affective Illness

There have been no reported studies of the effect of CCK in affective illness. However, as noted above, a number of studies on schizophrenic patients have recorded a reduction in anxiety and depression (Moroji et al. 1985; Nair et al. 1985). Albus et al. (1984), in an open study, noted significant improvement in a group of six schizophrenic patients with postpsychotic depression who were given two i.m. doses of ceruletide. There is therefore a need to study the effects of CCK and its analogues on both postpsychotic depression and depressive states in larger, more controlled trials.

5 Conclusion

CCK is a ubiquitous peptide found in high concentrations in human brain. The evidence for its classification as a neurotransmitter and for an effect on behaviour in animals is strong. There is evidence that CCK is reduced in limbic areas in schizophrenic brain. Some intriguing studies have suggested that peripherally administered CCK may have both antipsychotic properties and effects on mood. However, most recent controlled studies do not substantiate these claims. The lack of effect of exogenous CCK may be explained by pharmacodynamic factors and does not necessarily disprove the hypothesis that endogenous brain CCK plays a crucial role in normal and abnormal mood states.

References

Albus M, Ackenheil M, Munch U, Naber D (1984) Ceruletide: a new drug for the treatment of schizophrenic patients? Arch Gen Psychiatry 41:528

Baile CA, Della Ferra MA (1985) Central nervous system cholecystokinin and the control of feeding. Ann NY Acad Sci 448:424–430

Bloom DM, Nair NPV, Schwartz G (1983) CCK-8 in the treatment of chronic schizophrenia. Psychopharmacol Bull 19:361–363

Bogerts B, Meertz E, Schonfedt-Bausch E (1985) Basal ganglia and limbic system pathology in schizophrenia. Arch Gen Psychiatry 42:784–791

Carruthers B, Dawbarn D, de Quidt M, Emson PC, Hunter J, Reynolds GP (1984) Changes in neuropeptide content of amygdala in schizophrenia. Br J Pharmacol 81:190P

Chang RSL, Lotta VJ, Martin GE, Chen TB (1983) Increase in brain ^{125}I-cholecystokinin (CCK) receptor binding following chronic haloperidol treatment, intracisternal 6-hydroxydopamine or ventral tegmental lesions. Life Sci 32:871–878

Chase TN, Barone P, Bruno G, Cohen SL, Juncos J, Knight M, Ruggeri S, Steardo L, Tamminga CA (1985) Cholecystokinin-mediated synaptic function and the treatment of neuropsychiatric disease. Ann NY Acad Sci 448:553–561

Chun JJM, Nakamira MJ, Shatz CJ (1987) Transient cells of the developing mammalian telencephalon are peptide-immunoreactive neurons. Nature 325:617–620
Crawley JN (1985) Comparative distribution of cholecystokinin and other neuropeptides. Ann NY Acad Sci 448:1–8
Crow TJ (1980) Molecular pathology of schizophrenia: more than one disease process? Br Med J [Clin Res] 280:66–68
Crow TJ, Ferrier IN, Johnstone EC (1986) The two syndrome concept and the neuroendocrinology of schizophrenia. Psychiatr Clin North Am 9:99–113
De Witte P, Swanet E, Gewiss M, Goldman S, Roques B, Vanderhaeghen JJ (1985) Psychopharmacological profile of cholecystokinin using the self-stimulation and the drug discrimination paradigms. Ann NY Acad Sci 448:470–487
Dodd J, Kelly JS (1981) The actions of cholecystokinin and related peptides on pyramidal neurons of the mammalian hippocampus. Brain Res 205:337–350
Edwardson JA, McDermott JR (1982) Neurochemical pathology of brain peptides. Br Med Bull 38:259–264
Emson PC, Rehfeld JF, Langevin H, Rossor M (1980a) Reduction in cholecystokinin-like immunoreactivity in the basal ganglia in Huntington's disease. Brain Res 198:497–500
Emson PC, Lee CM, Rehfeld JF (1980b) Cholecystokinin octapeptide: vesicular localization and calcium-dependent release from rat brain in vitro. Life Sci 26:2157–2163
Farmery SM, Owen F, Poulter M, Crow TJ (1984) Reduced high affinity cholecystokinin binding in hippocampus and frontal cortex of schizophrenic patients. Life Sci 36:473–477
Fekete M, Kadar T, Penke B, Kovacs K, Telegdy G (1981) Influence of cholecystokinin octapeptide sulfate ester on brain monamine metabolism in rats. J Neural Transm 50:81–88
Ferrier IN, Crow TJ, Farmery SM, Roberts GW, Bowen F, Adrian TE, Bloom SR (1985) Reduced cholecystokinin levels in the limbic lobe in schizophrenia. Ann NY Acad Sci 448:495–506
Fuxe K, Agnati L, Benfenati F, Cimmino M, Algeri S, Hokfelt T, Mutt V (1981) Modulation by cholecystokinin of [^{3}H]spiroperidol binding in rat striatum: evidence for increased affinity and reduction in number of binding sites. Acta Physiol Scand 113:567–569
Geola FL, Hershman JM, Warwick R, Reeve JR, Walsh JH, Tourtellotte WW (1981) Regional distribution of cholecystokinin-like immunoreactivity in the human brain. J Clin Endocrinol Metab 53:270–275
Gerner RH, Yamada T (1982) Altered neuropeptide concentrations in cerebrospinal fluid of psychiatric patients. Brain Res 238:298–302
Greenwood RS, Godar SE, Reaves TA, Hayward JN (1981) Cholecystokinin in hippocampal pathways. J Comp Neurol 203:335–350
Handelmann GE, Meyer DK, Beinfeld MC, Oertel WH (1981) CCK-containing terminals in the hippocampus are derived from intrinsic neurons: an immunohistochemical and radioimmunological study. Brain Res 224:180–184
Hays SE, Goodwin FK, Paul SM (1981) Cholecystokinin receptors are decreased in basal ganglia and cerebral cortex of Huntington's disease. Brain Res 225:452–456
Hokfelt T, Skirboll L, Rehfeld JF, Goldstein M, Markey K, Dann O (1980a) A subpopulation of mesencephalic dopamine neurons projecting to limbic areas contains a cholecystokinin-like peptide: evidence from immunohistochemistry combined with retrograde tracing. Neuroscience 5:2093–2124
Hokfelt T, Rehfeld JF, Skirboll L, Ivemark B, Goldstein M, Markey K (1980b) Evidence for coexistence of dopamine and CCK in mesolimbic neurons. Nature 285:476–478
Hommer D, Skirboll L (1983) Cholecystokinin-like peptides potentiate apomorphine-induced inhibition of dopamine neurons. Eur J Pharmacol 91:151–152
Hommer DW, Pickar D, Crawley JN, Weingartner H, Paul SM (1985) The effects of cholecystokinin-like peptides in schizophrenics and normal human subjects. Ann NY Acad Sci 448:542–552

Itoh H, Tanoue S, Yagi G, Tateyama M, Kamisada M, Fujii Y, Takamiya M, Nakajima S (1982) Clinical study of the psychotropic effects of caerulein: an open clinical trial in chronic schizophrenic patients. Keio J Med 31:71–95

Kleinman JE, Iadarola M, Govani S, Hong J, Gillin JC, Wyatt RJ (1983) Post mortem measurements of neuropeptides in human brain. Psychopharmacol Bull 19:375–377

Kovelman JA, Scheibel AB (1984) A neurohistological correlate of schizophrenia. Biol Psychiatry 19:1601–1621

Kruse-Larsen C, Rehfeld JF (1979) Gastrin in human cerebrospinal fluid: lack of correlation with serum concentrations. Brain Res 176:189–191

Kudo Y (1983) The effect of ceruletide on chronic schizophrenia. Jpn J Clin Psychiatry 12:703–710

Lindvall O, Bjorklund A (1978) Organisation of catecholamine neurons in rat central nervous system. In: Iversen LL, Iversen SD, Snyder SH (eds) Handbook of psychopharmacology, vol 9. Plenum, New York, pp 139–231

Lostra F, Verbanck PMP, Gilles C, Mendlewicz J, Vanderhaeghen JJ (1985) Reduced cholcystokinin levels in cerebrospinal fluid of parkinsonian and schizophrenic patients. Effect of ceruletide in schizophrenia. Ann NY Acad Sci 448:507–517

Lucignani G, Porrino LJ, Tamminga CA (1984) Effects of systemically administered cholecystokinin octapeptide on local cerebral metabolism. Eur J Pharmacol 101:147–151

Meibach RC, Katzman R (1981) Origin, course and termination of dopaminergic substantia nigra neurons projecting to the amygdaloid complex in the cat. Neuroscience 6:2159–2171

Morley JE (1982) The ascent of cholecystokinin (CCK) from gut to brain. Life Sci 30:479–493

Moroji T, Watanabe N, Aoki N, Itoh S (1982) Antipsychotic effects of caerulein in chronic schizophrenia. Arch Gen Psychiatry 39:485

Moroji T, Itoh K, Itoh K (1985) Antipsychotic effects of cerulitide in chronic schizophrenia: an appraisal of the long-term, intermittend medication of cerulitide in chronic schizophrenia. Ann NY Acad Sci 448:518–534

Nair NP, Bloom DM, Nestoros JN (1982) Cholecystokinin appears to have antipsychotic properties. Biol Psychiatry 6:509–512

Nair NPV, Bloom D, Lal S, Rebonnel G, Schwartz G, Mosticyan S (1985) Clinical and neuroendocrine studies with cholecystokinin peptides. Ann NY Acad Sci 448:535–541

Passaro E, Bebas H, Oldendorf W, Yamada T (1982) Rapid appearance of intraventricularly administered neuropeptides in the peripheral circulation. Brain Res 241:335–340

Perry RH, Dockray GJ, Dimaline R, Perry EK, Blessed G, Tomlinson BE (1981) Neuropeptides in Alzheimer disease depression and schizophrenia. A post-mortem analysis of vasoactive intestinal peptide and cholecystokinin in cerebral cortex. J Neurol Sci 51:465–472

Price JL (1981) The efferent projects of the amygdaloid complex in the rat, cat and monkey. In: Ben-Ari Y (ed) The amygdaloid complex. Elsevier/North-Holland, Amsterdam, pp 121–132

Rehfeld JF (1978) Immunochemical studies on cholecystokinin. II. Distribution and molecular heterogeneity in the central nervous system and small intestine of man and hog. J Biol Chem 253:4022–4030

Reynolds GP (1983) Increased concentrations and lateral asymmetry of amygdala dopamine in schizophrenia. Nature 305:527–529

Roberts GW, Woodhams PL, Polak JM, Crow TJ (1982) Distribution of neuropeptides in the limbic system of the rat: the amygdaloid complex. Neuroscience 7:99–131

Roberts GW, Ferrier IN, Lee Y, Crow TJ, Johnstone EC, Owens DGC, Bacarese-Hamilton AJ, McGregor C, O'Shaughnessy D, Polak JM, Bloom SR (1983) Peptides, the limbic lobe and schizophrenia. Brain Res 288:199–211

Rossor MN, Emson PC (1985) Neuropeptides in degenerative disease of the central nervous system. Trends Neurosci 5:399–401

Skirboll LR, Grace AA, Hommer DW, Rehfeld J, Goldstein M, Hokfelt T, Bunney BS (1981) Peptide-monoamine coexistence: studies of actions of cholecystokinin-like peptide on the electrical activity of midbrain dopamine neurons. Neuroscience 6:2111–2124

Smith GP, Gibbs J (1985) The satiety effect of cholecystokinin: recent progress and current problems. Ann NY Acad Sci 448:417–423

Stacher G, Bauer J, Steinringer H (1979) Cholecystokinin decreases appetite and activation evoked by stimuli arising from the preparation of a meal in man. Physiol Behav 23:325–331

Studler JM, Javoy-Agid F, Cesselin F, Legrand JC, Agid Y (1982) CCK-8 immunoreactivity distribution in human brain: selective decrease in the substantia nigra from parkinsonian patients. Brain Res 243:176–179

Vanderhaeghen JJ, Lotstra F, De Mey J, Gilles C (1980) Immunohistochemical localization of cholecystokinin- and gastrin-like peptides in the brain and hypophysis of the rat. Proc Natl Acad Sci USA 77:1190–1194

Verbanck PMP, Lotstra F, Gilles C, Linkowski P, Mendlewicz J, Vanderhaeghen JJ (1984) Reduced cholecystokinin immunoreactivity in the cerebrospinal fluid of patients with psychiatric disorders. Life Sci 34:67–72

Zarbin MA, Innis RB, Wamsley JK, Snyder SH, Kuhar MJ (1983) Autoradiographic localization of cholecystokinin receptors in rodent brain. Neuroscience 3:877–906

Zetler G (1985) Neuropharmacological profile of cholecystokinin-like peptides. Ann NY Acad Sci 448:448–469

Opioid Peptides and Mood: Neuroendocrine Aspects *

H. Ågren [1] and L. Terenius [2]

Contents

1 Introduction

The connections between the abuse of opium and mood changes have been known for centuries. Opium preparations, indigenous to the Near and Middle East, were part of materia medica alrady in ancient Egypt and were indispensable for medical practice in classical Greek medicine and later throughout European history, one of the indications being the reduction of emotional distress. In China, opium was not in general use until the 16th century, and the first detailed account of opium smoking and addiction was given around 1800. The medical actions of opium smoking were "harmonizing the circulation of blood and pneuma, *fortifying the spirits,* and curing abdominal unbalances as well as all diseases of the stomach and gut" (Wu 1926).

As worldwide anxiolytics, opium and morphine were undisputed until the arrival of barbiturates and benzodiazepines. They fell into disuse when the addictive properties were recognized (Carlson and Simpson 1963). The idea that opiates could possess true antidepressant properties has been intuitively refused, but few studies have specifically addressed this issue. In the days when opium preparations were part of a very small group of psychoactive drugs depression was cer-

* The authors are supported by the Swedish Medical Research Council (grants 5095 to L. T. and 6604 to H. Å.) and the Fredrik and Ingrid Thuring Foundation

1 Department of Psychiatry, University Hospital, S-751 85 Uppsala, Sweden
2 Department of Pharmacology, Biomedical Centre, Box 591, S-751 24 Uppsala, Sweden

Current Topics in Neuroendocrinology, Vol. 8

tainly one of their many therapeutic indications, a practice that has survived from Kraepelin's era until quite recently in the prescription of morphine in severe melancholic morning anxiety (for one lingering practice, see Abse et al. 1982). Antimanic, antidepressant and antipanic effects of opiates have been described in uncontrolled studies (Gold et al. 1979) and reviewed by Gold et al. (1982). An early observation by Fink et al. (1970) suggested that a partial opioid agonist, cyclazocine, had antidepressant properties. The partial agonist buprenorphine has been shown in a double-blind study to possess antidepressant effects (Emrich et al. 1982). However, single doses of methadone and morphine were shown to lack clear antidepressant effects (Extein et al. 1981).

The discovery of endogenous morphine-like (opioid) peptides has strongly influenced our understanding of especially pain physiology and pharmacology, and, given the euphoriant properties of opiates, has also contributed to several powerful research hypotheses concerning the regulation of mood states. In fact, there are three different opioid peptide systems. Thus, enkephalins are formed from proenkephalin A, dynorphins and neoendorphins from proenkephalin B, and β-endorphin from proopiomelanocortin (POMC); POMC also contains the corticotropin sequence. The neuronal distribution of each system is unique (Cuello 1983).

Essentially four routes have been followed in elucidating the possible role opioid peptides may have in mood disorders:

1. Nonspecific blockade of opioid receptors by naloxone or naltrexone, studying the behavioral consequences.
2. Measuring various opioid activities in the cerebrospinal fluid (CSF) as well as plasma of patients.
3. Investigating the clinical effect of administrating endorphins to patients; β-endorphin has been a favorite tool.
4. A psychoendocrine approach has recently gained momentum, taking into consideration the fact that β-endorphin and corticotropin are both released from the same anterior pituitary cells by corticotropin-releasing factor (CRF) originating in the hypothalamus. Plasma β-endorphin has thus been studied as plasma corticotropin and cortisol and followed after dexamethasone challenge.

For the past 10 years, it has been natural to assume that the opioid peptides may have the potential to strengthen the feeling of well-being (Beluzzi and Stein 1977) and would be dysfunctional in states of subjective distress such as depression. In this chapter, we will review studies relevant to the above four points.

2 Naloxone Studies

In nonspecific attempts to investigate the role of opioid antagonism in mood regulation, a number of investigators have injected depressed patients systematically with naloxone. Table 1 lists six controlled studies and two pilot studies using naloxone and one recent controlled naltrexone study on depressed patients, manic patients, and healthy volunteers. Results are negative with the possible ex-

Table 1. Mood effects of administration of opiate antagonists in humans

Authors	Diagnostic criteria	Diagnosis	Number of trials	Age ± SEM or range	Type of study	Concomitant treatment	Dosage of active drug	Effects
Naloxone								
Terenius et al. (1977)		Depressive states	6	50 ± 5.8	Pilot	AD	0.4–0.8 mg t.i.d.s.c. for 6–12 days	Negative. Possibly worsening of two patients after end of injections
Grevert and Goldstein (1978)		Healthy volunteers	30	?	Double-blind	–	10 + 5 mg at 1 time	No effects on Profile of Mood States questionnaire
Davis et al. (1980)	RDC	Manic symptoms	10	41 ± 3.3	Double-blind	AD, NL, etc. allowed	20 mg s.c. at 1 time	Negative
Judd et al. (1980)	RDC	Mania Healthy volunteers	12 7	22–57 22–31	Double-blind	Li, NL allowed	20 mg i.v. at 2 times	Four of 12 patients showed short-term reduction of some symptoms. Non-specific subduing effect in both groups
Volavka et al. (1980)		Healthy volunteers	13	21–38	Double-blind	–	20 mg i.v. at 1 time	No subjective difference between naloxone and placebo
Schull and Kaplan (1981)		Healthy volunteers	14	18–55	Double-blind	–	10 mg/70 kg 2 times	Significant positive effects on ratings and tolerance time for ischemic pain, anxiety, and mood
Pickar et al. (1982a)	ICD-9	Mania	26	39 ± 2	Double-blind	Li, NL allowed in 12 cases	0.3 mg/kg s.c.	Negative
Papakostas et al. (1985)		Melancholia during ECT	5	38–63	Pilot	ECT	0.4, 1.2, and 4.0 mg 3 times	No effect on ECT-induced prolactin increase
Naltrexone								
Hatsukami et al. (1986)		Overweight male volunteers	56	37 ± 5 (naltrexone) 41 ± 8 (placebo)	Double-blind	–	300 mg/day p.o. for 8 weeks	No change in mood or cognitive function

AD, antidepressants; Li, lithium; NL, neuroleptics

ceptions of Judd et al. (1980; antimanic effects in 4/12 patients) and Schull and Kaplan (1981; some positive mood effects in healthy volunteers). It has not been possible to demonstrate clearly any simple link between opioid blockade and affective states.

3 CSF Studies

3.1 Radioreceptor Assays

Endorphins were initially defined as the material in human CSF with inherent affinity to opioid receptors in synaptic plasma membrane preparations and were measured by radioreceptor assay (RRA) (Terenius and Wahlström 1975; Naber et al. 1981 a, b). Using chromatographic fractionation, Terenius separated this opioid activity into fractions I and II. Comparisons of RRA and radioimmunoassay (RIA) methodologies have demonstrated RIA-determined β-endorphin *not* to form part of either one of these fractions, and the enkephalins to be confined to fraction II (Terenius et al. 1979). Fraction I contains proenkephalin-related peptides of intermediate sizes (10–20 amino acids). The USA National Institute of Mental Health (NIMH) group have measured total opioid activity, justified as a powerful strategy in studying the endogenous opioid system in clinical populations (Naber et al. 1981 b).

Table 2 lists the results from four CSF investigations using four different patient populations. The three Swedish studies show unambiguous increases in fraction I as well as fraction II when comparing depressed patients with healthy controls. The largest study comprises 98 patients with Research Diagnostic Criteria (RDC) major depressive disorder who were free from antidepressants and neuroleptics for at least 10 days before sampling (Ågren and Terenius 1985). Both opioid fractions were clearly significantly evelated compared to the 19 normals sampled earlier by Lindström et al. (1978) and analyzed in the same manner ($p < 0.0001$ for both). Furthermore, a difference in fraction I between unipolar and bipolar patients was detected, with significantly higher activity in unipolar depressives (Ågren et al. 1982). The NIMH investigators reported no difference in their single measure between 35 RDC depressives and 41 controls (Naber et al. 1981 b). These discordant results are most likely due to dissimilarities in the actual opioid content of the total biological receptor activity.

3.2 Radioimmunoassays

Four studies of β-endorphin-like immunoreactivity in CSF of depressed patients have failed to show differences from normal controls (see Table 3).

Loeb et al. (1984) reported significantly higher concentrations of met-enkephalin and leu-enkephalin in 12 patients with "endogenous depression" than in six controls ($p < 0.001$ and $p < 0.005$, respectively) (Table 4). This result accords well with the elevated opioid fraction II noted above, since enkephalins form part of this fraction.

Table 2. Radioreceptor assay studies of CSF endorphin concentrations

Authors	Diagnostic criteria	Diagnosis	Subdiagnosis (unipolars, bipolars)	Number Sex	Age ± SEM or range	Days off antidepressant medication	Analysis	Concentration ± SEM (pmol/ml)		Reported significantly different from normals
Terenius et al. (1977)		Mixed depressive states		5 F	50 ± 5.8	–	Fraction I	4.1 ± 0.7	[Met-enkephalin equivalents]	●
							Fraction II	9.4 ± 3.0		
		Manic depression								
Lindström et al. (1978)		Depressive phase	4 BP	3 M 1 F	30–60	–	Fraction I	2.8 ± 0.5	[Met-enkephalin equivalents]	●
							Fraction II	5.9 ± 2.1		●
		Manic phase					Fraction I	4.5 ± 1.5	[Met-enkephalin equivalents]	●
							Fraction II	6.6 ± 1.7		●
		Normal controls		19	–		Fraction I	1.0 ± 0.1	[Met-enkephalin equivalents]	
							Fraction II	2.5 ± 0.3		
Naber et al. (1981 a, b)	RDC	Major depressive disorder	?	16 M 25 F	41 ± 2.2	14	Total opioid activity	3.8 ± 0.39	[β-Endorphin equivalents]	
	–	Normal controls	–	27 M 14 F	31 ± 2.0	–	Total opioid activity	4.2 ± 0.33	[β-Endorphin equivalents]	
Ågren and Terenius (1985, revised[a])	RDC	Major depressive disorders	64 UP 34 BP	42 M 56 F	42 ± 1.3	10	Fraction I	1.88 ± 0.12	[Met-enkephalin equivalents]	●
							Fraction II	11.9 ± 1.43		●

[a] Arithmetic means and variation of both fractions recalculated from original data

Table 3. Radioimmunoassay studies of CSF β-endorphin reactivity

Authors	Diagnostic criteria	Diagnosis	Sub-diagnosis (unipolars, bipolars)	Number, Sex	Age ± SEM	Days off anti-depressant medication	CSF	Concentration ± SEM (fmol/ml)
Naber et al. (1981a, b)	RDC	Major depressive disorder		12 M 16 F		14	Lumbar	65 ± 6.0
	–	Normal controls	–	21 M 12 F	31 ± 2.3	–	Lumbar	59 ± 6.8
Gerner and Sharp (1982)	RDC	Primary major depression	19 UP 8 BP	12 M 15 F	18–66	7	Lumbar	29 ± 1.4
		Normal controls	–	4 M 5 F	18–36	–	Lumbar	30 ± 2.2
Black et al. (1986)	DSM-III	Major depression		7 M 11 F	37 ± 1.9	0	Ventricles	4.6 ± 0.4
		Hydrocephalus		9	?	?	Ventricles	6.1 ± 1.3
		Controls with back pain		6	?	?	Lumbar	6.2 ± 0.9
Facchinetti et al. (1986a)	DSM-III	Major depression	10 UP	2 M 8 F	35–57	?	Lumbar Lumbar	β-Endorphin 53 ± 7.3 β-Lipotropin 20 ± 3.6
		Normal controls	–	5 M 8 F	34–59	–	Lumbar Lumbar	β-Endorphin 67 ± 2.0 β-Lipotropin 28 ± 4.6

Table 4. Radioimmunoassay study of CSF enkephalin reactivity

Authors	Diagnostic criteria	Diagnosis	Sub-diagnosis (unipolars, bipolars)	Number, Sex	Age ± SEM or range	Days off anti-depressant medication	Analysis	Con-centration ± SEM (fmol/ml)	Reported significantly different from normals
Loeb et al. (1984)	?	Endogenous depression	12 UP	6 M 6 F	38 ± 3.5	15	Met-enkephalin Leu-enkephalin	62 ± 6.6 166 ± 36	● ●
		Normal controls	–	6	?	–	Met-enkephalin Leu-enkephalin	36 ± 6.4 59 ± 15	

4 Plasma Radioimmunoassay Studies

Seven studies listed in Table 5 have measured plasma β-endorphin alone or in conjunction with β-lipotropin using differently specific antibodies, comparing these measures in depressed patients with those in normals. Six indicated elevated activities (six increases in β-endorphin, four also in β-lipotropin); one study reported no difference. Matthews et al. (1982) found no difference between six patients with "endogenous" depression and five with "non-endogenous" depression.

Three groups have reported significant patient-normal group differences in composite measures. A series of studies by Brambilla and her collaborators (1981) reported plasma corticotropin, β-endorphin, and β-lipotropin immunoreactivities to be significantly intercorrelated in controls as well as in patients with primary depression, secondary depression, and schizophrenia. Further, Cohen et al. (1984) found a significantly lower ratio of plasma β-endorphin to cortisol in depressive patients, especially so in a group with minor and intermittent depression discussed in more detail below. The only study measuring plasma and CSF activities simultaneously (Facchinetti et al. 1986 a) reported no differences in CSF β-endorphin, β-lipotropin, or corticotropin, but found increased plasma immunoactivities of all three peptides. Intriguingly, the ratio of CSF to plasma activity was consistently and significantly lower for all.

5 Administration of Opioid Peptides to Humans

Following positive reports from pilot studies suggesting immediate antidepressant effects on injecting β-endorphin intravenously in five and six depressed patients respectively (Kline et al. 1977; Angst et al. 1979), one double-blind controlled study noted some positive mood response for a few hours but was unable to record any lasting effect, while a controlled study by the NIMH group was completely negative (see Table 6). A pilot study of the effect of infusing the met-enkephalin analogue FK 33-824 failed to record alteration in the mood of nine depressed patients (Jungkunz et al. 1983).

Des-tyrosine-γ-endorphin (DTγE), which is the (2-17) fragment of β-endorphin and a naturally occurring peptide, had been administered to schizophrenic patients with inconsistent results (Verhoeven et al. 1979; Tamminga et al. 1981). Interestingly, one placebo-controlled study by Chazot et al. (1985) on 19 RDC depressives in whom 1 mg DTγE was injected i.m. each evening for 7 days, reports clearly significant and rapid ameliorations of mood. No side effects could be discovered. This study clearly needs replication. The mechanism of action is unclear. This peptide lacks opioid activity and does not interfere with opiate action acutely; however, chronic treatment of rats increases the efficacy of morphine (Neil et al. 1981). Maybe, therefore, DTγE affects opioid receptor mechanisms.

Table 5. Radioimmunoassay studies of plasma β-endorphin and β-lipotropin reactivity

Author(s)	Diagnostic criteria	Diagnosis	Sub-diagnosis (unipolars, bipolars)	Number, Sex	Age ± SEM or range	Days off anti-depressant medication	Peptide	Concentration ± SEM (fmol/ml)	Reported significantly different from normals
Brambilla et al. (1981)	?	Primary affective disorder	9 UP	1 M 8 F	33–61	10	β-Endorphin β-Lipotropin	21.7 ± 5.1 19.2 ± 3.1	● ●
	?	Secondary affective disorder	–	1 M 6 F	23–50	10	β-Endorphin β-Lipotropin	52.5 ± 20.6 28.5 ± 3.5	● ●
	–	Normal controls matched		35		–	β-Endorphin β-Lipotropin	≈ 9 11.4 ± 0.8	
Risch et al. (1983)	RDC	Major depressive disorder	6 UP 6 BP	12	?	14	β-Endorphin	35.1 ± 2.7	●
		Normal controls matched for age and sex		12	?	–	β-Endorphin	18.6 ± 1.9	
Alexopoulos et al. (1983)	RDC	Primary major depressive disorder	10 UP	8	40–87	10	β-Endorphin	6.9 ± 1.9	
		Normal controls matched for age and sex		16		–	β-Endorphin	8.6 ± 1.9	
Cohen et al. (1984)	RDC	Major depressive disorder		5 M 12 F	?	4	β-Endorphin Cortisol Ratio: β-endorphin/cortisol	32.3 ± 1.4 21.4 ± 2.3 µg/100 ml 1.81 ± 0.23	 ● ●
	RDC	Minor and intermittent depressive disorder		1 M 10 F	?	4	β-Endorphin Cortisol Ratio:	23.7 ± 1.2 17.1 ± 1.2 µg/100 ml 1.49 ± 0.15	● ● ●

		Normal controls		8	?		β-Endorphin	31.8±1.3	
							Cortisol	12.5±0.6 μg/100 ml	
							Ratio: β-endorphin/cortisol	2.60±0.17	
Brambilla et al. (1986)	DSM-III	Primary affective disorders		7 M 42 F	27–74	10	β-Endorphin	18.4±1.9	●
							β-Lipotropin	16.4±1.2	●
	DSM-III	Secondary affective disorders		2 M 10 F	23–50	10	β-Endorphin	26.1±5.1	●
							β-Lipotropin	22.0±2.9	●
		Normal controls matched for sex and age		35		–	β-Endorphin	6.2±0.3	
							β-Lipotropin	6.1±0.3	
Genazzani et al. (1986)	DSM-III	Dysthymic disorder		9 M	27–60	10	β-Endorphin	22.1±0.8	●
							β-Lipotropin	13.7±0.3	●
		Normal controls		8	25–53	–	β-Endorphin	6.6±0.3	
							β-Lipotropin	6.2±0.2	
Facchinetti et al. (1986a)	DSM-III	Major depression	10 UP	2 M 8 F	35–57	?	β-Endorphin	18 ±2.3	●
							Ratio: CSF/plasma β-Endorphin	3.0±0.4	●
							β-Lipotropin	19 ±2.6	●
							Ratio: CSF/plasma β-Lipotropin	1.1±0.2	●
		Normal controls	–	5 M 8 F	34–59	–	β-Endorphin	10 ±0.4	
							Ratio: CSF/Plasma β-Endorphin	7.8±0.2	
							β-Lipotropin	8.6±0.4	
							Ratio: CSF/plasma β-Lipotropin	3.5±0.1	

Table 6. Mood effects of systemic injections of opioid peptides in humans

Authors	Diagnostic criteria	Diagnosis	Number of trials	Age ± SEM or range	Type of study	Concomitant medication	Dosage of active drug	Effects
β-Endorphin								
Kline et al. (1977)		Depressive states	2+3	59–63	Pilot	AD continued	Up to 7 mg i.v. at 2 times	Short-term impression positive
Angst et al. (1979)		Depressive states	6	34–56	Pilot	No AD for last 3 days	10 mg i. v. at 1 time	Increase of energy and mood for 2 h post injection, rapid relapse
Gerner et al. (1980)	RDC	Major depressive disorder	10	42±4.4	Double-blind	No drugs for last 8 to >60 days	10 mg i. v. at 2 times	Significant positive response 2–4 h post injection; no lasting effect
Pickar et al. (1981)	RDC	Major depressive disorder	4	47±6.6	Double-blind	No drugs for last 21 days	0.15 mg/kg, 4–15 mg, i. v., at 1 time	No significant changes in behavioral ratings
Met-enkephalin analogue FK 33–824								
Jungkunz et al. (1983)	ICD-8	Endogenous depression	9	55±4.1	Pilot	No AD for last 7 days	1.0 mg i. v. at two times	No significant changes
Des-tyrosine-γ-endorphin								
Chazot et al. (1985)	RDC	Major depressive disorder	19	43.3±3.0	Double-blind	No AD for last 10 days	1 mg i.m. in evening for 7 days	Rapid and significant antidepressant effect after 3 days

AD, antidepressants

6 Effects of Antidepressant Therapies

In animals, there is evidence that electroconvulsive stimulation activates brain opioid systems (Holaday et al. 1981). Clinical studies on electroconvulsive therapy (ECT) have only been directed at plasma measurements. Increase in plasma β-endorphin immunoreactivity in three patients undergoing ECT was reported by Emrich et al. (1979). Also in depressed patients, Alexopoulos et al. (1983) found each ECT to induce a transient threefold increase in RIA-determined plasma β-endorphin paralleling a similarly transient elevation of corticotropin, interpreted as a hypothalamic response to ECT. Similarly, Misiaszek et al. (1984) confirmed a major increase in plasma β-endorphin after ECT in seven depressed patients. These findings are not directly reconcilable with the evidence summarized above that plasma β-endorphin levels are elevated in depressed patients when compared to controls. The ECT elevations were acute and transient, not paralleling the ECT-induced mood elevation.

Carbamazepine exerted no effects on CSF total opioid activity (Post et al. 1981).

A study in rats provides evidence that acute activation of serotonin neurotransmission by the serotonin precursor 5-hydroxytryptophan, the serotonin reuptake blocker fluvoxamine, or the serotonin receptor agonist quipazine induces a marked decrease of β-endorphin immunoreactivity in the anterior pituitary and a concomitant striking rise in plasma β-endorphin (Petraglia et al. 1982). On the other hand, mianserin treatment of depressed patients reduces their plasma β-endorphin (Drago et al. 1982).

Reisine and Soubrié (1982) have reported on downregulated cortical opioid receptors (reduced ^{3}H-naloxone binding) resulting from chronic intraperitoneal administration of the tricyclic antidepressant desipramine to rats. The decrease in number of cortical opioid receptors reached a maximum after 3 weeks. Stengaard and Schou (1985) attempted to replicate this provocative finding and demonstrated a decreased binding of ^{3}H-met-enkephalin amide after desipramine treatment. The decreased binding was located in hippocampus and basal ganglia, but was *not* found in cerebral cortex. However, the sensitivity of individual cells, as judged from the responsiveness to iontophoretically applied morphine, does not seem to change (Scuvée-Moreau and Svensson 1982).

7 Diurnal and Other Rhythms

Diurnal variations in β-endorphin immunoreactivity in both CSF and plasma were observed in seven hydrocephalic patients being sampled every 1–2 h (Barreca et al. 1986). Both measures peaked at about 6–7 a.m. Similar diurnal variations in plasma β-endorphin were seen in normal controls, but not in patients with schizophrenia (Gil-Ad et al. 1986).

Following the female menstrual cycle, plasma β-endorphin did not vary significantly as for mean immunoreactivity, but the variance in the measure was largest during the ovulatory phase (Veith et al. 1984).

A seasonal variation in fraction I endorphins was described by Ågren and Terenius (1983), demonstrating a significant peak in late August for 62 unipolar depressed patients and a peak in late October in 30 bipolar patients. Seasonal studies on depression and on suicide have usually found peaks in spring or early summer, i.e., following a pattern not easily reconcilable with the fraction I rhythmicity.

8 Psychoendocrine Studies

One of the early findings of RRA CSF studies was the significant connections between CSF opioid activity and hyperactivity of the hypothalamus-pituitary-adrenocortical (HPA) axis. Two independent studies found positive correlations between CSF activity and urinary free cortisol excretion [opioid fraction I; $r=0.41$, $n=54$, $p=0.002$ (Ågren et al. 1982); total opioid activity: $r=0.47$, $n=21$, $p<0.05$ (Rubinow et al. 1981; Pickar et al. 1982b)]. The former study also noted significant and positive correlations between fraction I and postdexamethasone plasma cortisol levels. Additionally, the NIMH group found a positive trend correlation between RIA β-endorphin in CSF and urinary free cortisol. No correlations were seen in normal controls between either the RRA or RIA material in CSF and urinary free cortisol. Since β-endorphin is *not* part of fraction I, the combined information would suggest a coactivation of the HPA axis and the formation of neuronal opioid peptides in depression.

Since β-endorphin is formed in the corticotroph cells in the anterior pituitary in humans, the favored target for inquiry into opioid endocrinology has been the pituitary.

Morphine stimulates prolactin release in the anterior pituitary. Two studies on depressed patients found a blunted prolactin response to i.v. administration of morphine (Extein et al. 1980) and methadone (Judd et al. 1982).

Zis et al. (1985) showed that morphine given i.v. to unmedicated psychiatric patients as well as to healthy controls suppresses the secretion of cortisol. Depressed patients significantly more often escaped from this morphine suppression of cortisol than controls. The finding points to an inhibitory opioid control on the HPA axis, most likely in structures at or above the anterior pituitary.

Volavka et al. (1979) reported that i.v. naloxone given to healthy men increased both plasma cortisol and corticotropin, indicating corticotropin to be under inhibitory control of opioid peptides. In a repeat study on the same volunteers, naloxone elicited significant increases in corticotropin, cortisol, and luteotropin (Volavka et al. 1980).

A cholinergic mechanism in depressive β-endorphin hypersecretion was proposed by Risch and colleagues (Risch 1982; Risch et al. 1983), who demonstrated the physostigmin-stimulated release of β-endorphin immunoreactivity to be significantly greater in depressed patients than in controls. Baseline β-endorphin was elevated as well. The results were interpreted as indicating a muscarinic receptor supersensitivity in depression.

Cohen et al. (1984) found plasma cortisol to be elevated in 28 RDC depressives compared to controls, but, contrary to most other investigators, detected no difference in plasma β-endorphin immunoreactivity between controls and patients with RDC major depressive disorder. However, β-endorphin in patients with nonmajor depression (i.e., dysthymic and cyclothymic disorders) was significantly *lower* than in both controls and major depressives. The ratio of plasma β-endorphin to cortisol was significantly lower in both groups of depressed patients than in controls. The "adjustment" of β-endorphin to a value of cortisol set to unity thus facilitates the interpretation of results in showing a deranged balance in the formation of the two peptides in the anterior pituitary.

Facchinetti et al. (1986a) measured corticotropin, β-endorphin, and β-lipotropin in both CSF and plasma in 10 patients with major depression and 13 age-matched healthy controls. The plasma measures were all significantly elevated in the depressed group, but the CSF ones were not. No correlations were found between CSF and plasma data, adding further support to the notion of independent regulation of central and peripheral POMC-related peptides. The authors also argue that the noncorrelation reduces the possibility that peptides of pituitary origin, directly from the gland or through the peripheral circulation, could penetrate the CSF. This is clearly supported by the recent finding that the total brain kinetics of two metabolically stable ^{11}C-labeled met-enkephalin analogues in the rhesus monkey show very limited penetrance into the brain, with only 0.05 times the homogeneous dilution of the radioactive dose (Hartvig et al. 1986). Another study by Facchinetti et al. (1986b) indicates that obese patients hypersecrete β-endorphin and that this is only partially dependent on hypothalamic control.

Plasma β-endorphin analyses were included in a dexamethasone suppression test paradigm by Matthews et al. (1986). Plasma β-endorphin showed a diurnal rhythm similar to that of corticotropin (see above) and was suppressable with dexamethasone. While 33%–69% of melancholic patients were abnormal in post-dexamethasone cortisol levels, the corresponding percentage for β-endorphin was 50%–69%. Each given patient could show an abnormality in suppressability of cortisol, β-endorphin, or both. A larger number of melancholic patients were "detected" when both measures were taken into account than with only one predictor. Plasma corticotropin levels were not measured.

A divergent view has been submitted by Lin et al. (1986) in a study on 42 patients with RDC major depressive disorders, 76% of whom escaped plasma cortisol suppression by dexamethasone. The suppressors did not differ from the nonsuppressors in measures of plasma corticotropin or β-endorphin immunoreactivity. The authors hold that the disturbance involving the HPA axis is due to differences in adrenal secretion of cortisol but not in central nervous system pituitary activity. They emphasize that their methodology for β-endorphin measurement is superior to others.

9 Opioid Peptides in Mood Disorders: Markers or Mediators?

Reviewing the literature regarding opioid peptides and mood, it is common to find a lack of clear distinction between the peptides considered as diagnostic or prognostic markers, as opposed to mediators, directly involved in the regulation of mood. The secretion of β-endorphin into plasma (from the pituitary) has at times been taken as an index of CNS activity: this is clearly not valid. For instance, after pituitary ablation (Schlachter et al. 1983), plasma levels of β-endorphin become undetectable whereas CSF levels remain detectable.

The data reviewed in Table 5 clearly indicate that plasma β-endorphin is a useful neuroendocrine marker in affective disorders. β-Endorphin is here one product of POMC. Its target organs and potential significance as a hormone are unknown. It would be just as logical to measure corticotropin or, as commonly done, cortisol. Provided that the proportion between β-endorphin and corticotropin from the corticotroph would not be influenced by the depressive state (Cohen et al. 1984), it becomes a technical matter, rather than a principal choice between markers of the activity in the HPA axis.

Levels of peptides in CSF are likely to reflect activity in the CNS pathways. As already indicated, the blood-brain barrier restricts passage between plasma and brain. Besides, β-endorphin levels are higher in CSF than in plasma (e.g., Facchinetti et al. 1986a), emphasizing the likelihood that CSF β-endorphin originates in the CNS.

CSF levels of neuropeptides should reflect CNS activity more accurately than plasma levels. There is clearly little evidence for tonic opioid activity in pathways affecting mood, since naloxone hardly affects mood in healthy volunteers. However, the opioid system may interact with other signal systems, primarily monoaminergic.

Clinically, the most evident example of interaction is the efficacy of α_2-adrenoceptor stimulatory agents (such as clonidine) in inhibiting opiate abstinence (reviewed by Ågren 1986). Charney and Heninger (1986) have recently demonstrated that naloxone and the α_2-adrenoceptor antagonist yohimbine act synergistically in causing volunteers to experience more nervousness and anxiety symptoms as well as causing plasma cortisol levels to increase. These observations can be understood in terms of an opioidergic input acting on noradrenergic neurons.

Measurements of β-endorphin in CSF have consistently failed to show differences between patients and controls (Table 3). Peptides from proenkephalin A appear elevated on RIA (Table 4) and indirectly on RRA (Table 2), the major component of so-called fraction II endorphins being derived from proenkephalin A (Nyberg et al. 1983). Studies are completely lacking regarding the third opioid peptide system, proenkephalin B (involving dynorphins and neoendorphins).

Proenkephalin A and proenkephalin B have a very broad distribution throughout the neuraxis. For instance, they are clearly associated with sensory systems, notably pain pathways (Cuello 1983). It seems probable that the altered sensory perception frequently observed in depression could be secondary to increased opioid activity. In fact, increased fraction II levels have been found to relate to decreased pain sensitivity and selfmutilation in autistic children (Gillberg et al. 1985).

Although we have emphasized the need to distinguish markers of opioid neural systems and the opioid endocrine (pituitary) system, it is nevertheless clear that those systems are interconnected. Opioid neurons are in fact able to modulate the release of releasing hormones, including CRF, and thereby indirectly regulate secretion not only of corticotropin (and β-endorphin) but also of luteotropin, folliculotropin, and possibly vasopressin. Here, the opioid neurons appear tonically active as judged by naloxone effects (Howlett and Rees 1986). A feedback loop of potential significance is provided by cortisol, which exerts a negative feedback influence on the POMC system but a positive feedback on proenkephalin A (Civelli et al. 1983). Cortisol undoubtedly easily equilibrates over the blood-brain barrier. Thus the uncontrolled secretion of cortisol in depression may result in sustained activation of proenkephalin A biosynthesis. Indeed, positive correlations were observed between RRA activity in CSF and urinary free cortisol (Rubinow et al. 1981; Ågren et al. 1982).

References

Abse DW, Rheuban WJ, Akhtar S (1982) The poppy: therapeutic potential in cases of dementia with depression. Ann NY Acad Sci 398:79–82

Ågren H (1986) Clonidine treatment in the opiate withdrawal syndrome. A review of clinical trials of a theory. Acta Psychiat Scand 73[Suppl 327]:91–113

Ågren H, Terenius L (1983) Depression and CSF endorphins fraction I: seasonal variation and higher levels in unipolar than in bipolar patients. Psychiatry Res 10:303–311

Ågren H, Terenius L (1985) Hallucinations in patients with major depression. Interactions between CSF monoaminergic and endorphinergic indices. J Affect Disord 9:25–34

Ågren H, Terenius L, Wahlström A (1982) Depressive phenomenology and levels of CSF endorphins. Ann NY Acad Sci 398:388–398

Alexopoulos GS, Inturrisi CE, Lipman R, Frances R, Haycox J, Dougherty JH, Rossier J (1983) Plasma immunoreactive β-endorphin levels in depression: effects of electroconvulsive therapy. Arch Gen Psychiatry 40:181–183

Angst J, Autenrieth V, Brem F, Koukkou M, Meyer H, Stassen HH, Storck U (1979) Preliminary results of treatment with β-endorphin in depression. In: Usdin E, Bunney WE Jr, Kline NS (eds) Endorphins in mental health research. Oxford University Press, London, pp 518–528

Barreca T, Siani C, Franceschini R, Francaviglia N, Messina V, Perria C, Rolandi E (1986) Diurnal beta-endorphin changes in human cerebrospinal fluid. Life Sci 38:2263–2267

Beluzzi JD, Stein L (1977) Enkephalin may mediate euphoria and drive reward. Nature 266:566–568

Black PMcL, Ballentine HT Jr, Carr DB, Beal MF, Martin JB (1986) Beta-endorphin and somatostatin concentrations in the ventricular cerebrospinal fluid of patients with affective disorders. Biol Psychiatry 21:1075–1077

Brambilla F, Genazzani A, Facchinetti F, Parrini D, Petraglia F, Sacchetti E, Scarone S, Guastalla A, D'Antona N (1981) β-Endorphin and β-lipotropin plasma levels in chronic schizophrenia, primary affective disorders and secondary affective disorders. Psychoneuroendocrinology 6:321–330

Brambilla F, Petraglia F, Facchinetti F, Maggioni M, Genazzani AR (1986) Pituitary proopiomelanocortin peptides in mental disorders. In: van Ree JM, Matthysse S (eds) Progress in brain research, vol 65. Elsevier, Amsterdam, pp 193–206

Carlson ET, Simpson MM (1963) Opium as a tranquilizer. Am J Psychiatry 120:112–117

Charney DS, Heninger GR (1986) α_2-Adrenergic and opiate receptor blockade. Arch Gen Psychiatry 43:1037–1041

Chazot G, Claustrat B, Brun J, Olivier M (1985) Rapid antidepressant activity of destyr gamma endorphin: correlation with urinary melatonin. Biol Psychiatry 20:1026–1030

Civelli O, Birnberg N, Comb M, Douglass J, Lissitzky JC, Uhler M, Herbert E (1983) Regulation of opioid gene expression. Peptides 4:651–656

Cohen MR, Pickar D, Extein I, Gold MS, Sweeney DR (1984) Plasma cortisol and β-endorphin immunoreactivity in nonmajor and major depression. Am J Psychiatry 141:628–632

Cuello AC (1983) Central distribution of opioid peptides. Br Med Bull 39:11–16
Davis GC, Extein I, Reus VI, Hamilton W, Post RM, Goodwin FK, Bunney WE Jr (1980) Failure of naloxone to reduce manic symptoms. Am J Psychiatry 137:1583–1585
Drago F, Wiegant VM, Sapienza C, Aguglia E, Rapisarda V, Scapagnini U (1982) Mianserin reduces plasma levels of β-endorphin immunoreactivity in depressed patients. Experientia 38:685
Emrich HM, Höllt V, Kissling W, Fischler M, Laspe H, Heinemann H, von Zerssen D, Herz A (1979) Beta-endorphin-like immunoreactivity in cerebrospinal fluid and plasma of patients with schizophrenia and other neuropsychiatric disorders. Pharmakopsychiatria 12:269–276
Emrich HM, Vogt M, Herz A, Kissling W (1982) Antidepressant effects of buprenorphine. Lancet II:709
Extein I, Pottash ALC, Gold MS, Sweeney DR, Martin DM, Goodwin FK (1980) Deficient prolactin response to morphine in depressed patients. Am J Psychiatry 137:845–846
Extein I, Pickar D, Gold MS, Gold PW, Pottash ALC, Sweeney DR, Ross RJ, Rebard R, Martin D, Goodwin FK (1981) Methadone and morphine in depression. Psychopharmacol Bull 17:29–32
Facchinetti F, Petraglia F, Sances G, Garuti C, Tosca P, Nappi G, Genazzani AR (1986a) Dissociation between CSF and plasma B-endorphin in major depressive disorders: evidence for a different regulation. J Endocrinol Invest 9:11–14
Facchinetti F, Giovannini C, Barletta C, Petraglia F, Buzzetti R, Burla F, Lazzari R, Genazzani AR, Scavo D (1986b) Hyperendorphinemia in obesity and relationships to affective state. Physiol Behav 36:937–940
Fink M, von Simeon J, Itil TM, Freedman AM (1970) Clinical antidepressant activity of cyclazocine, a narcotic antagonist. Clin Pharmacol Ther 11:41–48
Genazzani AR, Petraglia F, Sinforiani E, Brambilla F, Facchinetti F, Nappi G (1986) Dysregulation of plasma pro-opiomelanocortin-related peptides in neurotic depression. Acta Endocrinol 112:1–6
Gerner RH, Sharp B (1982) CSF β-endorphin-immunoreactivity in normal schizophrenic, depressed, manic and anorectic subjects. Brain Res 237:244–247
Gerner RH, Catlin DH, Gorelick DA, Hui KK, Li CH (1980) β-Endorphin. Intravenous infusion causes behavioral change in psychiatric inpatients. Arch Gen Psychiatry 37:642–647
Gil-Ad I, Dickerman, Amdursky S, Laron Z (1986) Diurnal rhythm of plasma beta endorphin, cortisol and growth hormone in schizophrenics as compared to control subjects. Psychopharmacology 88:496–499
Gillberg C, Terenius L, Lönnerholm G (1985) Endorphin activity in childhood psychosis. Spinal fluid levels in 24 cases. Arch Gen Psychiatry 42:780–783
Gold MS, Pottash ALC, Sweeney DR, Kleber HD, Redmond E (1979) Rapid opiate detoxification: clinical evidence of antidepressant and antipanic effects of opiates. Am J Psychiatry 136:982–983
Gold MS, Pottash AC, Sweeney D, Martin D, Extein I (1982) Antimanic, antidepressant, and antipanic effects of opiates: clinical, neuroanatomical, and biochemical evidence. Ann NY Acad Sci 398:140–150
Grevert P, Goldstein A (1978) Endorphins: naloxone fails to alter experimental pain or mood in humans. Science 199:1093–1095
Hartvig P, Någren K, Lundberg PO, Muhr C, Terenius L, Lundqvist H, Lärkfors L, Långström B (1986) Kinetics of four ^{11}C-labelled enkephalin peptides in the brain, pituitary and plasma of Rhesus monkeys. Regul Pept 16:1–13
Hatsukami DK, Mitchell JE, Morley JE, Morgan SF, Levine AS (1986) Effect of naltrexone on mood and cognitive functioning among overweight men. Biol Psychiatry 21:293–300
Hirsch MD, O'Donohue TL (1986) Structural modifications of pro-opiomelanocortin-derived peptides alter their behavioral effects markedly. J Pharmacol Exp Ther 237:378–385
Holaday JW, Tortella FC, Belenky GL (1981) Electroconvulsive shock results in a functional activation of endorphin systems. Mod Probl Pharmacopsychiatry 17:142–157

Howlett TA, Rees LH (1986) Endogenous opioid peptides and hypothalamopituitary function. Annu Rev Physiol 48:527–536
Judd LL, Janowsky DS, Segal DS, Huey LY (1980) Naloxone-induced behavioral and physiological effects in normal and manic subjects. Arch Gen Psychiatry 37:583–586
Judd LL, Risch SC, Parker GC, Janowsky DS, Segal DS, Huey LY (1982) Blunted prolactin response – a neuroendocrine abnormality manifested by depressed patients. Arch Gen Psychiatry 39:1413–1416
Jungkunz G, Nedopil N, Rüther E (1983) Acute effects of the synthetic analogue of methionine enkephalin FK 33-824 on depressive symptoms. Pharmacopsychiatria 16:90–92
Kline NS, Li CH, Lehmann HE, Lajtha A, Laski E, Cooper T (1977) β-Endorphin-induced changes in schizophrenic and depressed patients. Arch Gen Psychiatry 34:1111–1113
Lin S-C, Maruta T, Newman DC, Kao PC (1986) Plasma levels of cortisol, corticotropin, and β-endorphin in patients with major depression. J Clin Psychiatry 47:413–414
Lindström LH, Widerlöv E, Gunne LM, Wahlström A, Terenius L (1978) Endorphins in human cerebrospinal fluid: clinical correlations to some psychotic states. Acta Psychiat Scand 57:153–164
Loeb C, Albano C, Serrato C (1984) Studio dei livelli liquorali della leucinencefalina e della metioninencefalina in pazienti con alterazioni comportamentali. Arch Suisses Neurol Neurochir Psychiatrie 134:29–40
Matthews J, Akil H, Greden J, Watson S (1982) Plasma measures of B-endorphin-like immunoreactivity in depressive and other psychiatric subjects. Life Sci 31:1867–1870
Matthews J, Akil H, Greden J, Charney D, Weinberg V, Rosenbaum A, Watson SJ (1986) β-Endorphin/β-lipotropin immunoreactivity in endogenous depression. Effect of dexamethasone. Arch Gen Psychiatry 43:374–381
Misiaszek J, Cork RC, Hameroff SR, Finley J, Weiss JL (1984) The effect of electroconvulsive therapy on plasma β-endorphin. Biol Psychiatry 19:451–455
Naber D, Pickar D, Post RM, van Kammen DP, Ballenger J, Rubinow D, Waters RN, Bunney WE Jr (1981 a) CSF opioid activity in psychiatric patients. In: Perris C, Struwe G, Jansson B (eds) Biological psychiatry. Elsevier, Amsterdam, pp 372–375
Naber D, Pickar D, Post RM, van Kammen DP, Waters RN, Ballenger JC, Goodwin FK, Bunney WE Jr (1981 b) Endogenous opioid activity and β-endorphin immunoreactivity in CSF of psychiatric patients and normal volunteers. Am J Psychiatry 138:1457–1462
Neil A, Lahti RA, Lindeberg G, Terenius L (1981) Des-tyrosine^1-γ-endorphin increases the antinociceptive efficacy of morphine in the rat. Eur J Pharmacol 76:93–95
Nyberg F, Wahlström A, Sjölund B, Terenius L (1983) Characterization of electrophoretically separable endorphins in human CSF. Brain Res 259:267–274
Papakostas YG, Stefanis CC, Markianos M, Papadimitriou GN (1985) Naloxone fails to block ECT-induced prolactin increase. Biol Psychiatry 20:1326–1327
Petraglia F, Peñalva A, Genazzani AR, Müller EE (1982) Stimulation of β-endorphin and β-lipotropin release from the anterior but not the neurointermediate pituitary lobe in the rat after acute administration of serotonin-acting drugs. Life Sci 31:2809–2817
Pickar D, Naber D, Post RM, van Kammen DP, Ballenger J, Kalin N, Bunney WE Jr (1981) Measurement of endorphins in CSF. Relationship to psychiatric diagnosis. Mod Probl Pharmacopsychiatry 17:246–272
Pickar D, Vartanian F, Bunney WE Jr, Maier HP, Gastpar MT, Prakash R, Sethi BB, Lideman R, Belyaev BS, Tsutsulkovskaja MYa, Jungkunz G, Nedopil N, Verhoeven W, van Praag H (1982 a) Short-term naloxone administration in schizophrenic and manic patients – a World Health Organization collaborative study. Arch Gen Psychiatry 39:313–319
Pickar D, Cohen MR, Naber D, Cohen RM (1982 b) Clinical studies of the endogenous opioid system. Biol Psychiatry 17:1243–1276
Post RM, Pickar D, Naber D, Ballenger JC, Uhde TW, Bunney WE Jr (1981) Effect of carbamazepine on CSF opioid activity: relationship to antidepressant response. Psychiatry Res 5:59–66
Reisine T, Soubrié P (1982) Loss of rat cerebral cortical opiate receptors following chronic desipramine treatment. Eur J Pharmacol 77:39–44

Risch SC (1982) β-Endorphin hypersecretion in depression: possible cholinergic mechanisms. Biol Psychiatry 17:1071–1079

Risch SC, Janowsky DS, Gillin JC (1983) Muscarinic supersensitivity of anterior pituitary ACTH and B-endorphin release in major depressive illness. Peptides 4:789–792

Rubinow DR, Post RM, Pickar D, Naber D, Ballenger JC, Gold PW, Bunney WE Jr (1981) Relationship between urinary free cortisol and CSF opioid binding activity in depressed patients and normal volunteers. Psychiatry Res 5:87–93

Schlachter LB, Wardlaw SL, Tindall GT, Frantz AG (1983) Persistence of β-endorphin in human cerebrospinal fluid after hypophysectomy. J Clin Endocrinol Metab 57:221–224

Schull J, Kaplan H (1981) Naloxone can alter experimental pain and mood in humans. Physiol Psychol 9:245–250

Scuvée-Moreau JJ, Svensson TH (1982) Sensitivity in vivo of central α_2- and opiate receptors after chronic treatment with various antidepressants. J Neural Transm 54:51–63

Stengaard-Pedersen K, Schou M (1985) Opioid peptides and receptors in relation to affective illness. Effects of desipramine and lithium on opioid receptors in rat brain. Acta Pharmacol Toxicol 56[Suppl 1]:170–179

Tamminga CA, Tighe PJ, Chase TN, Defraites EG, Schaffer MH (1981) Des-tyrosine-γ-endorphine administration in chronic schizophrenics. Arch Gen Psychiatry 38:167–168

Terenius L, Wahlström A (1975) Morphine-like ligand for opiate receptors in human CSF. Life Sci 16:1759–1764

Terenius L, Wahlström A, Ågren H (1977) Naloxone treatment in depression: clinical observations and effects on CSF endorphins and monoamine metabolites. Psychopharmacology 54:31–33

Terenius L, Wahlström A, Johansson C (1979) Endorphins in human cerebrospinal fluid and their measurement. In: Usdin E, Bunney WE Jr, Kline NS (eds) Endorphins and mental health research. Oxford University Press, New York

Veith JL, Anderson J, Slade SA, Thompson P, Laugel GR, Getzlaf S (1984) Plasma β-endorphin, pain thresholds and anxiety levels across the human menstrual cycle. Physiol Behav 32:1984

Verhoeven WMA, van Praag HM, van Ree JM, de Wied D (1979) Improvements in schizophrenic patients treated with des-tyr-gamma-endorphin (DTγE). Arch Gen Psychiatry 36:294–298

Volavka J, Cho D, Mallya A, Bauman J (1979) Naloxone increases ACTH and cortisol levels in man. N Engl J Med 18:1056–1057

Volavka J, Bauman J, Pevnick J, Reker D, James B, Cho D (1980) Short-term hormonal effects of naloxone in man. Psychoneuroendocrinology 5:225–234

Wu Jin-Xie (ed) (1926) Zhongguo Yixue Dacidian [Unabridged Encyclopedia of Chinese Medicine]. Commercial Press, Shanghai, p 4073

Young EA, Watson SJ, Akil H (1986) Pituitary regulation in endogenous depression. In: van Ree JM, Matthysse S (eds) Progress in brain research, vol 65. Elsevier, Amsterdam, pp 153–165

Zis AP, Haskett RF, Albala AA, Carroll BJ, Lohr NE (1985) Opioid regulation of hypothalamic-pituitary-adrenal function in depression. Arch Gen Psychiatry 42:383–386

The Neuroendocrinology of Anorexia Nervosa

M. J. Devlin [1] and B. T. Walsh [2]

Contents

1 General Considerations

The syndrome of anorexia nervosa, with its striking juxtaposition of psychological and physiological features, has, since the earliest clinical descriptions (Morton 1689; Gull 1874; Lasegue 1873) eluded a comprehensive and integrated understanding of etiology and pathophysiology. Clinical thinking about anorexia nervosa has largely proceeded along two somewhat divergent lines: the first, originating with Simmonds' report in the early years of this century of a wasting syndrome associated with pituitary lesions in a postpartum woman (Simmonds 1914), has directed itself toward the uncovering of a primary central nervous system abnormality which gives rise to anorectic behavior; the second, advanced by such clinicians as Bruch (1973) has concerned itself with the detailed description of patients and their typical family backgrounds, intrapsychic conflicts, and cognitive distortions. As work has continued, it has become increasingly clear that consideration of both psychological and physiological aspects of the illness is essential, and that there is likely to be no simple unifactorial etiology. To complicate matters further, a marked increase in the incidence of anorexia nervosa over the past decade (Herzog and Copeland 1985), along with the progression of the

1 Instructor in Clinical Psychiatry, College of Physicians & Surgeons, Columbia University, 722 West 168th Street, New York, NY 10032, USA, and Research Fellow, New York State Psychiatric Institute
2 Associate Professor of Clinical Psychiatry, College of Physicians & Surgeons, Columbia University and Research Psychiatrist, New York State Psychiatric Institute

Current Topics in Neuroendocrinology, Vol. 8

related syndrome of bulimia from psychiatric "zebra" to everyday occurrence on college campuses (Zuckerman et al. 1986), has not only underscored the importance of epidemiologic and sociocultural analysis in attempting to understanding eating disorders, but has added a rapidly increasing clinical incentive to the already compelling theoretical interest which anorexia nervosa has always inspired.

The neuroendocrine approach to anorexia nervosa has continued to generate tremendous interest, particularly in light of indications that the abnormal neuroendocrine profile of patients with this syndrome may be more than simply a concomitant of weight loss. Several recent reviews (Hudson and Hudson 1984; Weiner 1985; Herzog and Copeland 1985; Pirke and Ploog 1986) have performed the difficult task of compiling and synthesizing the available neuroendocrine data. Adding to the complexity of the situation is the consideration that an observed hormonal abnormality may lend itself to a number of different interpretations: for example, it may represent a preexisting abnormal state which favors the onset or maintenance of the syndrome, or it may be the result of undernutrition or weight reduction. Also to be kept in mind is the possibility that the syndrome of anorexia nervosa may in fact comprise a number of distinct subgroups with differing etiologic contributions or with varying expressions of underlying abnormalities.

One approach to analyzing the multiplicity of data and clinical impressions concerning patients with anorexia nervosa has been to attempt to draw analogies to other incompletely understood but possibly related states, such as forced semistarvation, bulimia, and depression, in hopes of contributing to the theoretical understanding of anorexia nervosa. While the temptation is to conceptualize anorexia nervosa as simply a variant of some more general disorder, this approach has so far yielded interesting insights and speculations, but no obvious category within which the condition fits comfortably.

In this review, we will first enumerate the neuroendocrine findings which have so far been reported along the hypothalamic/pituitary/adrenal, hypothalamic/pituitary/thyroid, and hypothalamic/pituitary/gonadal axes and review briefly other reported neuroendocrine abnormalities. We will then consider from a neuroendocrine perspective the various syndromes with which anorexia nervosa has been compared: (1) malnourished states in the absence of an eating disorder – forced semistarvation; (2) eating disorders in the absence of obvious malnutrition – normal weight bulimia; and (3) major depression, which overlaps with anorexia nervosa in some of its clinical features.

2 Neuroendocrine Findings in Anorexia Nervosa

2.1 Hypothalamic/Pituitary/Adrenal Axis

The normal hypothalamic/pituitary/adrenal (HPA) axis is illustrated in Fig. 1. Cortisol is normally secreted in discrete pulses by the adrenal gland in accordance with a circadian rhythm imposed at the CNS level, with maximal secretion in the early morning hours and minimal secretion during the evening. Although the

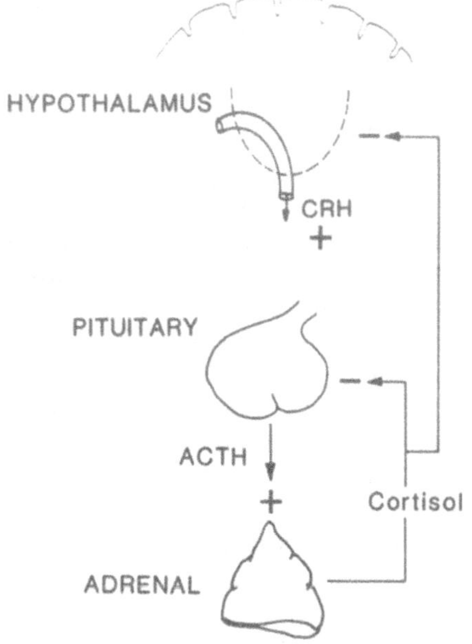

Fig. 1. The hypothalamic/pituitary/adrenal axis. (Modified from Reichlin 1985)

amount of cortisol secreted per day increases with normal maturation, the cortisol secretion rate (CSR) per body surface area remains constant (Kenny et al. 1966). The functional activity of the HPA axis is subject to regulation by a complex and multilevel control system. Corticotropin-releasing hormone (CRH) produced by hypothalamic neurons may be thought of as setting the pituitary's "adrenostat," i.e., as serving to modulate the effect of other inputs upon the production of ACTH by the pituitary corticotroph. These other inputs include: (1) closed-loop feedback systems such as the inhibitory feedback effect of cortisol at both the pituitary and hypothalamic levels and (2) external factors such as psychological and physical stress. Of particular relevance to the study of eating disorders is the responsiveness of the HPA axis to nutritional influences, such as insulin-induced hypoglycemia.

Studies in patients with anorexia nervosa have revealed a number of HPA abnormalities. First, absolute levels of cortisol are increased; in 24-h studies, mean cortisol levels are elevated while the typical diurnal variation is generally qualitatively preserved (Boyar et al. 1977a; Boyar 1977b; Doerr et al. 1980; Fichter et al. 1982), although attenuated in some patients (Brambilla et al. 1985). In addition, there have been reports of an increased number of adrenal secretory episodes as well as an increase in the total time spent in secretion (Doerr et al. 1980). Cortisol production rate (CPR), measured using radioactively labeled cortisol, was found by Walsh et al. (1978) to be comparable in underweight patients and normal weight controls; however, relative to body size, patients produce significantly more cortisol per day. This elevated CPR decreases with weight gain (Walsh et al. 1981). Additionally, urine free (or unconjugated) cortisol (UFC), a sensitive measure of integrated cortisol production, is increased in anorexia nervosa and decreases with clinical recovery (Walsh et al. 1978). Baseline adrenocorticotropin

(ACTH) levels are, by contrast, normal in anorexia nervosa (Gold et al. 1986b; Hotta et al. 1986).

Metabolism of cortisol, as gauged by half-life of plasma ^{14}C-cortisol and tetrahydrocortisol/tetrahydrocortisone ratio, is slowed, an effect which is probably related to decreased levels of triiodothyronine (Boyar et al. 1977a).

Dynamic testing of the HPA axis in anorexia nervosa has revealed normal HPA responsiveness to administration of metopirone which blocks the final step in cortisol synthesis (Marks and Bannister 1963); thus, the HPA axis is capable of responding appropriately to low levels of cortisol. However, the dexamethasone suppression test (DST), in which dexamethasone given at 11 p.m. normally suppresses morning and afternoon cortisol levels on the following day, has been shown to yield high rates of nonsuppression and early escape in patients with anorexia nervosa (Doerr et al. 1980; Gerner and Gwirtsman 1981), suggesting an inadequate compensatory feedback response to high levels of end product. This of course rests on the assumption of normal absorption and disposition of the administered dexamethasone (Weiner 1985), an assumption which has not withstood careful scrutiny in bulimic patients (Walsh et al. 1987a), although it has not yet been studied in anorexia nervosa. Doerr and colleagues (1980) report normalization of the DST in most patients following even incomplete refeeding.

Recent work by Gold and associates (1986b) has focused on responses of ACTH and cortisol to administered ovine CRH. They found that (1) the ACTH response to CRH was significantly blunted, which, given the negative feedback upon the pituitary of elevated cortisol levels found in anorexia nervosa, indicates appropriate corticotroph hyposensitivity, and (2) the cortisol response to CRH was normal, which, given the blunted ACTH response, suggests some increase in the adrenal sensitivity to ACTH. Their inference of the existence of a defect at or above the level of the hypothalamus in patients with anorexia nervosa is supported by observations that: (1) infusion of high levels of CRH in normal volunteers produces a 24-h cortisol secretion pattern similar to that seen in anorexia nervosa (Schulte et al. 1985); (2) CRH levels in the CSF of anorectics are elevated (Hotta et al. 1986). Thus, the observed elevated "adrenostat" setting in anorexia nervosa seems to occur at or proximal to the production of CRH. Whether this defect is truly primary or results from some other abnormality in patients with anorexia nervosa is as yet unknown; however, the abnormal ACTH response to CRH appears to revert to normal by 6 months following weight recovery.

Although the functional significance of these abnormalities is far from fully understood, certain evocative observations have been made. For example, Gold and associates (1986b) summarize findings that intracerebroventricular administration of CRH in animals produces certain behavioral and physiologic responses such as increased physical activity and decreased feeding which are reminiscent of those seen in anorexia nervosa as well as depression.

2.2 Hypothalamic/Pituitary/Thyroid Axis

The normal hypothalamic/pituitary/thyroid (HPT) axis is depicted in Fig. 2. The HPT axis is thought to have an important function in the regulation of body tem-

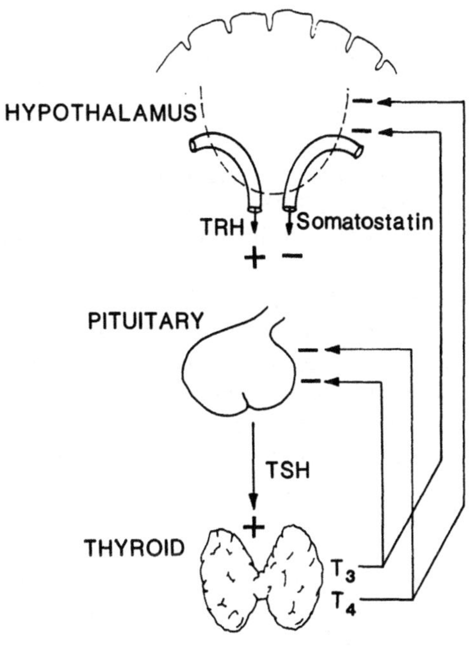

Fig. 2. The hypothalamic/pituitary/thyroid axis. (Modified from Reichlin 1985)

perature and energy expenditure. Thyroxine (T_4) produced by the thyroid is deiodinated peripherally to produce the much more potent (and probably clinically more important) 3,3′,5-triiodothyronine (T_3) as well as the relatively inactive 3,3′,5′-triiodothyronine (reverse T_3 or rT_3). As indicated, T_4 and T_3 exert negative feedback effects at both the pituitary and hypothalamic levels. Hypothalamic thyrotropin-releasing hormone (TRH) released into the hypophyseal portal system stimulates the release of thyroid-stimulating hormone (TSH) by pituitary thyrotrophs, while somatostatin inhibits TSH release. The HPT axis normally displays relatively short-term sensitivity to dietary amount and composition, with carbohydrate overfeeding inducing an increase in serum T_3 and decrease in rT_3 (Danforth et al. 1975) and fasting or consumption of an isocaloric carbohydrate-free weight-maintaining diet causing a decrease in T_3 and increase in rT_3 in normal weight subjects (Serog et al. 1982). In addition, chronically malnourished patients display a hypoactivity of the HPT axis with decreased T_3 (Chopra and Smith 1975) and both normal healthy subjects and obese subjects show a decreased T_3 and increased rT_3 during weight loss which partially normalizes with weight stabilization (Serog et al. 1982; Carlson et al. 1979).

The available data on the functioning of the HPT axis in anorexia nervosa have been comprehensively reviewed by Hudson and Hudson (1984). Levels of T_4 have generally been reported in the normal or low normal range, with T_3 levels distinctly lower than normal and rT_3 levels elevated. TSH levels have for the most part been found to be normal although there have been some conflicting reports. Pirke and associates (1985a) have found that T_3 levels, initially low in a group of patients with anorexia nervosa, were significantly increased after 10% weight gain but continued to rise with further weight gain toward normal.

Dynamic testing of the HPT axis via TRH stimulation has yielded somewhat discrepant results. Hudson and Hudson (1984), summarizing results of several studies report delayed TSH peaks in about half the anorectic patients and blunted peaks in a minority. Weiner (1985) estimates normal responses in 88% of patients with blunted or delayed responses in the remainder and suggests that this variability among anorectic patients may be mediated by growth hormone (GH), which is released by TRH in some patients with anorexia nervosa and which may serve to lessen the pituitary thyrotroph's response to TRH. One group reports that blunted and delayed TSH responses may persist even after weight recovery (Casper and Frohman 1982).

Several of the clinical features of anorexia nervosa, including low resting metabolic rate, decreased oxygen consumption, dry skin, bradycardia, disturbed thermoregulation, increased serum carotene and cholesterol, and altered metabolism of certain steroid hormones, are potentially attributable to the low T_3 syndrome (Weiner 1985). However, anorexia nervosa is distinguished from true hypothalamic hypothyroidism by a disproportionate decrease in T_3 as opposed to T_4, and by an increase in rT_3 levels (Herzog and Copeland 1985). Thus, most available data are consistent with the interpretation that HPT axis abnormalities in anorexia nervosa are secondary responses to malnutrition, rather than primary etiologic factors. However, the time course of these adaptive changes as well as the time course of their reversal with refeeding are as yet incompletely understood.

2.3 Hypothalamic/Pituitary/Gonadal Axis

The normal hypothalamic/pituitary/gonadal (HPG) axis is illustrated in Fig. 3. The gonadotropins, follicle-stimulating hormone (FSH) and luteinizing hormone (LH), are normally secreted in pulses by the pituitary, reflecting the pulsatile stimulation of pituitary gonadotrophs by hypothalamic gonadotropin-releasing hormone (GnRH). The functioning of the HPG axis is regulated by a complex feedback system involving positive and negative inputs at both the pituitary and hypothalamic levels, giving rise to normal menstrual cycling in postpubertal women. Prepubertal and pubertal women exhibit immature HPG axis functioning with distinct 24-h patterns of gonadotropin secretion (see Fig. 4). It has long been observed that normal menstrual functioning is related to weight and nutrition, with loss of menses occurring in various forms of starvation, possibly with a more specific relationship to fat content than to overall weight (Frisch and McArthur 1974; Frisch 1977). In the case of anorexia nervosa, loss of normal menstrual functioning is such a consistent feature of the illness that it has been proposed as a diagnostic criterion in the DSM-III-R (American Psychiatric Association 1987). In addition, HPG axis functioning can be affected by psychological stress and, independent of weight and fat content, by exercise (McArthur et al. 1980; Warren 1980), both of which are also frequently observed clinical features of anorexia nervosa.

The functioning of the HPG axis in anorexia nervosa has been particularly well reviewed by Weiner (1985). FSH and LH levels in postpubertal patients are

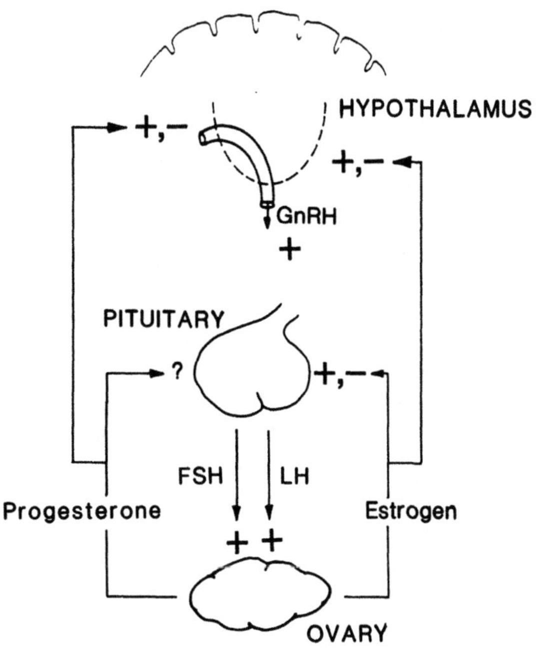

Fig. 3. The hypothalamic/pituitary/gonadal axis. (Modified from Reichlin 1985)

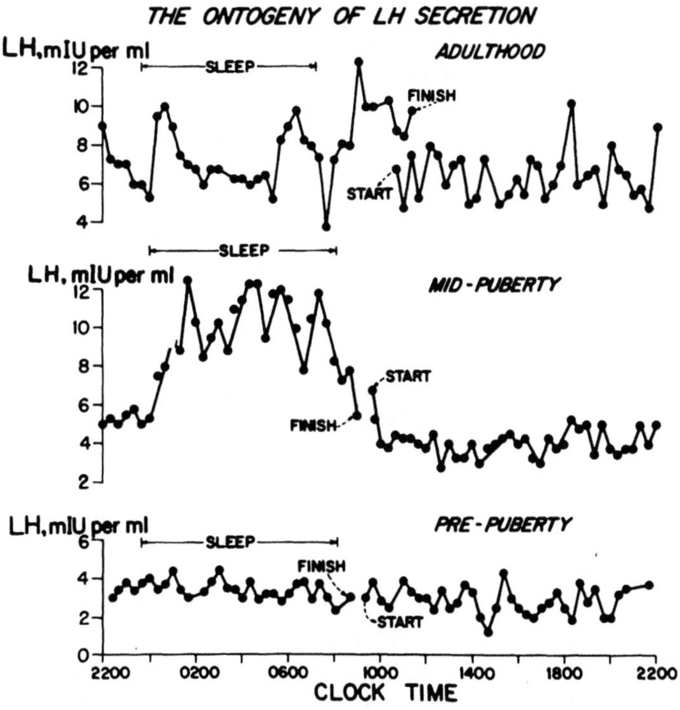

Fig. 4. The ontogeny of luteinizing hormone (*LH*) secretion. (Adapted from Boyar et al. 1972, 1974; Walsh 1980)

subnormal, as are resting levels of estradiol; other than anorexia nervosa, the only known situation in which this pattern is observed is normal prepuberty. The idea of a retrogression in anorexia nervosa to a prepubertal condition is supported by the further observations that 24-h LH secretion patterns in patients revert to pubertal or prepubertal patterns (Boyar et al. 1974), and that with weight recovery these can advance back toward adult patterns (Pirke et al. 1983). Dynamic testing of the HPG axis has elucidated to some extent the nature of the HPG axis disturbance in anorexia nervosa. Ethinyl estradiol normally produces an initial lowering followed by a rise in LH secretion, and clomiphene citrate normally blocks the tonic inhibitory feedback effects of estrogen, thus producing a rise in LH secretion: both agents fail to induce increased LH secretion in underweight patients with anorexia nervosa and, interestingly, in some weight-recovered patients as well (Wakeling et al. 1976, 1977). Although the LH and FSH response to simple GnRH stimulation have been variable, the pulsatile administration of GnRH mimicking normal hypothalamic GnRH secretion may, as summarized by Hudson and Hudson (1984), lead to normal gonadotropin responses and recovery of gonadal function in anorexia nervosa. These and other findings suggest that the HPG axis defect in anorexia nervosa is most likely due to a disturbance or failure of secretion of GnRH by the hypothalamus (Weiner 1983).

The functional significance of HPG axis abnormalities in anorexia nervosa is far from clear. Although the above abnormalities are largely similar to those seen in weight loss, the time course of the occurrence of these changes when compared to the clinical course of anorexia nervosa does not support a simple causal relationship: for example, some patients develop amenorrhea *before* significant weight loss occurs and do not resume normal menstrual functioning with weight recovery (summarized in Hudson and Hudson 1984). Those patients who remain amenorrheic following weight recovery reportedly show more preoccupation with food and weight than those who resume menstruating (Halmi and Falk 1983). These results are intriguing; however, others (Mrosovsky 1984; Hudson and Hudson 1984) provide the important caution that weight status is generally measured against a standard calculated ideal body weight which may not be the same as a person's individual weight set point; thus, a patient with anorexia nervosa who has regained ideal body weight or even premorbid weight may still be underweight relative to her weight set point. The striking coexistence of aspects of psychological and physiological sexual immaturity (Walsh 1980) in many of these patients is similarly evocative but also as yet poorly understood.

2.4 Hypothalamic/Pituitary (Growth Hormone)/Somatomedin Axis

The normal hypothalamic/pituitary/somatomedin (HPS) axis is illustrated in Fig. 5. As in the case of LH, the secretion of GH varies throughout the life cycle, with secretion only at night in prepubertal children, several secretory episodes per day in adolescents, and only one major daily episode in adults, occurring shortly after the initiation of sleep (Finkelstein et al. 1972). The growth-promoting effects of GH are thought to be mediated by growth factors such as somatomedin C which are secreted diffusely in multiple tissues. Somatomedin C exerts negative

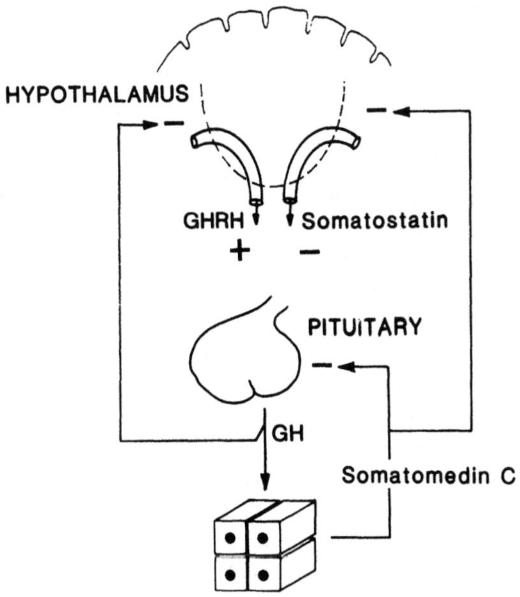

Fig. 5. The hypothalamic/pituitary/somatomedin axis. (Modified from Reichlin 1985)

feedback at both pituitary and hypothalamic levels, while CNS input is provided via growth hormone-releasing hormone (GHRH) and somatostatin, the latter of which also exerts its effects directly on a wide range of structures. The function of the HPS axis is known to be responsive to nutritional and weight factors, with blunting in obese subjects and sensitization in undernourished subjects. On a shorter time scale, GH release is known to be triggered by high protein intake and high carbohydrate meals (during the falling phase of blood glucose). Additional inputs relevant to anorexia nervosa are exercise and physical and emotional stress.

GH levels have been observed to be greatly elevated in many or most patients with anorexia nervosa (Pirke et al. 1985a; Brown et al. 1977; Casper et al. 1977; Vigersky and Loriaux 1977). Overall, these increased GH levels correlate more closely with current nutritional intake than with degree of weight loss (Garfinkel et al. 1975; Brown et al. 1977) and thus rapidly return to normal upon refeeding with as little as 10% weight gain (Pirke et al. 1985a). The observed elevation in GH is thought possibly to be related to the lessening of negative feedback by somatomedin C, the production of which is reduced in anorexia nervosa (Tanaka et al. 1985). As summarized by Weiner (1985), the GH response in anorexia nervosa to stimulation tests has been variable, with a paradoxical increase following oral glucose, reduced response to apomorphine, flattened response to L-dopa, attenuated response to insulin-induced hypoglycemia, normal response to arginine, and increased response to TRH. The functional significance of these is not entirely clear; what may be said is that in anorexia nervosa, as in starvation, alterations in GH levels appear to be relatively rapid in onset and in reversal.

2.5 Other Hypothalamic/Pituitary Systems

Other pituitary hormones which have been studied in patients with anorexia nervosa include prolactin and arginine-vasopressin. As summarized by Weiner (1985), resting prolactin levels in anorexia nervosa have consistently been reported in the normal range, and according to Pirke et al. (1985 a) they remain normal during weight recovery. However some studies have found alterations in circadian secretion patterns with flattening of the normal nighttime elevation occurring in very underweight patients (Kalucy et al. 1976; Darby et al. 1983). Dynamic studies of prolactin secretion have revealed: (1) a diminished prolactin response to TRH in a minority of patients (Vigersky and Loriaux 1977), a finding also associated with short-term starvation (Vinik et al. 1974); and (2) attenuated prolactin responses to insulin-induced hypoglycemia (Nakagawa et al. 1985). Thus, studies of prolactin in anorexia nervosa have provided some indications of hypothalamic dysfunction, possibly related to weight and nutrition.

Abnormalities in secretion of arginine-vasopressin in anorexia nervosa have been suggested by clinical observations of nocturia and polyuria and reports of impaired ability to concentrate or dilute urine in patients with anorexia nervosa (Mecklenburg et al. 1974; Russell and Bruce 1966). To further study the role of arginine-vasopressin in producing this phenomenon, Gold and associates (1983) examined both basal hormone levels and responses to hypertonic saline in underweight and weight-recovered patients with anorexia nervosa. They found reversal of the normal (less than 1.0) CSF/plasma ratio of arginine-vasopressin and abnormal osmoregulation in underweight patients which corrected very slowly with weight gain. The pathophysiologic significance of these findings is not fully understood.

Although several other CNS peptides, including gut-brain peptides (such as cholecystokinin and pancreatic peptides) and endogenous opioids, may play an important role in central homeostatic mechanisms which influence eating and drinking behavior, and neurotransmitters such as norepinephrine may be abnormally regulated in underweight and anorectic states, these areas of research are beyond the scope of this review.

3 Neuroendocrinology of Related Conditions

3.1 Anorexia Nervosa and Semistarvation

Patients with anorexia nervosa are in many ways prototypical semistarved patients; in fact, much of what is now known about semistarvation has come from studies of these patients. Beneath their distorted body images and pathological attitudes toward food and weight, they are conspicuously and undeniably undernourished. Thus, given that semistarvation is known to exert widespread effects upon neuroendocrine systems, the most parsimonious view of neuroendocrine abnormalities in anorexia nervosa would attribute them to semistarvation until demonstrated otherwise. A comparison of neuroendocrine abnormalities reported in anorexia nervosa and in other states of semistarvation is necessarily complicated at the outset by the difficulty in comparing studies performed in different places at different times using different methods and often very different groups of subjects. Nonetheless, the comparative approach has served to suggest areas in which the neuroendocrine profile of anorexia nervosa may not be entirely accounted for by nutritional factors. For example, the observed increase in cortisol production rate in patients with anorexia nervosa (Walsh et al. 1978) is at variance with findings of normal or reduced cortisol production in protein-calorie malnutrition (Alleyne and Young 1967; Smith et al. 1975). Other abnormalities of HPA functioning, such as reduced metabolic clearance of cortisol and high rates of DST nonsuppression, exist in both groups. Thus, any primary HPA abnormality in anorexia nervosa would be superimposed upon starvation-related HPA abnormalities, a complicated state of affairs which will require more careful study. A general underactivity of the HPT axis, as indicated by reduced T_3, elevated rT_3, and abnormal TSH response to TRH stimulation, is an overlapping feature of both anorexia nervosa and semistarvation. The HPS axis has been somewhat less extensively investigated; however, the paradoxical rise in GH following glucose loading in anorexia nervosa is paralleled by reports in malnutrition of minimal suppression (Smith et al. 1974) or paradoxical rise (Alvarez et al. 1972) in postglucose GH.

Another approach to the discrimination of nutrition-related abnormalities in anorexia nervosa is the longitudinal study of patients before, during, and after malnutrition. The most striking outcome of this approach is the finding that the amenorrhea which is a feature of both semistarvation and anorexia nervosa often precedes significant weight loss and lags behind weight recovery in patients with anorexia nervosa. This suggests that the relationship of this symptom to malnutrition is not a simple one and that factors such as body composition, weight fluctuation, and individual weight set point, as well as psychological and behavioral characteristics, may be important. Pirke and colleagues (1985 a) have studied the longitudinal course of abnormalities in T_3 and growth hormone during weight gain in anorexia nervosa and have found rapid normalization of growth hormone and somewhat slower normalization of T_3 with weight gain, suggesting a relationship of these abnormalities to weight and nutritional factors.

Most recently, Fichter and Pirke and associates have undertaken a systematic study of neuroendocrine function in experimentally starved but otherwise healthy subjects. They report that elevated plasma cortisol levels, increased numbers of cortisol secretory episodes, insufficient dexamethasone suppression, blunted TSH response to TRH, and elevated growth hormone levels can all be demonstrated in fasting individuals in the absence of anorexia nervosa (Fichter and Pirke 1984; Fichter et al. 1986). Such investigations which carefully separate nutritional and psychological factors should contribute importantly to an understanding of neuroendocrine abnormalities in anorexia nervosa.

3.2 Anorexia Nervosa and Bulimia

An alternative approach to the factoring out of nutritional components in the study of anorexia nervosa is to compare its neuroendocrine profile to that of bulimia, an eating disorder characterized by recurrent episodes of binge eating frequently followed by self-induced vomiting or other methods of purging in patients who are often of statistically normal weight. There are several indications that bulimia is related to anorexia nervosa, including: (1) psychological and cognitive similarities, such as a preoccupation with body size and a wish to be thinner; (2) significant comorbidity, with about half of the patients hospitalized for anorexia nervosa engaging simultaneously in binge eating, and one-third of patients of normal body weight who request help from eating disorder centers reporting a past history of anorexia nervosa (Fairburn and Cooper 1984; Casper et al. 1980; Garfinkel et al. 1980; Pyle et al. 1981). Thus, neuroendocrine disturbances observed in normal weight patients with bulimia as well as in patients with anorexia nervosa may represent a primary biological dysfunction in eating disorders rather than simply a concomitant of weight loss.

Several investigators have reported rates of cortisol nonsuppression by dexamethasone among normal weight patients with bulimia which are increased over normal (Hudson et al. 1982, 1983; Gwirtsman et al. 1983; Lindy et al. 1985) but not as high as those reported in anorexia nervosa. However, Walsh and colleagues (1987 b) have found that 24-h cortisol secretion patterns in patients with bulimia are normal, in contrast to the elevated 24-h cortisol secretion seen in anorexia nervosa. Recently, Walsh and colleagues (1987 a) have found that plasma dexamethasone levels in patients with bulimia undergoing DSTs are lower than in controls and inversely related to plasma cortisol levels following dexamethasone administration, thus suggesting that factors other than a disturbance of HPA function may contribute to positive DSTs in bulimia.

Studies of HPT axis functioning in bulimia have yielded somewhat discrepant results. Baseline T3 levels have variably been reported as normal in five of six patients with normal weight bulimia (Mitchell and Bantle 1983) and low in 15 patients with normal weight bulimia (Pirke et al. 1985 b). Gwirtsman and colleagues (1983) reported blunted TSH response to TRH in 8 of 10 patients with normal weight bulimia. HPG axis functioning in patients with bulimia, as gauged by clinical findings of abnormal menstrual functioning, is often but not invariably disturbed as it is in patients with anorexia nervosa. HPS axis findings in normal weight bulimia include failure of growth hormone suppression following glucose loading in three of six patients (Mitchell and Bantle 1983) and an abnormal increase in GH following TRH administration in some patients (Mitchell and Bantle 1983; Gwirtsman et al. 1983). Prolactin levels have been reported to be modestly elevated in three of six patients with normal weight bulimia all of whom also displayed menstrual irregularities (Mitchell and Bantle 1983). Taken together, these findings suggest the possibility that a non-nutritional factor contributes to the abnormal neuroendocrine profile of anorexia nervosa. However, the possibility that a more subtle nutritional abnormality may affect the neuroendocrine status of even normal weight patients with bulimia must also be considered.

3.3 Anorexia Nervosa and Depression

The suggestion that a non-nutritional factor may be contributing to the neuroendocrine profile of anorexia nervosa invites speculation as to the nature of such a factor. Although various psychopathological conditions, including anxiety disorders, addictive disorders, and obsessive compulsive disorder, have been compared in some of their features to anorexia nervosa, at this point the most evidence to suggest a link to another diagnosis exists in the case of depression. Lines of evidence in support of this link include the following: (1) findings of an increased frequency of major affective illness in the families of patients with anorexia nervosa (Theander 1970; Cantwell et al. 1977; Winokur et al. 1980); (2) some indications that antidepressant medications have some therapeutic effects in the treatment of anorexia nervosa (Halmi et al. 1986).

The comparison of neuroendocrine profiles of anorexia nervosa and depression provides some additional evidence for such a link. HPA axis abnormalities in depression which overlap with those seen in anorexia nervosa include, as summarized by Brown et al. (1984), elevated plasma cortisol, hypersecretion of cortisol, and inadequate cortisol suppression by dexamethasone in some patients. Gold and colleagues (1986a) report a blunted ACTH response to CRH similar to that seen an anorexia nervosa. As in anorexia nervosa, TRH stimulation produces a blunted TSH response in large numbers of depressed patients (summarized in Brown et al. 1984), as well as an elevated GH response (Maeda et al. 1975) not seen in normal controls. The HPG axis in depression has not yet been extensively studied; however, alteration in menstrual function is not thought to be a typical clinical feature of depression as it is of anorexia nervosa. Although these findings lead to intriguing speculations concerning possible etiologic or pathophysiologic relationships between the two conditions, it may be that similar neuroendocrine disturbances exist in anorexia nervosa and depression for different reasons. For example, the caution that alterations in caloric intake can produce marked neuroendocrine changes and thus should be very carefully controlled in such studies (Fichter et al. 1986) must be taken seriously.

In sum, although the neuroendocrinologic study of anorexia nervosa to date has contributed greatly to the characterization of this illness, its etiology and pathophysiologic mechanisms remain largely obscure. Nonetheless, the continued physiological and psychological investigation of this syndrome promises to contribute not only to a better understanding of anorexia nervosa but perhaps also to a deeper understanding of the interaction of mind, body, and culture in the production of and recovery from illness.

References

Alleyne GA, Young VH (1967) Adrenocorticol function in children with severe protein calorie malnutrition. Clin Sci 33:189–200

Alvarez LC, Dimas CO, Castro A, Rossman LG, Vanderlaan EF, Vanderlaan WP (1972) Growth hormone in malnutrition. J Clin Endocrinol Metab 34:400–409

American Psychiatric Association (1987) Diagnostic and Statistical Manual of Mental Disorders, Third Edition, Revised. American Psychiatric Association, Washington DC, p 67

Boyar RM, Finkelstein JW, Roffwarg H, Kapen S, Weitzman ED, Hellman L (1972) Synchronization of augmented luteinizing hormone secretion with sleep during puberty. N Engl J Med 287:582–586
Boyar RM, Katz J, Finkelstein JW, Kapen S, Weiner H, Weitzman ED, Hellman L (1974) Anorexia nervosa, immaturity of the 24-h luteinizing hormone secretory pattern. N Engl J Med 291:861–865
Boyar RM, Hellman LD, Roffwarg HP, Katz J, Zumoff B, O'Connor J, Bradlow HL, Fukushima DK (1977a) Cortisol secretion and metabolism in anorexia nervosa. N Engl J Med 296:190–193
Boyar RM (1977b) Anorexia nervosa: circadian rhythms of plasma hormones. N Engl J Med 296:1069
Brambilla F, Cavagnini F, Invitti C, Poterzio F, Lampertico M, Sali L, Maggioni M, Candolfi C, Panerai AE, Mueller EE (1985) Neuroendocrine and psychopathological measures in anorexia nervosa: resemblances to primary affective disorders. Psychiatry Res 16:165–176
Brown GM, Garfinkel PE, Jeuniewic N, Modolfsky H, Stancer HC (1977) Endocrine profiles in anorexia nervosa. In: Vigersky RA (ed) Anorexia nervosa. Raven, New York, pp 123–135
Brown GM, Graf P, Graf E (1984) Psychoendocrinology of depression. In: Shah WS, Donald AG (eds) Psychoneuroendocrine dysfunction. Plenum, New York, pp 343–356
Bruch H (1973) Eating disorders. Basic, New York
Cantwell DP, Sturzenberger S, Burroughs J, Salkin B, Green JK (1977) Anorexia nervosa. An affective disorder? Arch Gen Psychiatry 34:1087–1093
Carlson HE, Drenick EJ, Chopra IJ, Hershman JM (1979) Alterations in basal and TRH-stimulated serum levels of thyrotropin, prolactin, and thyroid hormones in starved obese men. J Clin Endocrinol Metab 45:707–713
Casper RC, Frohman LA (1982) Delayed TSH release in anorexia nervosa following injection of thyrotropin releasing hormone (TRH). Psychoneuroendocrinology 7:59–68
Casper RC, Davis JM, Pandey CN (1977) The effect of nutritional status and weight changes in hypothalamic function tests in anorexia nervosa. In: Vigersky RA (ed) Anorexia nervosa. Raven, New York, pp 137–147
Casper RC, Eckert ED, Halmi KA, Goldberg SC, Davis JM (1980) Bulimia. Its incidence and clinical importance in patients with anorexia nervosa. Arch Gen Psychiatry 37:1030–1035
Chopra IJ, Smith SR (1975) Circulating thyroid hormone and thyrotropin in adult patients with protein calorie malnutrition. J Clin Endocrinol Metab 40:221–227
Danforth E Jr, Desilets EJ, Horton ES (1975) Reciprocal changes in serum triiodothyroinine (T_3) induced by altering the carbohydrate content of the diet. Clin Res 23:573A (abstr)
Darby PL, Brown GM, Garfinkel PE (1983) Patterns of prolactin secretion in anorexia nervosa. In: Darby PL, Garfinkel PE, Garner DM (eds) Anorexia nervosa: recent developments in research. Liss, New York, pp 279–284
Doerr P, Fichter M, Pirke KM, Lund R (1980) Relationship between weight gain and hypothalamic pituitary adrenal function in patients with anorexia nervosa. J Steroid Biochem 13:529–542
Fairburn CG, Cooper PJ (1984) The clinical features of bulimia nervosa. Br J Psychiatry 144:238–246
Fichter MM, Pirke KM (1984) Hypothalamic pituitary function in starving healthy subjects. In: Pirke KM, Ploog D (eds) The psychobiology of anorexia nervosa. Springer, Berlin Heidelberg New York, pp 124–135
Fichter MM, Doerr P, Pirke KM, Lund R (1982) Behavior, attitude, nutrition, and endocrinology in anorexia nervosa: a longitudinal study in 24 patients. Acta Psychiatry Scand 66:429–444
Fichter MM, Pirke KM, Holsboer F (1986) Weight loss causes neuroendocrine disturbances: experimental study in healthy starving subjects. Psychiatry Res 17:61–72
Finkelstein JW, Roffwarg HP, Boyar RM, Kream J, Hellman L (1972) Age-related change in the twenty-four-hour spontaneous secretion of growth hormone. J Clin Endocrinol Metab 35:665–670

Frisch RE (1977) Food intake, fatness, and reproductive ability. In: Vigersky RA (ed) Anorexia nervosa. Raven, New York, pp 149–162
Frisch RE, McArthur JW (1974) Menstrual cycles: fatness as a determinent of minimum weight for height necessary for their maintenance or onset. Science 185:949–951
Garfinkel PE, Brown GM, Stancer HC, Moldofsky H (1975) Hypothalamic pituitary function in anorexia nervosa. Arch Gen Psychiatry 32:739–744
Garfinkel PE, Moldofsy H, Garner DM (1980) The heterogeneity of anorexia nervosa. Bulimia as a distinct subgroup. Arch Gen Psychiatry 37:1036–1040
Gerner RH, Gwirtsman HE (1981) Abnormalities of dexamethasone suppression test and urinary MHPG in anorexia nervosa. Am J Psychiatry 138:650–653
Gold PW, Kaye W, Robertson GL, Ebert M (1983) Abnormalities in plasma and cerebrospinal-fluid arginine vasopressin in patients with anorexia nervosa. N Engl J Med 308:1117–1123
Gold PW, Loriaux DL, Roy A, Kling MA, Calabrese JR, Kellner CH, Nieman LK, Post RM, Pickar D, Gallucci W, Paul S, Oldfield EH, Cutler GB, Chrousos GP (1986a) Response to corticotropin-releasing hormone in the hypercortisolism of depression and Cushing's disease: pathophysiology and diagnostic implications. N Engl J Med 314:1330–1335
Gold PW, Gwirtsman H, Avgerinos PC, Nieman LK, Gallucci WT, Kaye W, Jimerson D, Ebert M, Rittmaster M, Loriaux DL, Chrousos GP (1986b) Abnormal hypothalamic-pituitary-adrenal function in anorexia nervosa. N Engl J Med 314:1335–1342
Gull WW (1874) Anorexia nervosa (Apepsia hysterica, anorexia hysterica). Trans Clin Soc Lond 7:22–28
Gwirtsman HE, Roy-Byrne P, Yager J, Gerner RM (1983) Neuroendocrine abnormalities in bulimia. Am J Psychiatry 140:559–563
Halmi KA, Falk JR (1983) Behavioral and dietary discriminators of menstrual function in anorexia nervosa. In: Darby PL, Garfinkel PE, Garner DM, Coscina DV (eds) Anorexia nervosa: recent developments in research. Liss, New York, pp 323–329
Halmi KA, Eckert E, LaDu TJ, Cohen J (1986) Anorexia nervosa: treatment efficacy of cyproheptadine and amitriptyline. Arch Gen Psychiatry 43:177–181
Herzog DB, Copeland PM (1985) Eating disorders. N Engl J Med 313:295–303
Hotta M, Shibasaki T, Masuda A, Imaki T, Demura H, Ling N, Shizume K (1986) The response of plasma adrenocorticotropin and cortisol to corticotropin-releasing hormone (CRH) and cerebrospinal fluid immunoreactive CRH in anorexia nervosa patients. J Clin Endocrinol Metab 62:319–324
Hudson JI, Hudson MS (1984) Endocrine dysfunction in anorexia nervosa and bulimia: comparison with abnormalities in other psychiatric disorders and disturbances due to metabolic factors. Psychiatry Dev 4:237–272
Hudson JI, Laffer PS, Pope HG (1982) Bulimia related to affective disorder by family history and response to the dexamethasone suppression test. Am J Psychiatry 139:685–687
Hudson JI, Pope HG, Jonas JM, Laffer PS, Hudson MS, Melby JC (1983) Hypothalamic-pituitary-adrenal axis hyperactivity in bulimia. Psychiatry Res 8:111–117
Kalucy RS, Crisp AH, Chard T, McNeilly A, Chen CN, Lacey JH (1976) Nocturnal hormonal profiles in massive obesity, anorexia nervosa, and normal females. J Psychosom Res 20:595–604
Kenny FM, Preeaysombat C, Migeon CJ (1966) Cortisol production rate in normal infants, children and adults. Pediatrics 37:34–42
Lasegue C (1873) On hysterical anorexia. Med Times Gaz 2:265–266, 367–369
Lindy DC, Walsh BT, Roose SP, Gladis M, Glassman AH (1985) The dexamethasone suppression test in bulimia. Am J Psychiatry 142:1375-1376
Maeda K, Kato Y, Ohgo S, Chiharo K Yoshimoto Y, Yamaguchi N, Kuromaru S, Imuro H (1975) Growth hormone and prolactin release after injection of thyrotropin releasing hormone in patients with depression. J Clin Endocrinol Metab 40:501–505
Marks V, Bannister RG (1963) Pituitary and adrenal function in undernutrition with mental illness (including anorexia nervosa). Br J Psychiatry 109:480–484

McArthur JW, Bullen BA, Beitins IZ, Pagano M, Badger TM, Klibanski A (1980) Hypothalamic amenorrhea in runners of normal body composition. Endocrinol Res Commun 7:13–25
Mecklenburg RS, Loriaux DL, Thompson RH, Anderson AE, Lipsett MB (1974) Hyothalamic dysfunction in patients with anorexia nervosa. Medicine 53:147–159
Mitchell JE, Bantle JP (1983) Metabolic and endocrine investigations in women of normal weight with the bulimia syndrome. Biol Psychiatry 18:355–365
Morton R (1689) Phthisiologica: or a treatise of consumptions. London
Mrosovsky N (1984) Animal models: anorexia yes, nervosa no. In: Pirke KM, Ploog D (eds) The psychobiology of anorexia nervosa. Springer, Berlin Heidelberg New York, pp 22–34
Nakagawa K, Matsubara M, Obara T, Kubo M, Akikawa K (1985) Responses of pituitary and adrenal medulla to insulin-induced hypoglycemia in patients with anorexia nervosa. Endocrinol Jpn 32:719–724
Pirke KM, Ploog D (1986) Psychobiology of anorexia nervosa. In: Wurtman RJ, Wurtman JJ (eds) Nutrition and the brain. Raven, New York, pp 167–198
Pirke KM, Fichter MM, Warnhoff M, Dorsch G, Spyra B (1983) Hypothalamic regulation of gonadatropin secretion in anorexia nervosa and starvation. Int J Eat Dis 2:151–158
Pirke KM, Fichter MM, Pahl J (1985a) Noradrenaline, triiodothyronine, growth hormone and prolactin during weight gain in anorexia nervosa. Int J Eat Dis 4:499–509
Pirke KM, Pahl J, Schweiger U, Warnhoff M (1985b) Metabolic and endocrine indices of starvation in bulimia: a comparison with anorexia nervosa. Psychiatry Res 15:33–39
Pyle RL, Mitchell JE, Eckert ED (1981) Bulimia: a report of 34 cases. J Clin Psychiatry 42:60–64
Reichlin S (1985) Neuroendocrinology. In: Wilson JD, Foster DW (eds) Williams textbook of endocrinology. Saunders, New York
Russell GFM, Bruce JT (1966) Impaired water diuresis in patients with anorexia nervosa. Am J Med 40:38–48
Schulte HM, Chrousos GP, Gold PW, Booth JD, Oldfield EH, Cutler GB, Loriaux LD (1985) Continuous administration of synthetic ovine corticotropin releasing factor in man: physiological and pathological implications. J Clin Invest 75:1781–1785
Serog P, Apfelbaum M, Autissier N, Gaigts F, Brigant L, Ktorz A (1982) Effects of slimming and composition of diets on $V0_2$ and thyroid hormones in healthy subjects. Am J Clin Nutr 35:24–35
Simmonds M (1914) Ueber embolische Prozesse in der Hypophysis. Arch Pathol Anat 217:226
Smith SR, Edgar PJ, Pozefsky T, Chhetri MK, Prout TE (1974) Growth hormone in adults with protein-calorie malnutrition. J Clin Endocrinol Metab 39:53–62
Smith SR, Bledsoe T, Chhetri MK (1975) Cortisol metabolism and pituitary-adrenal axis in adults with protein-calorie malnutrition. J Clin Endocrinol Metab 40:43–52
Tanaka T, Maesaka H, Suwa S (1985) Changes in somatomedin activity in anorexia nervosa. Endocrinol Jpn 32:891–897
Theander S (1970) Anorexia nervosa. A psychiatric investigation of 94 female patients. Acta Psychiatry Scand 214:1–194
Vigersky RA, Loriaux DL (1977) Anorexia nervosa as a mode of hypothalamic dysfunction. In: Vigersky RA (ed) Anorexia nervosa. Raven, New York, pp 124–135
Vinik AI, Kalk WJ, McLaren H, Paul M (1974) Impaired prolactin response to synthetic thyrotropin releasing hormone after a 36 hour fast. Horm Metab Res 6:499–501
Wakeling A, Marshall JC, Beardwood CJ, De Souza VFA, Russell GFM (1976) The effects of clomiphene citrate on the hypothalamic pituitary gonadal axis in anorexia nervosa. Psychol Med 6:371–380
Wakeling A, De Souza V, Beardwood CJ (1977) Effects of administered estrogen in luteinizing hormone release in subjects with anorexia nervosa in acute and recovery stages. In: Vigersky RA (ed) Anorexia nervosa. Raven, New York, pp 199–209
Walsh BT (1980) The endocrinology of anorexia nervosa. Psychiatry Clin North Am 2:299–312

Walsh BT, Katz JL, Levin J, Kream J, Fukushima DK, Hellman LD, Weiner H, Zumoff B (1978) Adrenal activity in anorexia nervosa. Psychosom Med 40:499–506
Walsh BT, Katz JL, Levin J, Kream J, Fukushima DK, Weiner H, Zumoff B (1981) The production rate of cortisol declines during recovery from anorexia nervosa. J Clin Endocrinol Metab 53:203–205
Walsh BT, Lo ES, Cooper T, Lindy DC, Roose SP, Gladis M, Glassman AH (1987a) The DST and plasma dexamethasone levels in bulimia. Arch Gen Psychiatry (to be published)
Walsh BT, Roose SP, Katz JL, Dyrenfurth I, Wright L, Vande Wiele R, Glassman AH (1987b) Hypothalamic-pituitary-adrenal activity in anorexia nervosa and bulimia. Psychoneuroendocrinology (to be published)
Warren MP (1980) The effects of exercise on pubertal progression and reproductive function in girls. J Clin Endocrinol Metab 51:1150–1157
Weiner H (1983) Abiding problems in the psychoendocrinology of anorexia nervosa. In: Ross conference on medical research: understanding anorexia nervosa and bulimia. Ross Laboratories, Columbus
Weiner H (1985) The physiology of eating disorders. Int J Eat Dis 4:347–388
Winokur A, March V, Mendels J (1980) Primary affective disorder in relatives of patients with anorexia nervosa. Am J Psychiatry 137:695–698
Zuckerman DM, Colby A, Ware NC, Lazersen JS (1986) The prevalence of bulimia among college students. Am J Public Health 76:1135–1137

Effects of Peripheral Thyroid Hormones on the Central Nervous System: Relevance to Disorders of Mood

P. C. Whybrow and M. S. Bauer

Contents

1 Introduction

The association of mood symptoms with disorders of the thyroid has been evident since the earliest clinical descriptions of such disorders. From Caleb Parry's mention of "nervous affectations" (1825) and Graves' delineation of "symptoms which were supposed to be hysterical" (1940), to more recent longitudinal studies such as that of Hermann and Quarton (1965), thyroid hyperfunction has been shown to play an etiologic role in the production of many and varied mental symptoms. Similarly, since the early study of hypothyroidism by the Clinical Society of London (1888) and Asher's classic longitudinal study (1949), thyroid hypofunction has been identified as a causative agent in various sorts of psychopathology. This close relationship between thyroid economy and mental function has interested psychiatric researchers with diverse interests over at least the last 40 years (e.g., Gjessing et al. 1976; Lidz and Whitehorn 1949). More recently, this interest in the role of the thyroid axis in psychiatric disorders, particularly disorders of mood, has grown. Impetus for recent research stems from such clinical observations as the high prevalence of abnormal thyrotroph responsiveness to thyrotropin-releasing hormone (TRH) in unipolar and bipolar depression, and

Affective Disorders Program, Department of Psychiatry, University of Pennsylvania, Philadelphia, PA 19104, USA

Current Topics in Neuroendocrinology, Vol. 8

the efficacy of thyroid hormones as adjuvants in the treatment of unipolar and bipolar affective disorders.

However, the mechanism by which the hormones of the thyroid axis play a role in mental functioning is by no means clear. Despite intriguing clinical evidence, there are large gaps in our ability to relate these clinical findings to what is known from laboratory studies of the effects of thyroid hormones on the central nervous system. This chapter will review the clinical evidence that altered thyroid function is associated with mood disorders, and that such alterations may play a role in their pathogenesis and successful treatment. We will focus primarily on the evidence that altered levels of circulating or "peripheral" thyroid hormones – thyroxine (T_4), triiodothyronine (T_3), and reverse T_3 (rT_3) – whether altered due to endogenous mechanisms or due to exogenous administration, play such a role. We will also discuss those preclinical studies which may be important in understanding the role of these iodothyronines in the central nervous system in disorders of mood, and suggest directions for further clinical research.

2 Iodothyronines and Mood: Clinical Studies

2.1 Thyroid Disease and Mood

The occurrence of prominent mental symptoms in clinically identified cases of hypo- and hyperthyroidism has been an important point of departure for the study of the thyroid axis in psychopathology. This topic has recently been reviewed in depth elsewhere (Bauer et al. 1987). It is clear from studies over the last 100 years that cognitive or mood symptoms are present in virtually all persons who have clinical evidence of disordered thyroid economy, although the classic description of irritable anxiety in hyperthyroidism and retarded depression in hypothyroidism may be an oversimplication. Psychopathologic symptoms usually remit with correction of thyroid status. The severity of these mental symptoms may not parallel that of the somatic symptoms or the degree of abnormality of laboratory indices of peripheral thyroid function (Jain 1972). Furthermore, resolution of mental symptoms may lag behind the normalization of serum iodothyronine levels, and the time course of this resolution may also be much longer than that for somatic symptoms (Schon et al. 1961; Wallace et al. 1980).

To summarize, clinically significant alterations of thyroid function invariably lead to cognitive and mood pathology, with affective symptoms predominating, such as depression; subjective and objective impairment of thought processes, attention, and memory; anxiety; irritability; and occasionally psychosis and delirium. In all but the most severe cases, these changes are reversible with normalization of thyroid status, though often over a time course of several months after normalization of levels of circulating thyroid hormones.

Whether subtle thyroid abnormalities, insufficient to cause somatic symptoms, can produce cognitive and affective symptoms has not been comprehensively addressed by the endocrinologic literature. Two psychiatric studies have postulated that subclinical thyroid hypofunction may play a pathogenic role in the affective disorders. Gold et al. (1982) described grade II or III hypothyroidism (decreased T4 levels in grade I; normal serum iodothyronine levels with elevated serum levels of basal thyroid-stimulating hormone (TSH) in grade II; normal basal TSH with exaggerated response of TSH to TRH infusion in grade III) in 15 of 100 consecutive "depressed or anergic" patients admitted to their inpatient unit. Similarly, Cowdry et al. (1983) found a 92% incidence of grade II hypothyroidism in persons with rapid-cycling bipolar affective disorder. In both these studies, which focused primarily on pituitary thyrotroph function, levels of circulating iodothyronines were normal.

Since clinical observations indicate that most patients with unipolar and bipolar depression do not have obvious clinical stigmata of hypo- or hyperthyroidism, what then is the evidence that thyroid hormone factors may play a role in affective homeostasis?

2.2 Thyroid Hormone Augmentation of Antidepressant Treatment

In the early 1960s several papers stimulated interest in the role of thyroid hormones in depression. Carrier and Buday (1961) reported an increased toxicity to monoamine oxidase-inhibiting antidepressants (MAOIs) in hyperthyroid rats. Prange and Lipton (1962) demonstrated that the LD_{50} for imipramine was halved in mice made thyrotoxic by 4 days of treatment with dessicated thyroid. Extending these observations to the clinical sphere, Prange (1963) described the case of a woman in whom the addition of imipramine to an existing stable regimen of dessicated thyroid precipitated signs and symptoms of thyrotoxicosis, including an elevated basal metabolic rate, without change in serum butanol-extractable iodine (the then-current parameter for determination of circulating iodothyronine levels).

Interest in the psychopharmacologic effects of thyroid hormones was further stimulated by evidence that the addition of 25–75 μg T_3 daily to antidepressant treatment with imipramine results in a faster resolution of depressive symptoms, at least in women (Prange et al. 1969). This finding has been replicated (Wilson et al. 1970; Coppen et al. 1972) and also described using amitriptyline (Wheatley 1972), though not all studies have been positive (Feighner et al. 1972; Steiner et al. 1978). In addition, T_3 supplementation of tricyclic antidepressant treatment has been reported to induce remission in two-thirds of refractory depressives, both men and women (Goodwin et al. 1981; Earle 1970; Banki 1975, 1977; Prange et al. 1985). This ability of T_3 supplementation does not affect the disposition of tricyclics in animals (Breese et al. 1974), or alter blood levels of imipramine or its major active metabolite, desipramine, in patients (Garbutt et al. 1979). Taken together, these studies of antidepressant treatment response and toxicity suggest that thyroid hormones may act on systems which play a role in the pathogenesis or the resolution of depressive symptoms.

It appears, therefore, that manipulation of the thyroid axis can potentiate the response to tricyclic antidepressants, in a manner that does not depend on alteration of the pharmacokinetics of the antidepressant itself. It is unclear, however, what these fairly robust clinical findings tell us about the pathophysiologic role of the thyroid axis in disorders of mood. Several important questions which might further characterize this role remain unanswered: Does T_4 also potentiate the tricyclic antidepressant response? Can T_3 potentiate antidepressants with pharmacologic profiles unlike those of imipramine and amitriptyline (e.g., MAOIs or the serotonin uptake inhibitors)? What is the relationship of initial thyroid status to response to T_3 potentiation, or indeed to antidepressant treatment responsiveness in general? We are aware of no studies which address the first two questions, except the study of Feighner et al. (1972) which found no effect of T_3 when added to electroconvulsive therapy, although they also failed to find an effect of T_3 augmentation of imipramine therapy. There is scant data concerning whether response to thyroid axis manipulation is determined by preexisting thyroid status. Tsuitsui et al. (1979) described 10 of 11 refractory depressives who exhibited a blunted TSH response to TRH who responded to T_3 potentiation. On the other hand, Targum et al. (1984) showed that seven of 21 refractory depressives responded to T_3 or T_4 augmentation of antidepressant treatment; five of these seven had exaggerated TSH responses to TRH, yielding a significantly higher incidence of grade III hypothyroidism than in the 14 augmentation-resistant patients. However, Schwarcz et al. (1984) found no thyroid axis abnormalities, including normal TSH response to TRH, in four refractory depressives who were converted to responders with T_3 augmentation of desipramine treatment.

2.3 Rapid-Cycling Bipolar Affective Disorder

One specific subtype of affective disorder, rapid-cycling bipolar disorder, may shed further light on the role of thyroid function in the maintenance of psychiatric homeostasis. Persons with a rapid-cycling course of bipolar illness have by definition four or more affective episodes per year; many cycle continuously from mania or hypomania to depression (Dunner 1979). These persons have a high prevalence of present or past hypothyroidism (Cowdry et al. 1983) and tend to be particularly susceptible to lithium-induced hypothyroidism (Cho et al. 1979). Furthermore, treatment with "hypermetabolic" doses of T_4 (2–3 times the normal replacement dose) leads to resolution of these otherwise refractory symptoms, as indicated by one published case series (Stancer and Persad 1982) and several case reports (reviewed in Bauer and Whybrow 1986).

Our own data also indicate a high incidence of thyroid abnormalities in this population. Three of 18 rapid-cycling bipolar patients who were refractory to their current psychotropic medications had a blunted TSH response to TRH infusion; more striking, nine others had grade III hypothyroidism or had had a previous diagnosis of hypothyroidism and were taking suboptimal doses of replacement T_4 (unpublished observations). In a subset whom we followed longitudinally without medication changes, there appeared to be no alteration of peripheral thyroid indices or TSH response to TRH with affective state. This suggests that cycling of the thyroid system itself does not appear to drive the cyclic affective state, at least not according to the parameters which we are able to measure. Nevertheless, abnormalities of the thyroid axis, in particular hypothyroidism, are associated with a malignant course in bipolar illness.

We have thus far treated seven patients by adding T_4 to their existing regimen and induced remission of both mania and depression in six. Three points became clear during the course of this study. First, T_4 alone has little effect, and must be used in conjunction with standard antidepressant and/or anticycling medications. Second, it appears that some patients respond to replacement rather than hypermetabolic doses of T_4, i.e., doses of T_4 at which TSH decreases but circulating total T_4 remains within normal limits; however in such patients *free* thyroxine becomes elevated beyond the normal range. Third, it appears that the basic phenomenon which we have observed is that when rapid-cycling patients are refractory to standard treatments, augmenting their thyroid status leads to remission of their symptoms – strikingly similar to the ability of T_3 potentiation of tricyclics to induce remission in otherwise refractory depressives. This again suggests that a certain level of activity of the thyroid axis may be necessary for the maintenance or reestablishment of normal mood.

2.4 Serum Iodothyronines in Affective Disorders

If disorders of the thyroid, particularly hypothyroidism, can lead to significant mood symptoms, and if the addition of thyroid hormones can induce or speed the therapeutic response to antidepressant or anticycling treatments, it is logical to ask whether depression per se is associated with abnormal serum levels of iodothyronines. Numerous studies have addressed this issue (Table 1).

As can be seen from Table 1, there is no clear unanimity on this topic. Certainly there are methodologic issues which contribute to the variance of results. For instance, some of these investigations studied unipolar depressives, others bipolars, and others mixed groups; more important, some studied inpatients and others outpatients, who presumably differ on such important clinical features as severity or duration of depressive symptoms. Of further importance, these studies use various methods to determine normalcy, or lack thereof, in their patients; for example, some studies employed control groups and others simply compared their patients to an established normal range.

However, certain trends do emerge. Although eight studies found that iodothyronine levels did not distinguish their patients as abnormal, seven studies found evidence of increased total and/or free T_4 as the only abnormality, and one more found such evidence of hyperthyroxinemia accompanied by decreased levels of total and/or free T_3. Two studies found decreased levels of total and/or free T_4.

It does not appear that the affective disorders are commonly characterized by the so-called euthyroid sick syndrome or low T_3 syndrome, as are several medical illnesses (Wartofsky and Burman 1982). Only two studies found evidence of decreased total and/or free T_3 as the primary abnormality. Four of the more recent studies have measured rT_3, the apparently peripherally inactive metabolite of T_4 (DeGroot et al. 1984). In three of these studies, rT_3 levels were normal, while one study found higher levels in unipolar depressed and manic women than in bipolar depressed or control women. Two studies have reported that rT_3 levels fall with treatment, and one has reported an increase.

Table 1. Circulating thyroid hormone levels in affective disorders

Authors (year)	Diagnosis $(n)^a$	Thyroid indices		
		Increased	Decreased	No change
Board (1957)	"Depressive reaction", inpatient (33)			PBI
Whybrow (1972)	Unipolar depression (30)	FT_4I (6 of 30)		PBI, T_3U
Rybakowski and Sowinsky (1973)	Bipolar disorder (15: 6 manic, 9 depressed)		FT_4I, "absolute T_4"	T_4, T_3U, T_4 dialysis fraction
Kirkegaard (1975a)	Endogenous depression (15)	T_4, FT_4I		T_3U, T_3, FT_3I, TSH
Kirkegaard (1975b)	Endogenous depression (19)	T_4, FT_4I	T_3, FT_3I	T_3U, TSH
Kirkegaard (1977a)	"Depressed mood" (21)			T_4, T_3U, FT_4I, T_3, FT_3I, TSH
Kirkegaard (1977b)	Affective disorder (79)	T_4 (unipolar > bipolar or reactive)		T_3U, TF_4I, T_3, FT_3I, TSH
Rinieris (1978)	Primary affective disorder (40 female)		T_4, FT_4I	T_3U
Weeke and Weeke (1978)	Endogenous depression (19: 10 nonmedicated; 9 on antidepressants)		Circadian rhythm of FT_3, TSH blunted	T_4, FT_4, T_3
Gold (1982)	"Depressed or anergic" (100)	TSH (15 of 100)		T_4, T_3U, FT_4IT_3
Kirkegaard and Faber (1981)	Endogenous depression (80)	T_4, FT_4I	3,5-T_2	T_3, FT_3I, rT_3, 3′5′-T_2, 3,3′-T_2
Linnoila (1982)	Endogenous depression (16 female: 4 unipolar, 5 manic, 7 bipolar depressed)	T_4, rT_3		T_3
Sternbach (1983)	Major depression outpatient (44)			T_4 (decreased in 1 of 44), T_3U, TSH (increased in 2 of 44)
Calloway (1984)	Depression (68: 50 definite major, 9 probable major, 9 minor)	FT_4I (16 of 68)		T_4, T_3U,
Roy-Byrne (1984)	Primary affective disorder (50)			T_4, FT_4I, TBG, T_3, TSH
Joffe (1985)	Primary affective (11)		T_3, FT_3	T_4, FT_4, TBG

Table 1 (continued)

Authors (year)	Diagnosis (n)[a]	Thyroid indices		
		Increased	Decreased	No change
Nemeroff (1985)	Major depression (28)			T_4, T_3U, FT_4I, TSH
Orsulak (1985)	Major depression (32)		FT_3	FT_4
Unden (1986)	Major depression, some on medications (31)		TSH (24-h mean)	T_4, T_3, rT_3, TBG
Kirkegaard and Faber (1986)	Major depression (36)	FT_4		T_4, T_3, FT_3, rT_3, FrT_3, TSH

This table summarizes those studies which analyzed circulating iodothyronine and thyrotropin levels in various affective disorders; data on TSH response to TRH infusion are not included.
PBI = protein-bound iodine; $T_3U = T_3$ resin uptake; FT_4I = free T_4 (or index); FT_3I = free T_3 (or index); FrT_3 = free rT_3; TBG = thyroxine-binding globulin; $n,n,$-T_2, respective diiodothyronines; other abbreviations as per text.
[a] Diagnosis is as stated by the authors; those where no explicit basis (e.g., RDC, DSM-III) is given are in quotation marks.

Table 2. Longitudinal studies of circulating thyroid hormones in affective disorders

Authors (year)	Thyroid indices[a]		
	Increased	Decreased	No change
Board (1959)		PBI	
Whybrow (1972)		FT_4I	
Kirkegaard (1975a)		T_4, FT_4I	T_3, FT_3I, TSH
Kirkegaard (1975b)		T_4, FT_4I	T_3U, T_3, FT_3I, TSH
Kirkegaard (1977a)		T_4, FT_4I	T_3U, T_3, FT_3I, TSH
Kirkegaard and Faber (1981)		T_4, FT_4I, rT_3, 3,3′-T_2, 3′,5′-T_2	T_3, FT_3I, 3,5-T_2
Roy-Byrne (1984)	BH	T_4, T_3, FT_4	TBG
Unden (1986)	TSH	T_4, rT_3	T_3, TBG
Kirkegaard and Faber (1986)		T_4, FT_4I, rT_3, FrT_3	

Data derived from studies summarized in Table 1. Abbreviations as in Table 1
[a] Follow-up iodothyronine levels are considered increased or decreased relative to their initial values

Of the nine studies which followed their patients longitudinally (Table 2), all nine documented a significant decrease of total and/or free T_4 with treatment. Five of these nine studies which demonstrated a relative decrease in total and/or free T_4 had found that the initial levels were "normal." Thus, 11 of 20 studies present evidence that the occurrence of affective disorders is associated with relative hyperthyroxinemia. Of the seven studies which followed T_3 indices longitudinally, total T_3 decreased in only one.

It appears therefore that the most frequent abnormality of iodothyronine metabolism in the affective disorders is a relative increase of total and/or free T_4, without clear concomitant alterations in its metabolites T_3 or rT_3. That this increase in T_4 parameters is of clinical significance is suggested by the fact that longitudinal studies demonstrate a fall in these parameters with treatment, whether or not the initial levels are abnormally high.

Whether this increase in T_4 parameters is specific for the affective disorders is not clear. In studies of nonselected psychiatric inpatients as well (Table 3), increased total and/or free T_4 appears to be a frequent finding, at least in schizophrenia and the affective disorders. This is not a common finding in medical illnesses or undernutrition (Wartofsky and Burman 1982).

The ability of experimentally induced stress to elevate serum iodothyronine levels has not been well studied in humans (Rose 1980), although there is some evidence that chair-restraint stress in nonhuman primates leads to an acute rise in butanol-extractable iodide and TSH which normalizes over the ensuing 2–3 weeks (Mason 1975). The issue of whether the increase in T_4 parameters in the affective disorders is a nonspecific stress effect, or whether it reflects a process

Table 3. Circulating thyroid hormone levels in unselected psychiatric inpatients

Authors (year)	Diagnoses (*n*)	Thyroid indices: Increase	Decrease	No change
Nicholson (1976)	Unspecified (98 female)			T_4 (increased in 1 of 98; decreased in 7 of 98), T_3U, FT_4I
McLarty (1978)	Unspecified (1206)			T_3, T_4, TSH
Prange (1979)	Schizophrenic (17)		T_3	T_4, FT_4I, T_3U
Morley and Schafer (1982)	Nonselected (386)	FT_4I:[a] 21% schizophrenics 14.1% amphetamine users 5% affective disorders TSH: 6.4% bipolars		T_4, T_3U, T_3, FT_3I
Spratt (1982)	Nonselected (645)	T_4:[b] 41% schizophrenics 34% depressives FT_4I: 21% schizophrenics 21% depressives		

Abbreviations as in Table 1

[a] In 32 with increased FT_4I who were followed longitudinally, levels in 29 fell to within normal limits

[b] In a subset retested 1 week later, T_4 fell in all, to within normal limits in 19 of 22. FT_4 fell in 15 of 15. T_3 fell in 17 and rose in five of 22

which plays a particular pathogenic or compensatory role in the affective disorders can therefore not be resolved at present.

However, regardless of whether the effect is stress-related or not, the relative increase in T_4 may reflect a compensatory response of the organism to maintain affective homeostasis. Alternatively, the increase in T_4 parameters may be part of the mechanism by which the affective symptoms are produced. The T_3 potentiation studies reviewed above support the former hypothesis, since in those studies, and in the studies of T_4 augmentation in rapid-cycling bipolar illness, a modest augmentation of thyroid economy effect achieves the therapeutic response. An alternative explanation for these findings (Joffe et al. 1985) is that it is specifically elevated serum T_4 that in some way pathogenic, and the mechanism by which T_3 augmentation achieves its effect is by lowering circulating T_4, which does occur (Whybrow et al. 1972); however, the studies of treatment of rapid cycling bipolar illness do not support this hypothesis.

2.5 Prognostic Value of Iodothyronine Levels

One may therefore hypothesize that in the affective disorders, increased thyroid axis activity, whether due to endogenous mechanisms or to administration of exogenous thyroid hormone, plays a *compensatory, homeostatic* role in maintaining normal mental function. Evidence for this role is based on converging lines of support from clinical studies: (1) Clinically evident thyroid hypofunction is invariably accompanied by cognitive and affective symptoms similar to those found in depression, and these symptoms are reversible with normalization of thyroid status, albeit often over a period of weeks; (2) T_3 augmentation of tricyclic antidepressant treatment converts a high percentage of refractory depressives into responders and speeds the antidepressant response in women; (3) In the specific case of rapid-cycling bipolar disorder, the addition of T_4 to the existing psychopharmacologic regimen can induce remission of both manic and depressive symptoms when standard pharmacotherapy has failed; and (4) The most common thyroid hormone abnormality occurring in depression is an increase in total and/or free T_4, which decreases over the course of treatment.

The strongest evidence substantiating this hypothesis would be obtained by direct measurement of central nervous system thyroid status (e.g., by quantitative imaging of T_4, T_3, and their metabolites in the brain), or by measuring other specifically thyroid-related neurochemical or neurophysiologic parameters. There have been, to our knowledge, no studies which have successfully imaged thyroid hormones in the brain of man or animals in vivo. Similarly, there have been few studies of central nervous system parameters which are known or postulated to be thyroid hormone-determined. One example is the use of auditory evoked potential to differentiate hypo- and hyperthyroid patients from normals (Himelfarb et al. 1981). Regarding specifically central nervous system thyroid status in the affective disorders, we are aware of one report which has measured cerebrospinal fluid rT_3 and found higher levels in unipolar depressed than bipolar depressed or manic patients; however, no mention is made of other iodothyronine levels (Linnoila et al. 1983).

Without direct measurement of central nervous system thyroid status, one may also ask two related questions: (1) Does a relative increase in thyroid axis activity predict a favorable outcome in disorders of mood? (2) Are there *peripheral* indices of central nervous system thyroid function which are amenable to measurement?

If increased levels of circulating thyroid hormones to indeed play a compensatory role, one might predict a positive treatment response from the presence of increased total and/or free T_4. However, few studies have addressed this issue. In the study of T_3 potentiation of tricyclic antidepressant effect by Whybrow et al. (1972), in those patients who received only imipramine and not T_3, faster ankle reflex time and lower serum cholesterol levels at intake, two tissue indices of relatively increased thyroid hormone effect (Abraham et al. 1966; DeGroot et al. 1984), significantly predicted favorable clinical response, although total T_4 and free T_4 index did not. More recently, Roy-Byrne et al. (1984) studied 50 affectively ill unipolar and bipolar patients treated with carbamazepine and found that responders, whether depressed or manic, had significantly greater decreases in both total and free T_4 on carbamazepine than nonresponders, although pretreatment levels were similar in responders and nonresponders.

Most recent papers have focused on changes in the TSH response to TRH as a prognostic factor, usually in predicting *recurrence* of affective episodes rather than response during the index episode; some of these studies have also inspected concurrent iodothyronine data. Kirkegaard et al. (1975 b) longitudinally studied nine patients who relapsed within 6 months of recovery from the index episode of depression and ten patients who did not, and found that those who did not relapse had somewhat higher levels of total T_4 and free T_4 index; the difference was not, however, statistically significant. Similarly, this group found no significant difference in T_4 parameters at intake or with treatment between those who relapsed within 6 months of successful treatment with electroconvulsive therapy and those who did not. However, only the group who did not relapse showed a significant fall in these indices during treatment of the index episode, suggesting again that iodothyronine metabolism may play an important role in the pathogenesis of or recovery from affective illness (Kirkegaard and Faber 1986).

Thus one study (Whybrow et al. 1972) indicates that increased tissue measures of thyroid hormone effect correlates with favorable outcome; and two studies (Roy-Byrne et al. 1984; Kirkegaard and Faber 1986), although using dissimilar treatments, demonstrate that a relative heightened activity of the thyroid axis predicts favorable clinical response.

3 Iodothyronines and Central Nervous System Thyroid Status

3.1 Peripheral Indicators of Central Thyroid Status: TSH?

Aside from the single study including ankle reflex time as a measurement of thyroid status in depression (Whybrow et al. 1972), there have been to our knowledge no other studies using indicators of thyroid effect on neural function in the affective disorders. Instead, studies of peripheral correlates of central thyroid function have focused on non-neural parameters, particularly basal production by the pituitary gland of TSH, and increase in release after stimulation with TRH. The reasons for interest in this system have been varied, since Prange et al. (1972) and Kostin et al. (1972) first described the diminished TSH response to TRH infusion in depressed patients. These original observations have given rise to a rich and varied literature concerning the hypothalamic-pituitary-thyroid axis in the affective disorders. This literature has been extensively and thoughtfully reviewed elsewhere (Kirkegaard 1981; Loosen 1985).

It is true that as the thyroid gland becomes grossly underactive or overactive, as in hypo- or hyperthyroidism, iodothyronine-related factors predominate in the control of both basal and TRH-stimulated TSH production (Morley 1981). However, the relationship of TSH production or thyrotroph responsiveness to TRH to the modest alteration in iodothyronine levels found in the affective disorders is not well established (see Sect. 3.2), despite the fact that in normal subjects the TSH response to TRH is exquisitely sensitive to both small increases and decreases in T_3 and T_4 levels (Snyder and Utiger 1972; Saberi and Utiger 1975).

Furthermore, production of TSH by thyrotroph cells in vitro and in vivo is affected by numerous factors besides TRH (extensively reviewed by Morley 1981), and appears to be controlled in man, perhaps under physiologic conditions, by several nonthyroid factors, including glucocorticoids (Azukiwaza et al. 1979), gonadal steroids (Morley et al. 1981), somatostatin (Weeke et al. 1981), and dopamine (Kaptein et al. 1980). In the affective disorders in particular, there is evidence that basal TSH production and the TSH response to TRH are determined by nonthyroid and non-TRH factors such as somatostatin (Bissette et al. 1984), dopamine (Kirkegaard et al. 1977a), and serotonin (Gold et al. 1977).

In fact, the nihilistic view would suggest that it may have been an historical accident of sorts that the tripeptide pyroglutamyl-histidyl-prolylamide was named TRH, as the bulk of TRH in the brain is extrahypothalamic in most mammalian species in which it has been measured (Leppaluoto et al. 1978; Stanton et al. 1982), including man (Parker and Porter 1983). Similarly, receptors for TRH are present in significant concentrations outside the pituitary in mammals (Taylor and Burt 1982; Manaker et al. 1985), including man (Manaker et al. 1986). Further, TRH has many actions which bear no clear relationship to stimulation of the thyroid axis (Griffiths 1985).

3.2 TSH and Iodothyronines in Affective Illness

In the affective disorders, the bulk of the evidence indicates that circulating iodothyronine levels do not correlate with either basal or stimulated TSH production in the affective disorders (Weeke and Weeke 1978; Kirkegaard et al. 1975a; Gold et al. 1977; Winokur et al. 1983; Witschy et al. 1984; Kjellman et al. 1985; Sternbach et al. 1985).

However, others have demonstrated a correlation. Calloway et al. (1984) classified patients into three groups, depending on whether their TSH response was blunted, normal, or exaggerated, and found that free T_4 index decreased as TSH response increased. On the other hand, Takahashi et al. (1974) demonstrated that patients with a low free T_4 index had the most blunted TSH response. Loosen et al. (1983) showed that T_4 levels correlated positively with the TSH response to TRH, but only in depressed patients with a blunted TSH response, not in those with a normal response or in control subjects.

Regarding *change* of these thyroid indices during treatment, Kirkegaard and Faber (1981) found a weak negative correlation between the change in free T_4 index and the change in TSH response, and a weak positive correlation between free T_3 index and TSH response; however, this correlation did not persist when patients were divided into those who showed an increased TSH response after treatment and those whose TSH response showed no change. A subsequent study by these investigators did not demonstrate these correlations (Kirkegaard and Faber 1986). Unden et al. (1986) demonstrated a significant negative correlation between mean 24-h T_4 levels and TSH response in depressives, but not normals, although there was no correlation between the changes in these parameters with treatment. In the study by Roy-Byrne et al. (1984), while responders had significantly greater decreases in the T_4 parameters than nonresponders, there was no difference in the use of TSH between the two groups. Unfortunately, formal correlational and covariance analyses were not reported, which would have further substantiated their hypothesis that the affective disorders are associated with altered negative feedback at the level of the thyrotroph.

It may be argued that all these studies were done with TSH assays that were relatively insensitive to small changes in TSH, particularly at lower values. Certainly this issue should be reinvestigated now that monoclonal antibody assays with greater sensitivity and precision are available (Rattle et al. 1984).

This outline of the complexities of the control of the TSH production is not meant to leave the reader with the view that blunting or exaggeration of the TSH response to TRH is merely an interesting observation deriving from a hopelessly complex system, the study of which will yield little information on the pathophysiology of disorders of mood. Rather, several features of TSH production by the pituitary thyrotroph make that system potentially very valuable in the study of central nervous system thyroid function in the affective disorders, if it is understood as one level of an integrated brain-pituitary-thyroid-peripheral organ axis.

3.3 Central Nervous System Thyroid Status

As noted above, there are currently no good direct measures of central nervous system thyroid activity in vivo. Central nervous system iodothyronine concentrations are determined by at least two major factors – uptake by a saturable carrier which transports T_4 and T_3 from the circulation into the brain (Pardridge 1979), and the enzyme 5′-deiodiinase (5′D) which converts T_4 into the metabolically more active T_3. Two different forms of 5′D exist, based on affinity for T_4 and resistance to inhibition by propylthouracil (Visser et al. 1982). The form of 5′D, type II, which exists in the cerebral cortex is also found in the pituitary (Visser et al. 1983). The conversion of T_4 to T_3 in the pituitary is a major, perhaps *the* major determinant of TSH production (Larsen et al. 1979). Hence, it has been hypothesized that type II 5′D activity is the primary determinant of TSH production, and data in support of the hypothesis in man have been presented (Kaplan and Breitbart 1983). Besides the pituitary, type II 5′D has been demonstrated in rats only in brain and brown adipose tissue (Leonard et al. 1982), while the liver and kidney contain type I 5′D, which is the major pathway of production of circulating T_3 via conversion from T_4 (deGroot et al. 1984). TSH production may therefore provide an easily measurable output variable which reflects the activity of an enzyme system present in the central nervous system.

The existence of a separate pathway in the brain for T_4 deiodination, as well as the ability of the brain to maintain iodothyronine concentrations near normal despite significant hypo- or hyperthyroidism (Dratman et al. 1983), suggests that the adult central nervous system is not dependent primarily upon peripheral thyroid hormone levels, but rather has the ability to autoregulate thyroid status. There are several implications of this autoregulation of thyroid status in the central nervous system for understanding the role of thyroid axis in maintaining psychiatric homeostasis. Clearly the affective symptoms precipitated by clinically evident hypo- or hyperthyroidism may result from overcoming these homeostatic mechanisms by severe deficiency or excess of circulating thyroid hormones.

However, if one postulates the existence of an autoregulatory mechanism, it is also quite possible that such mechanisms may become disordered in some persons, allowing small alterations in the thyroid axis to produce significant clinical effects. This may be why persons with rapid-cycling affective disorder frequently have evidence of mild but definite thyroid hypofunction, and why it appears that persons with evidence of high-normal tissue thyroid effect or with a relative increase in T_4 parameters respond better to antidepressant therapy than those with evidence of less thyroid axis activation. Similarly, an aberration of the homeostatic mechanism in some patients with affective disorders may explain the ability of *small* doses of T_3, equal to the daily endogenous production of T_3 (deGroot et al. 1984), to augment the antidepressant response to tricyclics. In like fashion, the frequent finding in the affective disorders of initially elevated T_4 parameters which fall with treatment may reflect a compensatory response of the organism which allows delivery of more T_4 to a brain whose homeostatic mechanism has gone awry. Regarding treatment of rapid-cycling affective illness, our experience indicates that whereas some patients respond to doses of T_4 which lower TSH while maintaining normal serum levels of total T_4 and T_3, others appear to need serum T_4 levels in the range of 150%–200% of top-normal. In either case, increase in circulatory levels of free T_4 accompanied by greater tissue effect of the iodothyronines, as evidenced by a decrease in TSH production, appears to correlate with resolution of previously refractory affective symptoms.

4 Summary and Future Directions

The foregoing discussion has focused on several major aspects of the role of thyroid function in disorders of mood. First and foremost, the evidence clearly shows that alterations of thyroid status are associated with affective psychopathology. Second, these alterations may be hypothesized to reflect in some cases a pathogenic deficit (e.g., hypothyroidism in rapid-cycling bipolar disorder), and in other cases a compensatory response (e.g., a relative increase in T_4 parameters in depression). Third, there are currently no good direct clinical measures of brain thyroid status. Finally, the study of the control of TSH production by the pituitary thyrotroph may yield important information about central nervous system thyroid status, if it is understood within the framework of the entire thyroid axis.

Several areas of research are suggested by the studies reviewed herein. First, better delineation of the source and specificity of altered T_4 parameters in the affective disorders is necessary. It is quite notable, in light of the evidence that depression is associated with an alteration of specific T_4 parameters which appear to have prognostic significance, that direct measures of iodothyronine metabolism, i.e., synthesis, distribution, and clearance, have not been undertaken.

Second, further investigation of thyrotroph activity, which may reflect activity of the major metabolic pathway for iodothyronines in brain, is necessary. Particular emphasis should be given to the role of thyroid factors and their control in the affective disorders. Loosen et al. (1980) have already demonstrated that a single dose of thyroid hormone blunts the response of TSH to TRH infusion in normals but not in depressives, indicating that the feedback loop may be altered in depression, perhaps due to altered 5′D activity. Related to this, other indicators of altered peripheral 5′D activity in depression would be of great interest.

Third, imaging of thyroid hormones in vivo in depressed patients will greatly clarify the role which thyroid function plays in psychiatric homeostasis.

Fourth, there is evidence from certain studies (e.g., Cho et al. 1979; Cowdry et al. 1983) and from our own observations that subtle thyroid hypofunction may be associated with at least increased severity of bipolar affective illness. Further exploration of this possible association, as well as determination of whether thyroid illness is a risk factor for the occurrence of affective illness, will therefore complement the pathophysiologic directions of research outlined. We know of no genetic or epidemiologic studies to date of the co-occurrence of thyroid dysfunction and unipolar or bipolar affective illness.

Finally, further definition of the parameters of response to thyroid hormone augmentation of treatment for both depressed and rapid-cycling patients will allow further characterization or the manner in which the thyroid axis plays a role in the response to antidepressant treatment and therefore also in the pathogenesis of affective illness. The specific characteristics of thyroid hormone effects will help to determine whether these thyroid effects on mood are due to interaction of thyroid hormones with a particular neurochemical system, for example the β-adrenergic system (Whybrow and Prange 1981).

Alternatively – or in addition – to such neurotransmitter – or neuropeptide – related effects of the iodothyronines, the endogenous increase in circulating T_4 levels or the administration of exogenous thyroid hormone may compensate for a primary defect in the processing of thyroid hormones by the central nervous system.

References

Abraham S, Atkinson M, Roscoe D (1966) Value of ankle-jerk timing in the assessment of thyroid function. Br Med J 2:830–833

Asher R (1949) Myxoedematous madness. Br Med J 22:555–562

Azukiwaza M, Mori S, Ohta H, Matsumura S, Yoshimoto H, Uozuari T, Miyai K, Kamahara Y (1979) Effect of a single dose of glucocorticoid on the diurnal variations of TSH, thyroxine, 3,5,3′-triiodothyronine, 3,3′,5′-triiodothyronine and cortisol in normal men. Endocrinol Jpn 26:719–723

Banki C (1975) Triiodothyronine in the treatment of depression. Orv Hetil 116:2543–2547

Banki C (1977) Cerebrospinal fluid amine metabolites after combined amitriptyline-triiodothyronine treatment of depressed women. Eur J Clin Pharmacol 11:311–315

Bauer M, Whybrow P (1986) The effect of changing thyroid function on cyclic affective illness in a human subject. Am J Psychiatry 143:633–636

Bauer M, Droba M, Whybrow P (1987) Disorders of thyroid and parathyroid. In: Loosen P, Nemeroff C (eds) Handbook of clinical psychoneuroendocrinology. Guilford, New York

Bissette G, Wallens H, Widerlou E, Karlsson I, Eklundt K, Loosen P, Nemeroff G (1984) Reduction of CSF concentration of SRIF-LI in dementia, major depression, and schizophrenia. Abstracts of annual meeting of Society for Neuroscience, Anaheim

Board F, Wadeson R, Persky H (1957) Depressive affect and endocrine function. Arch Neurol Psychiatry 78:612–620

Breese G, Prange A, Lipton M (1974) Pharmacologic studies of thyroid-imipramine interactions in animals. In: Prange A (ed) Thyroid axis, drugs, and behavior. Raven, New York

Calloway S, Donlan R, Fonagy P, DeSouza V, Wakeling A (1984) Endocrine changes and clinical profiles in depression: II. the thyrotropin-releasing hormone test. Psychol Med 14:759–765

Carrier R, Buday P (1961) Augmentation of toxicity of monoamine oxidase inhibitors by thyroid feeding. Nature 191:1107

Cho J, Bone S, Dunner D, Colt E, Fieve R (1979) The effect of lithium treatment on thyroid function in patients with primary affective disorder. Am J Psychiatry 136:115–116

Clinical Society of London (1881) Report on myxedema. Trans Clin Soc Lond [Suppl] 21

Coppen A, Whybrow P, Noguera R, Maggs R, Prange A (1972) The comparative antidepressant vlaue of L-tryptophan and imipramine with and without attempted potentiation by liothyronine. Arch Gen Psychiatry 26:234–241

Cowdry R, Wehr T, Zis A, Goodwin F (1983) Thyroid abnormalities associated with rapid-cycling bipolar illness. Arch Gen Psychiatry 46:414–420

DeGroot L, Larsen P, Refetoff S, Stanbury J (1986) The thyroid and its diseases, 5th edn. Wiley, New York

Dratman M, Crutchfield F, Gordon J, Jennings A (1983) Iodothyronine homeostasis in rat brain during hypo- and hyperthyroidism. Am J Physiol 245:185–193

Dunner D (1979) Rapid-cycling bipolar manic-depressive illness. Psychiatr Clin North Am 2:461–467

Earle B (1970) Thyroid hormone and tricyclic antidepressants in resistant depressions. Am J Psychiatry 126:1667–1669

Feighner J, King L, Schuckit M, Croughan J, Briscoe W (1972) Hormonal potentiation of imipramine and ECT in primary depression. Am J Psychiatry 128:1230–1238

Garbutt J, Malekpaur B, Brunswick D, Jennalagadda M, Jolliff L, Podolak R, Wilson I, Prange A (1979) Effects of triiodothyronine on drug levels and cardiac function in depressed patients treated with imipramine. Am J Psychiatry 136:980–982

Gjessing R, Jenner F (eds) (1976) Somatology of periodic catatonia. Pergamon, Oxford

Gold M, Pottash A, Extein I (1982) Symptomless autoimmune thyroiditis in depression. Psychiatry Res 7:261–269

Gold P, Goodman F, Wehr T, Rebar R (1977) Pituitary thyrotropin response to thyrotropin-releasing hormone in affective illness: relationship to spinal fluid amine metabolites. Am J Psychiatry 134:1028–1031
Goodwin F, Prange A, Post R, Mascattola G, Lipton M (1981) L-Triiodothyronine converts tricyclic antidepressant nonresponders to responders. Am J Psychiatry 139:334–338
Graves R (1940) Clinical lectures. Med Classics 5:33–36
Griffiths E (1985) Thyrotropin-releasing hormone: endocrine and central effects. Psychoneuroendocrinology 10:225–235
Herman H, Quarton RV (1965) Psychological changes and psychogenesis in thyroid hormone disorders. J Clin Endocrinol Metab 25:527–538
Himelfarb M, Lakretz, Gold S, Shanon E (1981) Auditory brain stem responses in thyroid dysfunction. J Laryngol Otol 95:670–686
Jain V (1972) A psychiatric study of hypothyroidism. Psychiatr Clin 5:121–130
Joffe R, Blank D, Post R, Uhde T (1985) Decreased triiodothyronines in depression: a preliminary report. Biol Psychiatry 20:922–925
Kaplan M, Breitbart R (1983) Conversion of thyroxine to triiodothyronine on the anterior pituitary gland and the influence of this process on thyroid status. Horm Metab Res [Suppl] 13:79–85
Kaptein E, Kletzky O, Spencer C, Nicoloff J (1980) Effects of prolonged dopamine infusion on anterior pituitary function in normal males. J Clin Endocrinol Metab 51:488–495
Kirkegaard C (1981) The thyrotropin response to thyrotropin-releasing hormone in endogenous depression. Psychoneuroendocrinology 6:189–212
Kirkegaard C, Faber J (1981) Altered serum levels of thyroxine, triiodothyronines and diiodothyronines in endogenous depression. Acta Endocrinol 96:199–207
Kirkegaard C, Faber J (1986) Influence of free thyroid hormone levels on the TSH response to TRH in endogenous depression. Psychoneuroendocrinology 11:491–497
Kirkegaard C, Norlem N, Lauridsen U, Bjorum N, Christiansen C (1975a) Protirelin stimulation test and thyroid function during treatment of depression. Arch Gen Psychiatry 32:1115–1118
Kirkegaard C, Norlem N, Lauridsen U, Bjorum N (1975b) Prognostic value of thyrotropin-releasing hormone stimulation test in endogenous depression. Acta Psychiatry Scand 52:170–177
Kirkegaard C, Bjorum N, Cohn D, Faber J, Lauridsen U, Nerup J (1977a) Studies of the influence of biogenic amines and psychoactive drugs on the prognostic value of the TRH stimulation test in endogenous depression. Psychoneuroendocrinology 2:131–136
Kirkegaard C, Bjorum N, Cohn D, Lauridsen U (1977b) Thyrotropin-releasing hormone (TRH) stimulation test in manic-depressive illness. Arch Gen Psychiatry 35:1017–1021
Kjellman B, Ljudggren J, Beck-Friis J, Wetterberg L (1985) Effect of TRH on TSH and prolactin levels in affective disorders. Psychiatry Res 14:353–363
Kostin A, Ehrensing R, Schalch D, Anderson M (1972) Improvement in mental depression with decreased thyrotropin response after administration of thyrotropin-releasing hormone. Lancet 2:740–742
Larsen P, Dick T, Markovitz B, Kaplan M, Gard T (1979) Inhibition of intrapituitary thyroxine to 3,5,3′-triiodothyronine conversion prevents the acute suppression of thyrotropin release by thyroxine in hypothyroid rats. J Clin Invest 64:117–128
Leonard J, Mellen S, Larsen P (1982) Thyroxine 5′-deiodinase activity in brown adipose tissue. Endocrinology 112:1153–1157
Leppaluoto J, Koivusalo F, Krama R (1978) Thyrotropin-releasing factor: distribution in neural and gastrointestinal tissues. Acta Physiol Scand 104:175–179
Lidz T, Whitehorn J (1949) Psychiatric problems in a thyroid clinic. J Am Med Assoc 139:698–701
Linnoila M, Lamberg B, Potter W, Gold P, Goodwin F (1982) High reverse T3 levels in manic and unipolar depressed women. Psychiatry Res 6:271–279

Linnoila M, Cowdry R, Lamberg B, Makineu T, Rubinow D (1983) CSF triiodothyronine (rT3) levels in patients with affective disorders. Biol Psychiatry 18:1489–1492
Loosen P (1985) The TRH-induced TSH response in psychiatric patients: a possible neuroendocrine marker. Psychoneuroendocrinology 10:237–260
Loosen P, Prange A (1980) TRH: a useful tool for psychoneuroendocrine. Invest Psychoneuroendocrinol 5:63–80
Loosen P, Wilson I, Prange A (1980) Endocrine and behavioral changes after TRH: alteration by pretreatment with thyroid hormones. J Affect Dis 2:267–278
Loosen P, Kistler K, Prange A (1983) The use of the TRH-induced TSH response as an independent variable. Am J Psychiatry 140:700–703
Manaker S, Winokur A, Rostene W, Rainbow T (1985) Autoradiographic localization of thyrotropin-releasing hormone receptors in the rat central nervous system. J Neurosci 5:167–174
Manaker S, Eicken A, Winokur A, Rhodes C, Rambo T (1986) Autoradiographic localization of thyrotropin-releasing hormone (TRH) receptors in human brain. Neurology 36:641–646
Mason J (1975) Emotion as reflected in patterns of endocrine integration. In: Levi L (ed) Emotions – their parameters and measurement. Raven, New York
McLarty D, Ratcliffe W, Ratcliffe J, Shimmins J, Goldberg A (1978) A study of thyroid function in psychiatric in-patients. Br J Psychiatry 133:211–218
Morley J (1981) Neuroendocrine control of thyrotropin secretion. Endocrinol Rev 2:396–436
Morley J, Schafer R (1982) Thyroid function screening in new psychiatric admissions. Arch Intern Med 142:591–593
Morley J, Sawin T, Carlson H, Longcope C, Hershman J (1981) The relationship of androgen to the thyrotropin and prolactin responses to thyrotropin-releasing hormone in hypogonadal and normal men. J Clin Endocrinol Metab 41:1161–1167
Nemeroff C, Simon J, Haggerty J, Evans D (1985) Antithyroid antibodies in depressed patients. Am J Psychiatry 142:840–843
Nicholson G, Liebling L, Hall R (1976) Thyroid dysfunction in female psychiatric patients. Br J Psychiatry 129:236–238
Orsulak P, Crowley G, Schlesser M, Giles D, Fairchild C, Rush A (1985) Free triiodothyronine (T3) and thyroxine (T4) in a group of unipolar depressed patients and normal subjects. Biol Psychiatry 20:1047–1054
Pardridge W (1979) Carrier-mediated transport of thyroid hormones through the rat blood-brain carriers: primary role of albumin-bound hormone. Endocrinology 105:605–612
Parker C, Porter J (1983) Regional localization and subcellular localization of thyrotropin-releasing hormone in the adult human brain. J Neurochem 40:1614–1622
Parry C (1825) Collected works, vol I. London, p 478
Prange A (1963) Paroxysmal auricular tachycardia apparently resulting from combined thyroid-imipramine treatment. Arch J Psychiatry 119:994–995
Prange A, Lipton M (1962) Enhancement of imipramine mortality in hyperthyroid mice. Nature 196:588–589
Prange A, Wilson I, Rabon A, Lipton M (1969) Enhancement of imipramine antidepressant activity by thyroid hormone. Am J Psychiatry 126:457–469
Prange A, Wilson I, Lara P, Alltop L, Breese G (1972) Effects of thyrotropin-releasing hormone in depression. Lancet II:999–1002
Prange A, Loosen P, Wilson I, Meltzer H, Fang V (1979) Behavioral and endocrine responses of schizophrenic patients to TRH (protirelin). Arch Gen Psychiatry 36:1086–1093
Prange A, Loosen D, Wilson I, Lipton M (1985) The therapeutic use of hormones of the thyroid axis in depression. In: Post R, Ballenger J (eds) The neurobiology of mood disorders. Williams and Wilkins, Baltimore
Rattle S, Purnell D, Williams P, Siddle K, Forrest G (1984) New separation method for monoclonal radioimmunometric assays and its application to assays for thyrotropin and human choriogonadotropin. Clin Chem 30:1457–1462

Rinieris P, Christodoulou G, Souvatzoglou A, Koutras D, Stefanis C (1978) Free-thyroxine index in mania and depression. Compr Psychiatry 19:561–564
Rose R (1980) Endocrine responses to stressful psychological events. Psychiatric Clin North Am 3:251–276
Roy-Byrne P, Joffe R, Uhde T, Post R (1984) Carbamazepine and thyroid function in affectively ill patients. Arch Gen Psychiatry 41:1150–1153
Rybakowski J, Sowinski J (1973) Free-thyroxine index and absolute free-thyroxine in affective disorders. Lancet I:889
Saberi M, Utiger R (1975) Augmentation of thyrotropin responses to thyrotropin-releasing hormone following small decreases in serum thyroid hormone concentrations. J Clin Endocrinol Metab 40:435–441
Schon M, Sutherland A, Rawson R (1961) Hormones and neuroses – the psychological effects of thyroid deficiency. Proc Third World Cong Psychiatry 2:835–839
Schwarcz G, Halaris A, Baxter L, Escobar J, Thompson M, Young M (1984) Normal thyroid function in desipramine nonresponders converted to responders by addition of L-triiodothyronine. Am J Psychiatry 141:1614–1616
Snyder P, Utiger R (1972) Inhibition of thyrotropin response to thyrotropin-releasing hormone by small quantities of thyroid hormones. J Clin Invest 51:2077–2084
Spratt D, Pont A, Miller M, McDougall I, Bayer M, McLaughlin W (1982) Hyperthyroxinemia in patients with acute psychiatric disorders. Am J Med 73:41–48
Stancer H, Persad E (1982) Treatment of intractable rapid-cycling manic-depressive disorder with levothyroxine. Arch Gen Psychiatry 39:311–312
Stanton T, Winokur A, Beckman A (1982) Seasonal variation in thyrotropin-releasing hormone (TRH) content of different brain regions and the pineal in the mammalian hibernator, *Citellus literalis*. Regul Peptides 3:135–144
Steiner M, Radquan M, Elizur A, Blum J, Atsman A, Davidson S (1978) Failure of L-triiodothyronine to potentiate tricyclic antidepressant response. Curr Ther Res 23:655–659
Sternbach H, Gwirtsman H, Gerner R, Heishman I, Pekary E (1985) The THR stimulation test and reverse T_3 in depression. J Aff Disorders 8:267–270
Sternbach H, Gold M, Pottash A, Extein I (1983) Thyroid failure and protirelin (thyrotropin-releasing hormone) test abnormalities in depressed outpatients. JAMA 249:1618–1620
Takahashi S, Kondo H, Yoshimara M, Ochi Y (1974) Thyrotropin responses to TRH in depressive illness: relation to clinical subtypes and prolonged duration of depressive episode. Folia Psychiatry Neurol Jpn 28:355–365
Targum S, Greenberg R, Harmon R, Kessler K, Salerian A, Fram D (1984) Thyroid hormone and the TRH stimulation test in refractory depression. J Clin Psychiatry 45:345–346
Taylor R, Burt D (1982) Species differences in the regional distribution of receptor binding for thyrotropin-releasing hormone. J Neurochem 38:1649–1656
Tsuitsui S, Yamazaki Y, Namba T, Tsushinia M (1979) Combined therapy of T3 and antidepressants in depression. J Int Med Res 7:138–146
Unden F, Ljunggren J, Kjellman B, Beck-Friis J, Wetterberg L (1986) Twenty-four-hour serum levels of T4 and T3 in relation to decreased TSH serum levels and decreased TSH response to TRH in affective disorders. Acta Psychiatry Scand 73:358–364
Visser T, Leonard J, Kaplan M, Larsen P (1982) Kinetic evidence suggesting two mechanisms for iodothyronine 5′-deiodination in rat cerebral cortex. Proc Natl Acad Sci USA 79:5080–5084
Visser T, Leonard J, Kaplan M, Larsen P (1983) Evidence for two pathways of iodothyronine 5′-deiodination in rat pituitary that differ in kinetics, propylthiouracil sensitivity, and response to hypothyroidism. J Clin Invest 71:992–999
Wallace J, MacCrimmon D, Goldberg W (1980) Acute hyperthyroidism: cognitive and emotional correlates. J Abnorm Psychol 89:519–527
Wartofsky L, Burman K (1982) Alterations in thyroid function in patients with systemic illness: the "euthyroid sick syndrome". Endocrinol Rev 3:164–217
Weeke A, Weeke J (1978) Disturbed circadian variation of serum thyrotropin in patients with endogenous depression. Acta Psychiatry Scand 57:281–289

Weeke J, Christiansen S, Hansen A, Laurberg P, Lundbaek K (1980) Somatostatin and the 24h levels of TSH, T3, T4, and reverse T3 in normals, diabetics, and patients treated for myxedema. Acta Endocrinol 94:30–35
Wheatley D (1972) Potentiation of amitriptyline by thyroid hormone. Arch Gen Psychiatry 26:229–233
Whybrow P, Prange A (1981) A hypothesis of thyroid-catecholamine-receptor interactions: its relevance to affective illness. Arch Gen Psychiatry 38:106–113
Whybrow P, Coppen A, Prange A, Noguera A, Bailey J (1972) Thyroid function and the response to liothyronine in depression. Arch Gen Psychiatry 26:242–245
Wilson I, Prange A, McClane T, Rabon A, Lipton M (1970) Thyroid-hormone enhancement of imipramine in nonretarded depression. N Engl J Med 282:1063–1067
Winokur A, Amsterdam J, Oler J, Mendels J, Snyder P, Caroff S, Brunswick D (1983) Multiple hormonal responses to protirelin (TRH) in depressed patients. Arch Gen Psychiatry 40:525–531
Witschy J, Schlesser N, Fulton C, Orsulak P, Giles D, Fairchild C, Crowley G, Rush A (1984) TRH-induced prolactin release is blunted in females with endogenous unipolar major depression. Psychiatry Res 12:321–331

Subject Index

GPSR Compliance

The European Union's (EU) General Product Safety Regulation (GPSR) is a set of rules that requires consumer products to be safe and our obligations to ensure this.

If you have any concerns about our products, you can contact us on ProductSafety@springernature.com

In case Publisher is established outside the EU, the EU authorized representative is:

Springer Nature Customer Service Center GmbH
Europaplatz 3
69115 Heidelberg, Germany

Batch number: 10838062

Printed by Printforce, the Netherlands